THE HANDBOOK OF NATURE

THE HANDBOOK
OF NATURE

Frank R. Spellman
and
Joni Price-Bayer

GOVERNMENT INSTITUTES
AN IMPRINT OF

THE SCARECROW PRESS, INC.
Lanham • Toronto • Plymouth, UK
2012

 **Government Institutes**

Published by Government Institutes
An imprint of The Scarecrow Press, Inc.
A wholly owned subsidiary of The Rowman & Littlefield Publishing Group, Inc.
4501 Forbes Boulevard, Suite 200, Lanham, Maryland 20706
www.govinstpress.com

Eastover Road, Plymouth PL6 7PY, United Kingdom

British Library Cataloguing in Publication Information Available

Library of Congress Cataloging-in-Publication Data

Spellman, Frank R.
  The handbook of nature / Frank R. Spellman and Joni Price-Bayer.
    p. cm.
  Includes bibliographical references and index.
  ISBN 978-1-60590-773-4 (cloth : alk. paper) — ISBN 978-1-60590-774-1 (electronic)
    1. Natural history—Popular works. 2. Natural history—Handbooks, manuals, etc.
  3. Biology—Popular works. 4. Biology—Handbooks, manuals, etc. 5. Earth sciences—Popular works. 6. Earth sciences—Handbooks, manuals, etc.
  I. Price-Bayer, Joni, 1964–   II. Title.

  QH45.5.S63 2012
  508—dc23
                                                            2011042042

∞™ The paper used in this publication meets the minimum requirements of American National Standard for Information Sciences—Permanence of Paper for Printed Library Materials, ANSI/NISO Z39.48-1992.

Printed in the United States of America

For
Joan-Carol Bensen
and
Tracy L. Teeter

Contents

Preface xv

Part I What Is Nature?

1 Introduction 3
 Nature's Jekyll and Hyde Personality 3
 Nature Gives, Takes Away, and . . . Gives Back 9
  The Warm Fire 10
  From Moonscape to Landscape 16
  "If the mountain goes, I'm going with it." 24
  Glacial Lake Missoula and Dry Falls 32
 What Is Nature? 33
 References and Recommended Reading 34

Part II Earth's Structure

2 Composition of Earth 41
 Planet Earth 41
 Earth Products, Beauty, and Divisions 42
 Earth Geology 43
 Structure of Earth 44
 Earth's Geological Processes 44
  Rock Cycle (aka Geologic Cycle) 45

The Lithosphere		46
Soil		46
Soil Formation		48
References and Recommended Reading		50
3 Reshaping Earth		52
Introduction		52
Endogenic Forces		52
Volcanism		53
Earthquakes		62
Plate Tectonics		66
Exogenic Forces		68
Weathering		68
Glaciation		72
Wind Erosion		80
Mass Wasting		83
Desertification		86
Water Flow (Streams): Transport of Material (Load)		87
References and Recommended Reading		89

Part III The Atmosphere: Earth's Breath

4 The Atmosphere		95
Introduction		95
Earth's Atmosphere Is Born		95
Earth's Thin Skin		98
Troposphere		98
Stratosphere		99
A Jekyll and Hyde View of the Atmosphere		100
Atmospheric Particulate Matter		100
Moisture in the Atmosphere		102
Cloud Formation		102
Major Cloud Types		103
Summary		105
References and Recommended Reading		106
5 Atmospheric Water Phenomena		107
Introduction		107
Precipitation		110
Coalescence		110
Bergeron Process		110
Types of Precipitation		112
Convectional Precipitation		112
Orographic Precipitation		112

Frontal Precipitation	113
Colored Rain	113
Cloudless Rain	114
Evapotranspiration	114
Evaporation	115
Transpiration	115
Evapotranspiration: The Process	117
Albino Rainbow	117
References and Recommended Reading	121
6 Atmosphere in Motion	122
Introduction	122
Global Air Movement	122
Global Wind Patterns	124
Earth's Atmosphere in Motion	125
Causes of Air Motion	126
Local and World Air Circulation	131
References and Recommended Reading	134
7 Weather and Climate	135
Introduction	136
What's the Difference between Weather and Climate?	136
El Niño–Southern Oscillation	137
The Sun: The Weather Generator	138
Air Masses	140
Thermal Inversions and Air Pollution	141
Climate	142
Climate Divisions	143
Climate Subdivisions	144
Microclimates	146
Microclimates near the Ground	147
Microclimates over Open Land Areas	149
Microclimates in Woodlands or Forested Areas	149
Microclimates in Valleys and Hillside Regions	149
Microclimates in Urban Areas	150
Microclimates in Seaside Locations	150
References and Recommended Reading	152

Part IV Water: Earth's Blood

8 Water on Earth	155
To the Reader	155
Introduction	156
Water Is Special, Strange, and Different	157

Setting the Stage   163
Historical Perspective   165
    So Necessary, but so Easy to Abuse   167
References and Recommended Reading   167
9   All about Water   168
Introduction   168
Characteristics of Water   171
    Inflammable Air + Vital Air = Water   172
    Just Two H's and One O   173
    Somewhere between 0° and 105°   173
Water's Physical Properties   174
Capillary Action   176
Water Cycle   176
    Specific Water Movements   178
Sources of Water   180
Portable Water Source   181
    Potable Water   181
Surface Water   181
    Location! Location! Location!   181
    How Readily Available Is Potable Water?   182
Groundwater Supply   186
    Groundwater   186
    Groundwater Quality   189
GUDISW   189
References and Recommended Reading   191
10 Earth's Oceans and Their Margins   192
Oceans   192
Ocean Floors   194
Ocean Tides, Currents, and Waves   195
    Tides   195
    Currents   196
    Waves   197
Coastal Erosion, Transportation, and Deposition   198
    Wave Erosion   198
    Marine Transportation   199
    Marine Deposition   199
References and Recommended Reading   200
11 Running Waters   201
Stream Genesis and Structure   201
Streams   203
Characteristics of Stream Channels   204
    Stream Profiles   204
    Sinuosity   205

Bars, Riffles, and Pools   206
Floodplain   206
Water Flow in a Stream   207
Stream Water Discharge   208
Transport of Material (Load)   209
Groundwater   211
Geologic Activity of Groundwater   214
References and Recommended Reading   215
12 Still Waters   216
Lentic Habitat   216
Classification of Lakes   219
Classification Based on Eutrophication   220
Special Types of Lakes   220
Impoundments (Shut-Ins)   221
References and Recommended Reading   222

Part V Earth's Life Forms

13 Biosphere   225
Introduction   225
What Is Life?   226
Life: An Afterthought   227
Levels of Organization   228
The Scientific Method   229
How Is the Scientific Method Used?   230
Case Study: Scientific Method—*Pfiesteria* and Fish Health   233
Why Is the Scientific Method Important?   235
References and Recommended Reading   236
14 Ecosystems   237
Introduction   237
Ecosystems   240
Types of Ecosystems   242
Biogeochemical Cycles   243
Nutrient Cycles   243
Water Cycle   247
Water Reservoirs, Fluxes, and Residence Times   249
Carbon Cycle   250
Nitrogen Cycle   252
Phosphorus Cycle   256
Sulfur Cycle   258
Energy Flow   259
Food Chain Efficiency   263
Ecological Pyramids   264

Relationships in Living Communities 266
Productivity 267
References and Recommended Reading 270
15 The Fundamental Unit of Life 273
Introduction 273
Types of Cells 276
Cell Structure 278
Intercellular Junctions (Animal Cells) 283
References and Recommended Reading 284
16 Biological Diversity 285
Introduction 285
Classification 286
Kingdoms of Life 287
Bacteria 288
Viruses 291
Protists 292
Protozoa 293
Algae 297
Slime and Water Molds 302
Fungi 303
Plants: The Great Starch Producers 306
The Plant Kingdom 307
The Plant Cell 309
Vascular Plants 309
Leaves 310
Chlorophyll/Chloroplast 311
Photosynthesis 311
Roots 312
Growth in Vascular Plants 312
Plant Hormones 313
Tropisms: Plant Behavior 313
Photoperiodism 314
Plant Reproduction 315
Animals 315
Invertebrates 316
Vertebrates 319
Human Evolution 319
References and Recommended Reading 320
17 Animal Groups 321
Birds 321
Songbirds 323
Waterfowl Populations 325

Geese   326
Ducks   327
Shorebirds   331
Seabirds   332
Colonial-nesting Waterbirds   335
Raptors   336
Wild Turkeys   337
Mourning Doves   338
Common Ravens   339
Mississippi Sandhill Cranes   340
Piping Plovers   341
Red-cockaded Woodpeckers   341
Southwestern Willow Flycatchers   342
Mammals   343
Marine Mammals   344
Indiana Bats   345
Gray Wolves   346
North American Black Bears   347
Grizzly Bears   348
Black-footed Ferrets   350
American Badgers   352
Northeastern White-tailed Deer   353
North American Elk   354
Reptiles and Amphibians   354
Turtles   356
Marine Turtles   357
Amphibians   359
American Alligators in Florida   361
Native Ranid Frogs   362
Desert Tortoises   362
Fringe-toed Lizards   363
Tarahumara Frog   365
Fishes   365
Freshwater Fishes of the Contiguous United States   367
Managed Populations: Loss of Genetic Integrity through Stocking   368
Colorado River Basin Fishes   369
Cutthroat Trout: Glacier National Park   370
White Sturgeon   372
Invertebrates   373
Diversity of Insects   376
Grasshoppers   376
Lepidoptera: Butterflies and Moths   377

Aquatic Insects and Biotic Indices 380
Insect Macroinvertebrates 384
Other Invertebrates 394
Plants 396
Fungi 397
Microfungi 398
Macrofungi 399
Lichens 401
Bryophytes 401
Native Vascular Plants 403
References and Recommended Reading 403

Part VI Wilderness

18 Land Organism 423
What Is Wilderness? 423
U.S. Forest Resources 425
Forested Wetlands 428
Fire Regimes within Fire-adapted Ecosystems 430
Vegetation Change in National Parks 433
Whitebark Ecosystem 436
Wisconsin Oak Savanna 437
References and Recommended Reading 438

Glossary 443

Index 453

About the Authors 463

Preface

Mᴏꜱᴛ ᴘᴇᴏᴘʟᴇ ʜᴀᴠᴇ one particular passion that sustains them in the outdoors; we have several. For some, it's the experience of carrying a backpack through an alpine meadow sprinkled with a host of colors of countless wildflowers. For others, it's the vision of fly-fishing for trout on a clear river perfectly mirroring the surrounding tree-lined mountain landscape. Still others are in their element, their glory, their habitat(s) where wind and leaf-fall drop quaking aspen leaves that look like golden doubloons in the sunlight. Still others just like the stillness of a frosty cold walk in the woods. A few like this same walk but prefer the silence broken by the chirp of a solitary bird or the bugling of a bull elk or the sighing of the wind through tall pines.

For us, we like all of the above . . . and more. We like the outdoors because we love nature. We love the land and its inhabitants, all shaped and balanced by the forces of nature. The outdoor environment is a wonderland of shadows, textures, color, habitats, and experiences. We have trekked Mother Nature's landscapes from mountain, alpine tundra, subalpine, montane forest, dune field, sand sheets and grasslands, sabkha, riparian and wetland regions, and many others. In all we are overwhelmed with the incomparable handiwork of Mother Nature. And overwhelming is a good way to describe the works of Mother Nature. No one human being can ingest and digest all that she has to offer. We can't count the number of times we have walked the woods and/or deserts and climbed alpine meadows and great sand dunes and have wondered the "how" of it all. Where did all the splendor come from, how did great sea cliffs form; how big are the biggest trees; how deep is the deepest lake; how were lakes and rivers and streams and rivulets and springs formed; why is grass tall in some areas and nonexistent in others?

These are some of the questions we have spent a lifetime trying to answer. Based on our experience, we have found that some of the easiest questions to ask are the most difficult to answer, but we do not avoid "difficult" topics (you might say it is not our "nature" to avoid difficulty)—if a process is essential to a true understanding of the dynamics of earth and life, then it is included, along with the best, plain English, non-gibberish (avoiding all forms of gobbledygook), explanation science has to offer in this text.

The Handbook of Nature has been created to give specialists and non-specialists a real understanding of the way our planet and its living organisms function. It provides scientific (natural) answers to questions that arise when looking at the world around us or studying—answers that strip away the mystery and cut through to the basic foundational processes.

This handbook is for people who want to refresh their memories and/or learn about nature without taking a formal course. From our Earth and its four environmental mediums—air, water, soil, and biota—through evolution, ecosystems, biomes, biogeochemical cycles, wilderness, on the extraordinary diversity of form and behavior of today's living species and challenges, *The Handbook of Nature* provides an all-inclusive broad sweep of both earth and life sciences, employing several illustrations to translate detailed technical information into terms that everyone can follow and readily reference. It can serve as a classroom supplement, tutorial aid, self-teaching guide, home-schooling reference, a nature-lover's handbook, and as a comprehensive desk reference for any environmental professional in the field or laboratory.

Again, in regard to professionals and based on personal experience, we have found that many professionals do have some background in nature studies (Don't we all? How can anyone avoid nature?), but many of these folks need to stay current. (Remember that there is nothing static about nature—and the same can be said [hopefully] about our understanding of nature.) Moreover, many practitioners are specialists (e.g., engineers, environmental scientists, environmental health professionals, medical professionals, epidemiologists, water quality technicians, toxicologists, environmental scientists, occupational health and safety professionals, industrial hygienists, teachers, speech pathologists, and so forth). Herein lies the problem—that is, specialization. Our view is that practitioners of any blend, flavor, or type should absolutely know the ins and outs of their field of specialization but in doing so should also be generalists with a wide range of knowledge—not just specialists whose knowledge may be too narrowly focused. Again, based on personal experience, our students, for example, who have generalized their education (spread out their exposure to include disciplines in several aspects of general studies), versus those who narrowly specialize, are afforded many more opportunities to broaden their view of the world and increase their chances for upward mobility in employment ventures. They have had a much better opportunity to ascend to upper management positions because with on-the-job experience and their generalized education—their holistic view of the world and life around them—they become well-rounded individuals, ready to face just about any challenge that may arise. Is this not the standard characteristic of a successful person?

The Handbook of Nature fills the gap between general introductory nature texts and the more advanced environmental science books used in graduate courses. The handbook fills this gap by surveying and covering the basics of natural studies. This book is both a technical and non-technical survey for those with or without a background in natural studies—presented in reader-friendly written style.

The study of nature is multidisciplinary in that it incorporates aspects of biology, chemistry, physics, ecology, geology, meteorology, pedology, sociology, and many other fields. Books on the subject are typically geared toward professionals in these fields. This makes undertaking a study of nature daunting to those without this specific background. However, this complexity also indicates nature study's broad scope of impact. Because nature affects us, sometimes in profound double-edged ways such as on the one edge that presents incomparable aesthetics and beauty and on the other edge that presents such events as volcanism, earthquakes, tornadoes, floods, border wars, social/cultural norms, and so forth. It is important to understand some basic concepts of the discipline.

Along with basic natural scientific principles, the handbook provides a clear, concise presentation of the consequences of the anthropogenic interactions with the environment we inhabit. Even if you are not tied to a desk, the handbook provides you, the naturalist or non-naturalist, with the jargon, concepts, and key concerns of nature and nature in action.

This reference book is compiled in an accessible, user-friendly format, unique in that it explains scientific concepts in the most basic way possible. Organized in dynamic themes that reflect the approach of modern science, *The Handbook of Nature* draws together common processes and activities rather than dividing and classifying. The beginning naturalist will quickly have foundation for the future. The more experienced may find their knowledge suddenly snapped into focus with a new and solid foundation under that which is already known.

Frank R. Spellman
Joni Price-Bayer
Norfolk, Virginia

I

WHAT IS NATURE?

1

Introduction

Nature never did betray
The Heart that Loved her.

—William Wordsworth (1770–1850),
Lines Composed a Few Miles Above Tintern Abbey, 1798

Nature's Jekyll and Hyde Personality

WE HAD NOT WALKED any part of Mt. Fremont Lookout Trail for more than several years. Though we had never walked its entire 5.7 mile length and elevation gain of approximately 800 feet with some up and down gain/loss, from Sunrise in Mt. Rainier National Park to Mt. Fremont Lookout at once, over the course of several years a long time ago, we had in piecemeal fashion covered most of it. But we had moved out of easy reach of the Trail, and for years had only our memories of it.

For us, the lure of sojourning the Mt. Fremont Trail had always been more than just an excuse to get away from it all—whatever "it" happened to be at the time. Notwithstanding the alpine environment which can produce unexpected hazardous conditions including wind, rain, cold, whiteouts, swarms of mosquitoes, as well as dehydrating extreme sunlight and heat (in the past we had suffered all these in the same day) the draw, the magnetism of the Trail was more—much more to us than that, though we have always found its magic difficult to define. Maybe it was a combination of elements—recollections, pleasant memories, ephemeral surprises found and never forgotten. Memories waking from the miles-deep sleep of earned exhaustion to the awareness of peace . . . inhaling deep draughts of cool, clean mountain air, breathing through nostrils tickled with the pungency of pure, sweet Ponderosa, Western white, and Whitebark pine . . . eardrums soothed by the light tattoo of fresh rain pattering against taut nylon . . . watching darkness lifted, then suddenly replaced with cloud-filtered daylight, spellbound by the sudden, ordinary miracle of a new morning . . . anticipating our expected adventure and realizing the pure, unadulterated treasure of pristine wilderness we momentarily owned, with minds not weighed down by the everyday mundanities of existence. That is what we took away from our Trail experiences, years ago, what we remembered about living on the trail, on our untroubled sojourn through one of the last pure wilderness areas left in the United States. Those memories were magnets. They drew us inexorably to the Trail—back again and again.

But, of course, the Trail for us had another drawing card—the Natural World and all its glory. The Trail defined that for us. The flora that surrounds you on the Trail literally encapsulates you as it does in any dense forest—in addition to pine, there are the Bigleaf Maple, Bitter Cherry, Douglas-Fir, Noble Fir, Pacific Dogwood, Sitka Spruce, wild Crabapple, Red Alders, and many others—and brings you fully into its own world, shutting out all the other worlds of your life. For a brief span of time, along the Trail, the office was gone, cities, traffic, the buzz and grind of work melted away into forest. But this forest was different; its floral inhabitants created the difference. Not only the thickets of rhododendrons and azaleas (in memory, always in full bloom), the other forest growths drew us there: the magnificent trees—that wild assortment of incomparable beauty that stood as if for forever—that was the Trail.

But there is more to this trail, to this life experience, to this temporary trek in paradise—much more. Trekking the Trail is poetry in action; specifically, William Wordsworth comes to mind, recall "Daffodils" (1804):

> I wander'd lonely as a cloud
> That floats on high o'er vales and hills,
> When all at once I saw a crow,
> A host, of golden daffodils;
> Beside the lake, beneath the trees,
> Fluttering and dancing in the breeze.

Did we stumble into a host of golden daffodils? No; not exactly. Instead, what we hiked into is one or more of the pristine mountain meadows that adjoin and where we transect (i.e., where we take in, photograph, count, or just enjoy a host of incomparable wildflowers) a multitude of blooms of incomparable beauty. These meadows, during late July and early August are indeed awash in wildflowers—a short but glorious flower season. The names are nearly as colorful as the blossoms themselves: Indian paintbrush, glacier lily, Cascade aster, Cascade catchfly, elegant Jacob's ladder, yellow-dot saxifrage, alpine lupine, tongue-leaf Rainiera, cliff paintbrush, alpine golden daisy, and white-mountain heather.

Along with the large variety of tall trees and high alpine meadows awash in wildflowers, it is difficult, virtually impossible to ignore the conspicuous and dynamic glaciers. When we look up at those massive sheets of ice, we first think how static, rigid, unchanging and well-entrenched they are. All these terms are unsuitable, of course, because they certainly do not fit these particular massive sheets of ice, covering up to 39 square miles of the mountain's surface, they are dynamic—not at rest; no, never—eroding the mountain's volcanic cone and providing the primary sources of stream flow for several rivers, including some that provide water for hydroelectric power and irrigation.

Then there is the never to be overlooked main feature, of course: Mt. Rainier. From just about anywhere on the Mt. Fremont Lookout Trail whether you look to your right, your left, over your shoulder behind you, in front of you, downward or upward there it is; it stands out in unrivaled prominence.

As lifelong students of the physical sciences, we know that the matchless beauty presented by the one half to one million year old volcano can't be ignored; moreover, we can't forget that the mountain is classified as an episodically active volcano. The current summit, Columbia Crest, lies at 14,411 feet (4,392 m) above sea level, on the rim of the recent lava cone. Simply, Mt. Rainier is the elevated jewel of the Cascade Range.

Upon reaching Mt. Fremont Lookout, at 7,181 ft of elevation, and after recognizing, absorbing, and appreciating the beauty of the place, almost invariably, sooner or later, a cold chill comes over us. No, this is not the frisson or cold chill that Huron (2009) describes as "a pleasant tingling feeling" sometimes associated with the flexing of hair follicles resulting in goose bumps accompanied by a cold sensation, and sometimes producing a shiver or shudder. No, our chill is different, not driven by temperature, menopause, or anxiety but rather is an emotionally triggered response when we are deeply affected by recognition.

Recognition? Yes; recognition that the trees, flowers, snow, ice, glaciers, pure mountain air, and the mountain itself mask the underlying truth: Potentially Mt. Rainier is a very dangerous volcano. This is the case even though the most recent recorded eruption was between 1820 and 1854. Moreover, consider that approximately 5,000 years ago the mountain likely stood at more than 16,000 feet (4,900 m) before a major debris avalanche and the resulting mudflows (lahars) lowered it to its present height (Scott and Vallance 1993), discharging mudflow all the way to Tacoma and south Seattle. Large debris avalanches and lahars in the past were not that unusual on Rainier because of the large amount of glacial ice present.

The question is what about the present? The past is full of physical catastrophes and other calamities and there is no question we need to learn from the past, but we live in the present and hope for an extended future. One thing seems certain; when Mt. Rainier erupts again the effects will be much more devastating than Mt. St. Helens (see Figures 1 and 2). With the amount of ice and the shape of Mount Rainier (see Figure 3), lahars will be the biggest danger, swallowing up numerous small cities built on deposits of great mudflow and further affecting the lives of many people (Zellers 2010).

Again, Mt. Rainier is considered to be an active volcano, but as of 2010 there was no evidence of an imminent eruption. However, an eruption could be devastating for all areas surrounding the volcano. Mount Rainier is currently listed as a Decade volcano, or one of the 16 volcanoes with the greatest likelihood of causing great loss of life and property if eruptive activity resumes (USGS 2008).

FIGURE 1.   Mt. St. Helens as it appeared in July 2010
Photo by Frank R. Spellman

FIGURE 2. Inside the crater of Mt. St. Helens. Photo taken from a hovering helicopter. The frozen ice and snow to the right side of the crater indicate the cooler side while the stream, steam, and bare rock on the left indicate the hot, active section of the crater. A water stream flows from this spot down the lateral blast area into a larger iron oxide laden stream below; when measured, it typically indicates a temperature level of 185–190°F.

Photo by Frank R. Spellman

FIGURE 3. Mt. Rainier as it appeared in July 2010
Photo by Frank R. Spellman

So, each time we stand at Mt. Fremont Lookout and take in the majesty and beauty of our surroundings we realize that the volcano's beauty is deceptive— Mount Rainier is classified as one of our Nation's volcanoes. It has been the source of countless eruptions and volcanic mudflows that have surged down the valley on its flanks and buried broad areas that are now densely populated (USGS 2008).

Mount Rainier's natural, normal, and unpredictable cycling from a show of beauty to acts of enormous calamity via eruptions and mudslides points out an important aspect about nature that we often do not think about and contemplate; that is, Nature can present a full array of natural beauty in just about anything and everything we gaze upon or can imagine, but at the same time can be brutal in its heavy-handedness in reshaping the planet to its own liking. If we personify Nature in terms of human aspects, we can say that Nature basically possesses and often displays a Dr. Jekyll and Mr. Hyde split personality: In the sense of serene beauty, good on the one side and, in the sense of the bad, potential brutal destruction on the other.

Nowhere is this contrast in personality so readily observed than in the occurrence of and result of natural disaster. Thankfully, natural disasters are not frequent occurrences (they do not occur on a daily basis). However, when they do occur, humankind stands up and takes notice or is bulldozed over and becomes only a memory. We especially take notice of such events whenever they occur in our own backyards. We all know that natural disasters affect the environment, and can lead to financial, environmental, and/or human losses. The resulting loss, of course, depends on our resilience; on our capacity to support or resist the disaster (Bankoff et al., 2003).

Keep in mind that the key word in the statement above—to make the statement correct—is "observed." Natural disasters occur only when hazards meet vulnerability—a natural disaster will never result in areas without vulnerability, where there is no one to observe or be affected, e.g., a volcanic eruption in uninhabited areas is a natural event and not always a natural disaster (Wisner et al., 2004). This is simply a rehash of the old adage: "If a tree falls in a forest and no one is around to hear it, does it make a sound?" In other words, there must be observation (via our senses) to perceive knowledge of reality. We have to be in the forest to actually hear the tree fall. We have to observe or be affected in some way by an avalanche, earthquake, lahars, volcanic eruption, flood, limnic eruption, tsunami, blizzards, cyclonic storm, drought, hailstorm, fire, heat wave, tornado, epidemic, and/or famine for such an event to be considered a natural disaster. Otherwise, such occurrences are simply events; again, we can say that they are "natural events," because they are.

Did You Know?

Decade volcanoes are so designated if they exhibit more than one volcanic hazard: people living near the Decade volcanoes may experience tephra (volcanic rock) fall, pyroclastic flows, lava flows, lahars, volcanic edifice instability and lava dome collapse; a Decade volcano shows recent geological activity, and is located in a populated area.

The following 16 volcanoes are listed as Decade volcanoes:

1. Avachinsky-Koryaksky, Kamchatka
2. Colima Volcano, Mexico
3. Mount Etna, Italy
4. Galeras Volcano, Colombia
5. Mauna Loa, Hawaii
6. Merapi Volcano, Indonesia
7. Niragongo Volcano, Democratic Republic of the Congo
8. Mount Rainier, Washington
9. Sakurajima Volcano, Greece
10. Santa Maria/Santiaguito Volcano, Guatemala
11. Santorini Volcano, Greece
12. Taal Volcano, Philippines
13. Teide Volcano, Canary Islands, Spain
14. Ulawun Volcano, Papua New Guinea
15. Unzen Volcano, Japan
16. Vesuvius Volcano, Italy

We spend a lot of time and money attempting to understand "natural" and natural disasters. And this is the way it should be. We live on Earth and thus we should try to understand Nature and why natural events occur. We need to ask: What are the mechanisms that trigger such events? Because we live on Earth and no place is immune from natural occurrences, we need to know and to understand as much as possible about Nature's mechanisms and/or Nature's way.

The bottom line: In summary of a critical point: A natural disaster is an event that affects life, limb, and property of humans. On the other hand, a natural event is an occurrence that has no apparent effect on humankind.

Nature Gives, Takes Away, and . . . Gives Back

One realization that those of us who have spent a lifetime studying nature come to recognize, whether it be about a natural disaster (where humans and their property are damaged) or a natural event (no human impact), is that nature gives, takes, and—if not interfered with—gives back, sometimes in a manner better than the original. Many call this natural phenomenon nature's restorative power or ability. In the real-world cases that follow, we present a study of a naturally caused forest fire incident and its aftermath; rift volcanism on the Snake River Plain and its remarkable nature-aided transformation from an outer-space landscape to a thriving eastern Idaho terrain; Mt. St. Helens after its historic 1980 eruption and its present status; and finally Glacial Lake Missoula, which when the ice melted released water that roared across the Idaho panhandle region into Washington State and carved the scablands regions of eastern Washington today, not only leaving channeled scablands but also one of Earth's highest waterfalls, all bone dry.

The Warm Fire

One of our all-time favorite places to visit and explore is Jacob Lake, Arizona. It is a small unincorporated community on the Kaibab (pronounced kie-bab) Plateau in Coconino County, at the junction of U.S. Route 89A and State Route 67. Known as the "Gateway to the North Rim of the Grand Canyon" and named after the Mormon explorer Jacob Hamblin, it is the starting point of route 67, the only paved road leading to the North Rim of the Grand Canyon, 44 miles to the south.

The quaint settlement of Jacob Lake, with its inn, cabins, restaurant, lunch counter, gift shop, bakery, horse riding center, campground, general store, vast forest area, and a U.S. Forest Service visitors center, is one of those out of the way and often looked-forward-to stopping off places that we always enjoy visiting. And while it is the North Rim of the Grand Canyon with its Bright Angle Point Trail, Cliff Springs Trail, and the much longer North Kaibab Trail (to the Canyon floor and 14 miles across it to the South Rim) that are the main draws to this particular area, it is the 44 mile scenic drive through the priceless heritage forest from Jacob Lake to the North Rim that we have come to appreciate. Along this paved highway, situated at roughly 8000 feet in a large ponderosa pine forest, giving way in places to aspen, spruce, and fir, the trees, grazing meadows, and wildlife are breathtaking and soothing to the eye at the same time (see Figure 4). Home to the endangered Kaibab Squirrel, this much-traveled road is also home to mule deer, coyotes, porcupines, bobcats, numerous bird species, horned lizards, mountain lions and, as shown in Figures 4 and 5, buffalo.

In regard to scenery, the drive along Route 67 is different today from what it was in the past. Actually, the drive has been different since 2006. Nature stepped in and made a few changes along the way. On the afternoon of June 8, 2006,

FIGURE 4.   Buffalo herd in Kaibab Forest ponderosa pine fringed-meadow along Route 67 from Jacob Lake, Arizona, to North Rim, Grand Canyon

FIGURE 5.   Buffalo and wallow along Route 67 near Jacob Lake, Arizona

a lightning storm swept across the Kaibab Plateau. One of the high voltage, sky/earth exchanges set a tree on fire south of Jacob Lake. Local fire crews could have responded immediately, but that is not what happened. Instead, the National Forest Service decided to let the fire burn. They named it the Warm Fire, and designated it a Wildland Use Fire (WUF). While "wildland" can have several different definitions, for our purposes we define it as a natural environment that has not been significantly modified by human activity; a last truly wild natural area.

U.S. Forest Service management chose to let the naturally ignited fire burn in the **national forest** because they understood that fire plays a critical role in wildlands by recycling nutrients, regenerating plants, and by reducing high concentrations of fuels that contribute to disastrous wildland fires. U.S. land managers recognize the role that controlled wildland fire plays in ecosystems and through careful planning, can manage naturally occurring fires, such as Warm Fire, for resource benefits. It is important to point out that human-caused fires are never used for resource benefit and are always declared as unwanted wildland fire to be suppressed using the most cost-effective means to protect lives, property, and the environment (NIFC 2010).

As mentioned, wildland fire use is the management of naturally ignited wildland fire to accomplish resource management objectives for specific areas. There are three primary objectives for allowing wildland fire use:

- Provide for the health and safety of firefighters and the public.
- Maintain the natural ecosystems of a given area and allow fire to play its natural role in those ecosystems.
- Reduce the risks and consequences of unwanted fire.

In regard to the above objectives related to the Warm Fire, the old idiom "The best laid plans of mice and men oft go astray" is apropos. Why? Consider the following account compiled by our investigation (and regulatory agency reports) of this incident.

According to various private, public, and U.S. Forest Service accounts, after lightning struck and ignited one of the trees, the Warm Fire smoldered in the decaying leaves and branches (duff) covering the forest floor and crept around for a few days. At first glance all appeared well under control. But then the wind picked up and the Warm Fire jumped Route 67 on June 15. It jumped it again two days later and consumed 750 acres.

Three days later, on the 18th, the Warm Fire had grown to over 3,000 acres. Then overnight it doubled to over 6,000 acres. By the 22nd the Warm Fire had grown to over 10,000 acres. No one panicked. The mindset? Let it burn! And burn it did. On June 23 strong winds came blowing hard out of the west and the Warm Fire blew up, and overnight grew to over 15,000 acres. On the 24th, the Kaibab National Forest fire service came to the conclusion that enough was enough and it was time for intervention, for action. By the time action was taken on June 26th, the Warm Fire had grown to 30,000 acres. By the next day it was 60,000 acres, the doubling mainly due to the back fires set by the professional firefighters (aka smoke-eaters) fighting to contain the fire. These same professionals, using abundant ground and air resources contained, controlled, and mostly extinguished the Warm Fire by July 4th. The cost to Kaibab Forest was immeasurable. Many critics feel that the fire was not a cleanser but instead was a destroyer.

In February 2007 the Kaibab Ranger District posted a rather frank report on their website entitled Warm Fire Assessment: Post-Fire Conditions and management Considerations on the ecological damages inflicted on the Kaibab Forest by the Warm Fire. According to the Ranger District, the following types of damage have and will continue to occur:

- **Financial**—Total cost/loss of the Warm Fire is estimated at $70 million.
- **Soil Erosion**—Intense burn denuded vast acreages. Rain and snow runoff incised deeper channels and accelerated soil loss. Known archaeological sites have been eroded away and partially lost.
- **Water Quality**—Ash and sediment have been transported in Marble and Kanab creeks to the Colorado River.
- **Forest Vegetation**—More than 60% of the Warm Burn experienced 100% **mortality** to the trees.
- **Fire Hazard**—Surface and fine fuels were partially consumed, but the addition of dead snags increased coarse fuels.
- **Wildlife**—A large portion of the Mexican spotted owl **habitat** was destroyed by the Warm Fire. Moreover, habitat for the northern goshawk was destroyed.
- **Heritage Sites**—Many historic, cultural sites were damaged or destroyed by the Warm Fire.

After the Warm Fire, many private organizations, think tanks, and environmental thinkers jumped all over the National Fire Service for letting the Warm Fire cause the damage that it did. Nature burns in forest areas may be at odds with images of Smokey the Bear and blackened forests; the effects of fire, however, are temporary. During several walk-arounds of hundreds of acres scorched by the Warm Fire along Route 67, shown in Figures 6a–d, we have concluded that life in the form of various wildlife species and the obvious growth of tree seedlings and other undergrowth are apparent throughout—the burned area is recovering.

FIGURE 6a.   Kaibab Forest section off Route 67 Warm Fire burn area, July 2008

Photo by Frank R. Spellman

FIGURE 6b.   Kaibab Forest section off Route 67 Warm Fire burn area, July 2008

Photo by Frank R. Spellman

FIGURE 6c. Kaibab Forest section of Warm Fire burn area south of Jacob Lake, Arizona, July 2009

Photo by Frank R. Spellman

FIGURE 6d. Kaibab Forest section of Warm Fire burn area south of Jacob Lake, Arizona, July 2009

Photo by Frank R. Spellman

Instead of pointing the finger of blame toward those who made the decision to let Warm Fire burn, the emphasis should be on gathering lessons learned and making sure they are disseminated to those who might face a similar situation in the future. The rapid recovery of vegetation, the apparent ability for most species of wildlife, especially mule deer, to use the Warm Fire burn areas and high-quality habitat provided during post-fire recovery suggest that, in this instance, fire actually enhanced the habitat for most plants and animals in the Kaibab Plateau.

From Moonscape to Landscape

Many of us who were around during the 1960s remember with pride when the U.S. landed a man on the moon in July of 1969. We also took pride in the other lunar landings and the voyages of the Apollo 11–17 space vehicles until the program ended in December 1972. To say that landing a man on the moon was one of humankind's greatest all-time achievements is to make a gross understatement.

Most of us who were around during that heady timeframe and watched the television reports of the lunar landings have fond memories of the spectacular lunar-landing accomplishments. However, there is a fly in the lunar-ointment, so to speak. Shortly after the moon landings, different moon landing conspiracy theories claimed that some or all elements of the Apollo project and the associated moon landings were staged by NASA, the media, and members of other organizations. After the Apollo program ended, a number of related accounts espousing a belief that landings were faked in some fashion have been advanced by various individuals and groups.

One of the major points of contention is if the landings did not take place on the lunar surface, where did they take place? Where else on Earth, other than a humongous Hollywood movie set, could a fake lunar landing be staged? Where could such an event be portrayed to make it look real—to make it appear that the landing did indeed take place on the surface of the moon?

Well, there is one such place that we know of. It is a physical site located in the United States, one that is perfect in the minds of the conspiracy theorists; it is located in southeastern Idaho. The place? Craters of the Moon National Monument and Preserve. It is located in the Snake River Plain in central Idaho between two small cities, at an average elevation of 5,900 feet (1,800 m) above sea level (USGS 2010). The protected area features volcanic landscape and represents one to the best preserved flood basalt areas in the continental United States.

The Monument and Preserve encompasses three major lava fields and about 400 square miles (1,036 km²) of sagebrush steppe grasslands to cover a total area of more than 1,000 square miles (2800 km²). All three lava fields lie along the Great Rift of Idaho, with some of the best examples of open rift cracks in the world, including the deepest known on Earth at 800 feet (240 m). The Monument and Preserve contains excellent examples of almost every variety of basaltic lava, as well as lava tubes and many other volcanic features.

During our many hours of hiking several miles within the Craters of the Moon National Monument and Preserve, two salient points became readily clear to us. In the first place, as shown in Figures 7a–e, the doubting Thomases, those moon-landing conspiracy theorists at least got one small part of their outrageous argument correct: Craters of the Moon National Monument and Preserve is indeed moonscape-like terrain. The various types or forms of lava flows (basaltic and rhyolitic) and formations (cinder cones, spatter cones, and tunnels) make this conclusion rather obvious.

FIGURE 7a.   Cinder dome. Craters of the Moon National Monument, Idaho
Photo by Frank R. Spellman

FIGURE 7b.   Lava field. Craters of the Moon National Monument, Idaho
Photo by Frank R. Spellman

FIGURE 7c.   Squeeze up lava field. Craters of the Moon National Monument, Idaho
Photo by Frank R. Spellman

FIGURE 7d.   Ash mounds. Craters of the Moon National Monument, Idaho
Photo by Frank R. Spellman

FIGURE 7e.   Lava tube/cave. Craters of the Moon National Monument, Idaho
Photo by Frank R. Spellman

Did You Know?

"The Devil's Vomit" is how one Oregon-bound pioneer described his encounter with Craters of the Moon. Hundreds of pioneers traveled through there on the Goodale's Cutoff (a spur of the Oregon Trail) in the 1850's and 1860's.

For us the other significant eye-catcher is the plant life. Although much of the Craters of the Moon National Monument is covered by young lava flows (around 2000 years old), it supports a surprising diversity of plant communities. Uniquely adapted plants and a variety of abundant vegetation can be found here (see Figures 8a–f).

Although we could not count and catalog each and every one of them, there are over 660 different types of plants (taxa) that have been identified in the Monument. Vegetation in different successional stages (*Note:* Plant succession is discussed in greater detail later in the text.) can be found on lava flows, in cinder areas, on kipukas (areas that have older vegetation than the surrounding

FIGURE 8a.   Plant life and scrub trees taking root in lava bed at Craters of the Moon National Monument, Idaho

Photo by Frank R. Spellman

FIGURE 8b.   Sagebrush and grass growing in lava rock fragments. Craters of the Moon National Monument, Idaho

Photo by Frank R. Spellman

FIGURE 8c.   Grass spouting in cinder-ash area of Craters of the Moon National Monument, Idaho

Photo by Frank R. Spellman

FIGURE 8d.   Plant life taking root in fresh (2,000 year old) lava flow. Craters of the Moon National Monument, Idaho

Photo by Frank R. Spellman

FIGURE 8e.   New and old plant and tree life in lava flow area of Craters of the Moon National Monument, Idaho

Photo by Frank R. Spellman

FIGURE 8f.   Single flower bunch growing on cinder cone surface in Craters of the Moon National Monument, Idaho

Photo by Frank R. Spellman

areas; this is the result of their having had lava flows cover the surrounding areas, missing the kipukas and sparing its vegetation), and in mountain and riparian areas. Many unique plants have developed ways to adapt and to survive the extreme conditions they face here.

The National Park Service (2010) points out that the types and density of vegetation vary considerably and depend on such factors as geology, availability of soil and water, aspect (aspect generally refers to the horizontal direction to which a slope faces, which can make a significant influence on its microclimate; Bennie et al. 2006), air temperature, and exposure to wind. The density of vegetation on lava flows depends primarily on the amount of soil available. Although lava flow surfaces support only lichens, vascular plants are able to grow in depressions on those surfaces. When basalt rock is very young, the only soil available is whatever blows into cracks and fractures. As soil develops within these cracks over time, vegetation can begin to grow. The depth of crevices, cracks, and depressions fixes the amount of soil and moisture that can be held. The size of the crack also determines the types of plants that will grow and what degree of protection they will have from harsh climatic conditions such as extreme air temperatures and exposures to high winds.

Cinder cones support three different plant communities: shrub, limber pine (*Pinus flexilis*) and/or juniper pine, and cinder garden (characterized by a cinder surface and lower total plant cover). These communities are determined primarily by aspect and by succession.

Some kipukas in the monument have been protected from alteration by areas of rough lava and represent rare examples of undisturbed shrub steppe habitats. Dominant kipukas vegetation includes three-tip sagebrush/Idaho fescue, big sagebrush, bluebunch wheatgrass, and needlegrass.

Did You Know?

Moonscape? . . . At first glance the landscape of Craters of the Moon appears to be devoid of life. Look deeper and you will observe a rich diversity of life including more than 660 types of plants and over 280 animal species.

Nature has stepped in with its many tools in hand and is changing moonscape to landscape to habitat for many animals.

Craters of the Moon National Monument & Preserve is, at the time of this writing (2010), the home of songbirds, rodents, marmots, chipmunks, ground squirrels, mule deer, the Great Basin pocket mouse, coyotes, porcupines, mountain cottontails, jackrabbits, and several others. This high desert ecosystem is continually changing. Weather, climatic shifts, and geologic processes, including volcanic eruptions, will continue to shape the landscape as they have for thousands of years. Human settlement of the Snake River Plain brought about rapid environmental change, and as a result native animals such as big horn sheep, bison, and the grizzly bear have disappeared from the area. More recently, human-caused factors such as air pollutants and invasive introduced species have had an increasing impact on nature resources within the remote lava fields of the Great Rift. Fortunately, the remote and undeveloped landscape of this area provides an ideal place to study how various environmental factors affect desert ecosystems, and to help people predict the changes that might take place in the future.

"If the mountain goes, I'm going with it."[1]

As owner and proprietor of the Lodge, he knew the special place the surrounding area held in the hearts and minds of those who called it "home." In

---

1. Adapted from U.S. Forest Service *A Visitor's Guide to Mount St. Helens National Volcanic Monument* 2010. Retrieved from www.fs.fed.us/gpnt/mshnvm.

the past, the Cowlitz and Klickitat Native Americans lived around the area and neighboring mountains in the summer. Medicinal plants, roots, beargrass, berries, fish, and game were abundant in the mountains. Huckleberries grew in old burns and were heavily harvested. Many Native American children spent long hours chasing birds away from drying berries. Huckleberry season meant days of work followed by evenings of reminiscing with friends and relatives only seen at these special times of the year.

Euro-Americans initially came to try their hand at trapping, mining, and logging, but found the real wealth of the area was in the landscape itself. Like nourishment for the soul, days, weeks, or entire summers on the lake's shore refreshed youth and adults weary of war, the Great Depression, or the fast pace of city life. Who remembers the first look at the area as they cruised the lake or noted the passing of the Forest Service boat? How many exulted in their ascent to the summit or rambled joyfully through the Mount Margaret backcountry? Which campers wrote a letter home to gain their way into Sunday dinner at the YMCA Camp, or enjoyed a home-baked pie at Harmony Lodge? This was life at Spirit Lake and Mt. St. Helens that he, Harry R. Truman, his 16 cats, and 56 other souls knew before they perished May 18, 1980, when nature brought the Mountain to a short-lived but violent expression of its inexorable dynamism—a release of 24 megatons of thermal energy (7 by blast, the rest through release of heat; USGS 1997). In one of his final interviews with the press, Harry R. Truman stated that "if the mountain goes, I'm going with it" . . . and he did; he is forever entombed in the place he knew and loved more than life itself, covered by more than 150 feet of volcanic rubble from the 1980 Mt. St. Helens blast.

Personal Reflections: Crisis and Response[2]

On July 1, 2010, while standing at the top of Johnston Ridge Observation Point (named for geologist David A. Johnston, who died when the Volcano erupted), my backpack at foot, I looked out at a panoramic view of the north face of Mt. St. Helens' crater (see Figures 9a–b), the pumice plain and a large portion of the blast zone (see Figure 9c), as well as a partial view of Spirit Lake. To my right I looked westward toward the north branch of the Toutle River valley. In all directions I could see, except for the volcano itself, that many things were different. For instance, the physical landforms had changed in dramatic fashion from that 1984 May 18th morning when the avalanche of

---

2. This section is based on the research, recollections, and reflections of co-author Frank R. Spellman, who studied the aftermath of the Mt. St. Helens eruption during visits and hikes throughout the blast area in 1980, 1988, 1990, 1996, 2002, and 2010. Many of the recovery observations are also based on data contained in U.S. Army Corp of Engineers, Portland Oregon, *Mount St. Helens Recovery* (1990).

rock, mud, and ice released by the earthquake roared down the mountain, turned west and surged 17 miles down the North Fork Toutle River valley— one of the largest landslides in history. Another part of the slide pushed north across the valley, overtopped the ridge (the one I was now standing on) and flowed down South Coldwater canyon. The eastern part rammed into Spirit Lake, raising the lake level about 200 feet and blocking its outlet with debris hundreds of feet deep. Massive mudflows choked the Toutle and Cowlitz rivers and brought shipping to a halt on the Columbia River. The Corps, responsible for navigation and flood control, quickly became involved and raised levees and roads and removed debris and cleared blocked creeks between Castle Rock and Longview. Later, the Army Corps of Engineers built a Sediment Retention Structure (SRS), an 1800 ft long, 184 ft high, and 70 ft wide dam. A conventional dam can stop the flow of water completely. The SRS does not stop water; its works with nature to stop sediment, slowing down the flow of water so that the sediment drops out. Instead of traveling downstream and settling in the river channels where it can cause flooding and impede navigation, sediment builds behind the SRS in a single large manageable deposit.

FIGURE 9a.   North face of Mt. St. Helens showing the crater's lateral blast area
Photo by Frank R. Spellman, taken July 1, 2010, at the summit of Johnston Observation Point

FIGURE 9b.   Aerial view of inside of Mt. St. Helen's crater

Aerial photo taken by Frank R. Spellman from helicopter

FIGURE 9c.   Pumice plain in front of lateral blast zone of Mt. St. Helens

Photo taken at summit of Johnston Observation Point by Frank R. Spellman

On that July day in 2010, I also noticed that the land at the foot of the volcano and for as far as I could see, in all directions, was growing green again
(see Figure 9d). Shade, a rarity after the blast, was apparent now, along with
deer and elk herds that had returned, while above, young eagles and other
birds soared.

Another significant change was the blown down trees. "Where are they?" I
asked myself. Where did they go? One of the most enduring memories I have,
at least for a few years after the eruption, is the vast maze of trees, about 4
billion board feet (9.4 million m³) of timber, that were strewn across the landscape, staked like giant jackstraws by the powerful lateral blast (see Figures
9e–f). Today's blown down forest is difficult to see for reasons only nature
could bring about. The weight of the snow packs each year pushes the trunks
into the ground and soaks the trees, while the summer sun bakes them. The
intense wetting and drying has led to the rapid deterioration of the blown
down trees. But there is more, nature has many tools at her disposal. The old
saying that nothing is wasted in nature is true, as evidenced by the disappearing blown down forest. The question is, are the trees really gone, or have they
been transformed? Well, except for a lingering trunk and fallen deadfall here

FIGURE 9d.   New growth timber and shrubs in blast zone of North Toutle River valley
Aerial photo by Frank R. Spellman.

and there, their original physical presence is mostly gone, transformed to something else. Along with its wetting and drying tools, nature employed a veritable army of **decomposers** including **bacteria, molds,** and **fungi.** Collectively these provided food for themselves by extracting chemicals from the dead trees, using these to produce energy. The decomposers then produced waste of their own. In turn, this also decomposes, eventually returning nutrients to the soil. These nutrients are then taken up by the roots of living plants enabling them to grow and develop, so that organic material is naturally cycled. Again, virtually nothing goes to waste in nature. This is an endless cycle; more simply and importantly, it is nature's way.

Not all is decay and decomposition around Mt. St. Helens and the Toutle River valley. Windblown seeds caught by the moist, nutrient-rich tree trunks nourished seedlings and have created a veil of low-lying grasses and shrubs. The array of plants and animals that survived the eruption also made it easier for other life to colonize decimated areas. Today, the blown-down forest is hidden by an odd assortment of silver fir, huckleberry, and other pre-eruption forest survivors, beside sun-loving weeds and shrubs.

FIGURE 9e.   Demarcation line where blast zone and safe zone meet 14 miles from Mt. St. Helens
Aerial photo by Frank R. Spellman

FIGURE 9f.   Blast tree remnants near summit of Johnston Observation Point at Mt. St. Helens National Monument

Photo by Frank R. Spellman

One of the surprises, even for this scientist who has studied such phenomena for years, is the Mt. St. Helens landslide deposit. This area is living testimony to nature's ability to recreate habitats. The distinctive irregular surface of the landslide entrapped runoff from rain and snowmelt, forming almost 150 new ponds and wetlands. The massive landslide also blocked two creeks, creating Coldwater and Castle lakes. These new habitats powered the rapid resurgence of an array of birds, amphibians, insects, mammals, and plants—with the accompanying reborn ecosystem teeming with flying, swimming, chewing, croaking, chorusing, and singing life. The lush Sitka and red alder forests that spread from these new aquatic epicenters have, as mentioned, created a rare commodity at Mt. St. Helens, shade. As I stood there at Johnston Observation Point, at the trailheads of Eruption hiking trail #201 and Boundary hiking trail #1, it was obvious that the landslide hosts the most diverse and productive ecosystem in the National Volcanic Monument.

The last feature to catch my eye that day was off to the southeast, the target area of my subsequent hike, Spirit Lake (see Figure 9g). The eruption's impact to Spirit Lake was so severe that 30 years later many still believe that the lake was destroyed. In reality all life was destroyed, but Spirit Lake's surface area doubled! The lake's once cold, clear waters were transformed into primordial soup, a warm nutrient-rich broth of heavy metals and organic matter covered by floating log mat. Bacterial populations exploded in these ideal conditions and began to cleanse the lake. Within five years Spirit Lake had gone though a remarkable metamorphosis, returning to near pre-eruption conditions. Eight years later, rainbow trout were discovered, although scientists suspect they were illegally introduced. Today, Spirit Lake is really two lakes in one. Much of the lake is deep, cold, and biologically unproductive like many high elevation lakes in the Cascades, and, conversely, the shallow warmer waters in the southern and northwestern reaches host vibrant ecosystems.

That is how I viewed Mt. St. Helens that day. I remember as I picked up my backpack and strapped it on to start my downhill trek from Johnston Observation Point to Spirit Lake the woman observer who turned to me and said: "Aren't you a bit afraid to hike down there closer to that mountain . . . what if it blows up again?" I paused for a moment and then I simply replied with that refrain that was lodged in my thought-process at that moment of time: "Well, if the mountain goes, I'm going with it."

FIGURE 9g.   Spirit Lake as it appeared in July 2010
Photo by Frank R. Spellman

Glacial Lake Missoula and Dry Falls

Nearly twenty thousand years ago, during the last Ice Age, glaciers moved south and one ice sheet plugged the Clark Fork of the Columbia River, which kept water from being drained from Montana. Consequently, a significant portion of western Montana flooded, forming gigantic Lake Missoula. Eventually, enough pressure accumulated on the ice dam that it gave way. It is generally accepted that this process of ice-damming of the Clark Fork, refilling of Lake Missoula, and subsequent cataclysmic flooding happened dozens of times over the years of the last Ice Age (Alt 2001; Bjornstad 2006).

In regard to these cataclysmic flooding events, in central Washington on the opposite side of the Upper Grand Coulee from the Columbia River, and at the head of the Lower Grand Coulee, exists a three and a half mile-long scalloped precipice known as Dry Falls (see Figures 10 and 11). Ten times the size of Niagara, Dry Falls is thought to be the greatest known waterfall that ever existed. Geologists speculate that during the last ice age catastrophic flooding channeled water at 65 miles per hour through the Upper Grand Coulee and over this 400-foot (120 m) rock face. By comparison, Niagara, one mile wide with a drop of only 165 feet, would be dwarfed by Dry Falls. At this time, it is

FIGURE 10.   Dry Falls, Washington
Photo by Frank R. Spellman

FIGURE 11.   Dry Falls, Washington
Photo by Frank R. Spellman

estimated that the flow of the falls was ten times the current of all the rivers in the world combined. Today, flow rate is extinct; the falls are desiccated.

What Is Nature?

In attempting to define nature, some would choose actions over words. Thus, they could use our examples discussed and illustrated in the preceding sections. In those sections, we described nature at work, doing its natural, normal thing; that is, nature using a few of her tools, volcanism, lava flows, fire, and flooding, to sculpt and change the landscape of Earth. More importantly, in describing how nature uses these and other tools (earthquakes, droughts, extreme heat and cold, hurricanes, landslides, and so forth) to tear down and reshape or build up a landscape, we also pointed out that with time nature is the great restorer. For example, within a short time span, nature begins work converting a freshly strewn lava field with the aid of wind-carried soil particles and deposits in tiny crevices where just enough water and accompanying temperature changes/ranges produce primary growth that, again, with enough time can turn a barren lava-strewn field into a gigantic grassland steppe of tree-filled forest. This is nature's way; it's a normal cycle.

In our opinion, using actions over words to describe nature, to explain what nature is, to explain what it is all about, is the easy way out. To make our point, consider the word "nature." Describe it, if you can. Can you? Simply, what is nature? When we asked more than 100 individuals (college students) to tell us what they thought nature was, we ended up with a lot of blank looks and eventually more than 100 different answers. Why?

We truly did not understand the dilemma and frustration in trying to define or describe nature until we tried to do so ourselves. We quickly understood that nature is "something" that is regarded as reified in the sense that nature describes that "something" that is clearly in the abstract as a material or concrete thing. What? Ok, using common-speak about nature, it is probably better to consider our place in it; this is best described by T. E. Lawrence in the following: "The living knew themselves just sentient puppets on God's stage." Still confused, shaking your head, building a headache? Well, remember that the term nature can be used as a philosophical concept, as the personification of nature as a maternal figure, and can be used to describe innate behavior (i.e., the character or essence of a human or another living organism). Consider some of our attempts to define nature:

- Nature is the sum of all natural forces.
- Nature is equivalent to the natural world.
- Nature is a phenomenon of the natural world.
- Nature is life.
- Nature is all matter and energy of which all these things are composed.

In reviewing the definitions we came up with, we just scratched our heads and almost gave up on trying to define it. But then we turned to those who know or knew nature better than any scientist ever could, ever would, the poets. Finally, we settled on Ralph Waldo Emerson's definition: "Nature is a mutable cloud which is always and never the same."

References and Recommended Reading

Alexander, David. "Nature's Impartiality, Man's Inhumanity: Reflections on Terrorism and World Crisis in a Context of Historical Disaster." *The Journal of Disaster Studies, Policy and Management* 26, no. 1, March 2002, 1–9.

Alt, D., 2001. *Glacial Lake Missoula & Its Humongous Floods*. Missoula, MT: Mountain Press Publishing Company.

Anderson, D.L. 1989. *Theory of the Earth*. Boston: Blackwell Publications.

Bankoff, G., Frerks, G., and Hilhorts, D., eds. 2003. *Mapping Vulnerability: Disasters, Development and People*. London: Earthscan Publishers Ltd.

Bennie, J., Hill, M.O., Baxter, R., and Huntley, B. 2006. Influence of slope and aspect on long-term vegetation change in British Chalk Grasslands. *Journal of Ecology* 94, no. 2: 355–368.

Berkeley, G. 1998. *A Treatise Concerning the Principles of Human Knowledge*, 1734, section 45. New York: Oxford University Press.

Bjornstad, B. 2006. *On the Trail of the Ice Age Floods: A Geological Guide to the Mid-Columbia Basin*. Sandpoint, ID: Keokee Books.

Burton, I., Kates, R., and White, G. *The Environment as Hazard*, 2nd ed. New York: Guilford Press, 1993.

Cannon, Terry. "Vulnerability Analysis and the Explanation of 'Natural' Disasters." Chapter 2 (pp. 13–30) in *Disasters, Development and Environment*, A. Varley, ed. London: Wiley, 1994.

Dombrowsky, W.R. "Again and Again—Is a Disaster What We Call a 'Disaster'." Chapter 3 in *What Is a Disaster*, E.L. Quarantelli, ed. London: Routledge. 1998.

Duran, L.R. (Center for Co-ordination for Disaster Prevention in Central America). 1999. "The Social Impact of Disaster." Pp. 16–18 in Ingleton.

Dynes, R.R. 1993. "Disaster Reduction: The Importance of Adequate Assumptions about Social Organization." *Sociological Spectrum* 13: 175–192.

Dynes, R.R. 1997. *The Lisbon Earthquake in 1755: Contested Meanings In The First Modern Disaster*. Newark: University of Delaware, Department of Sociology and Criminal Justice, Disaster Research Center, Preliminary Paper 255. Downloadable from: http://www.udel.edu/DRC/preliminary/255.pdf (32 pages).

Erikson, Kai. *A New Species of Trouble—The Human Experience of Modern Disasters*. New York and London: W.W. Norton & Company, 1989.

Flint, R.R., and Skinner, B.J. 1977. *Physical Geology*. New York: John Wiley & Sons.

Gilbert, C. 1998. "Studying Disaster—Changes in the Main Conceptual Tools." Chapter 2 in *What Is a Disaster?* E.L. Quarantelli, ed. London and New York: Routledge.

Hasten, M. "Foster Says State Saved By 'Divine Intervention'." *Lafayette* (LA) *Advertiser*. October 4, 2002.

Hooke, W. (National Science and Technology Council). 1999. "Progress and Challenges in Reducing Losses From Natural Disasters," pp. 280–283 in Ingleton.

Horlick-Jones, T., and Peters, G. "Measuring Disaster Trends Part One: Some Observations on the Bradford Fatality Scale." *Disaster Management* 3, no. 3, 1991: 144–148.

Huron, D. 2009. *Frisson: Thrills for Chills*. PDF Document (http://www.cogsci.msu.edu/DSS/ 2008-2009/Huron/HuronFrisson.pdf).

ISDR (International Strategy for Disaster Reduction). 2002. *Living with Risk—A Global Review of Disaster Reduction Initiatives* (Preliminary Version). Geneva, Switzerland: United Nations ISDR. Downloaded from: http://wwwunisdr.org/unisdr/Globalreport/htm.

Martinet, Michael E. "Book Review: Acts of God." Pp. 9–10 in *IAEM Bulletin* 19, no. 4, April 2002.

McEntire, David A. (forthcoming). *Disaster Response Operations and Management*. FEMA Emergency Management Higher Education Project Instructor Guide. FEMA: Emmitsburg, MD. Downloadable from (when available): http://training. fema.gov/EMIWeb/edu/aem_courses.htm.

Mileti, Dennis. 2001. "Risk Assessment." Presentation given April 1, 2001, *National Symposium on Mitigating Severe Weather Impacts—Design for Disaster Reduction*, Tulsa Convention Center, Tulsa, OK, March 31–April 5.

National Park Service (2010). *Craters of the Moon National Monument & Preserve: Plants*. Retrieved 09/28/10 @ http://www.nps.gov/crmo/naturescience/plants.htm.

NIFC 2010. *Wildland Fire Use*. Retrieved 09/26/2010 @ http://www.nifc.gov/fuels/ overview /fireTreament.html.

Press, F., and Siever, R. 1974. *Earth*. San Francisco: W.H. Freeman.

Quarantelli, E.L. 1987. "What Should We Study? Questions and Suggestions for Researchers About the Concept of Disasters." *International Journal of Mass Emergencies and Disasters* 5, no. 1 (March): 7–32.

Scott, K.M., and Balance, J.W. 1993. *History of Landslides and Debris Flows at Mount Rainier*. Open-File Report 93–111. USGS.

Smith, Keith. 1996. *Environmental Hazards: Assessing Risk and Reducing Disaster*. London: Routledge.

Smith, P.J., ed. 1986. *The Earth*. New York: Macmillan.

Steinberg, T. 2000. *Acts of God—The Unnatural History of Natural Disaster in America*. Oxford: Oxford University Press.

Tierney, K. J., Lindell, M.K., and Perry, R.W. 2001. *Facing the Unexpected: Disaster Preparedness and Response in the United States*. Washington, DC: Joseph Henry Press.

Tobin, G. A., and Montz, B.E.. 1997. *Natural Hazards: Explanation and Integration*. New York and London: Guilford Press.

Toft, B. 1992. "The Failure of Hindsight." *Disaster Prevention and Management* 1, no. 3: 48–60.

Twigg, J. (Benfield Greg Hazard Research Centre, University College London). 2001. "Physician, Heal Thyself? The Politics of Disaster Mitigation." Disaster Management Working Paper #1. http://bghrc.com.

USGS. 1997. Summary of Mt St Helens Eruption. Retrieved 09/29/10 @ http://vulcan. wr.usgs.gov/Volcanoes/MSH/Publications/FS070-97/framework.html.

USGS. 2008. Mount Rainier—Learning to Live with Volcanic Risk. Retrieved from (http://pubs.usgs.gov/fs/2008/30627) *Fact Sheet 034-02*.

USGS. 2010. *Craters of the Moon Volcanic Field, Idaho*. Retrieved 09/27/10 @ http:// Vulcan.wr. USGS.gov/Volcanoes/Idaho/CratersMoon/description_craters_moon.html.

Von Kotze, A. "A New Concept of Risk?" Pp. 33–40 in *Risk, Sustainable Development and Disasters*, A. Holloway, ed. Cape Town, South Africa: Periperi Publications, 1999.

White, G.F., Kates, R.W., and Burton, I. "Knowing Better and Losing Even More: The Use of Knowledge in Hazards Management." *Environmental Hazards*, 3, nos. 3–4, (September/December 2001): 81–92.

Wisner, B. "Same Old Story." *Guardian*, February 1, 2001.

Wisner, B., Blaikie, P., Cannon, T., and Davis, I. 2004. *At Risk—Natural Hazards, People's Vulnerability and Disasters*. Wiltshire: Routledge.

Witt, J. Lee. 2001 (April 3). Keynote Presentation, Session III, *National Symposium on Mitigating Severe Weather Impacts—Design for Disaster Reduction*, Tulsa Convention Center, Tulsa OK, March 31 to April 4, 2001.

Yeats, R.S. 2001. *Living With Earthquakes in California: A Survivor's Guide*. Corvallis: Oregon State University.

Zellers, C. 2010. Mt. Rainier trumps Mount St. Helens in Danger Level. Accessed 09/24/10 @ http://www.emporia.edu/earthsci.student/zellers1/rainir.htm).

II

EARTH'S STRUCTURE

Adapt or perish, now as ever, is Nature's inexorable imperative.

—H.G. Wells, *Mind at the End of Its Tether*, 1945

2
Composition of Earth

How strange that Nature does not knock, and yet does not intrude!
—Emily Dickinson (1830–1886), *Letter to Mrs. J.S. Cooper*

Every soil-atom seems to yield enthusiastic obedience to law—bowlders and mud grains moving to music as harmoniously as the far-whiling planets.

—John Muir (1874)

Planet Earth

EARTH (or, "the earth") is the only planet presently known to support life. The earth has a radius of about 6370 km, although it is about 22 km larger at equator than at poles. The circumference of the earth is about 24,874 miles, and the surface area comprises roughly 197 million square miles. About 29% are surface lands. The remaining 71% of the earth's surface is covered by water.

The earth, like the other planets in our solar system, revolves around the sun within its own orbit and period of revolution. The earth also rotates on its own axis. The earth rotates from west to east and makes one complete rotation each day. It is this rotating motion that gives use to the alternating periods of daylight and darkness which we know as day and night.

The earth also precesses (or wobbles) as it rotates on its axis, much as a top wobbles as it spins. This single wobble has to do with the fact that the earth's

axis is tilted at an angle of 23½ degrees. The tilting of the axis is also responsible for the seasons. Over the years, various hypotheses have been put forward in attempts to determine what is the force, or excitation mechanism that
propels the wobble, such as atmospheric phenomena, continental water storage (changes in snow cover, river runoff, lake levels, or reservoir capacity),
and interaction at the boundary of Earth's core and its surrounding mantle,
and earthquakes. In an explanation by NASA (2000), the principal cause of
the wobble is fluctuating pressure on the bottom of the ocean, caused by temperature and salinity changes and wind-driven changes in the circulation of
the oceans.

The earth revolves around the sun in a slightly elliptical orbit approximately
once every 365¼ days. During this solar year, the earth travels at a speed of more
than 60,000 miles per hour, and on the average it remains about 93 million
miles from the sun.

Earth Products, Beauty, and Divisions

Without thinking about it (without any great level of geological knowledge),
there are some things about Earth and the natural environment that we all
know. We know about the products of soils which have been formed by
weathered rock, oil formed from the remains of prehistoric plants and animals, and the beauty and value of precious stones. These are the basics; they
represent only a small fraction of the useful materials with which the earth
provides us.

Think about how impossible it would be for modern industry to have developed as we know it today without other earth products. Mineral resources such
as coal, iron, lead, and petroleum derived from the earth have been made readily available through the application of basic geology, geological engineering,
Global Positioning System (GPS), and Generic Mapping Tools (GMT).

In addition to useful products, earth also provides us with areas of excep-
tional beauty. One need not be a geologist or naturalist to marvel at the breathtaking vastness of the Grand Canyon, the enriching warmth of the sandstone
formations and the very color of the earth at Zion National Park, Bryce Canyon,
the Arches National Park, Monument Valley, the mystery of Luray Caverns,
and the natural wonders of Yosemite or Yellowstone. All of these and many
more places of beauty and intrigue are the results of geologic processes that are
dynamic (still at work today). They are the same geologic processes which began
to shape the earth more than 4.54 billion years ago!

Earth's materials and beauty are not unique but instead are universal to any
of its four divisions. These divisions include (or inter-connected "geo-spheres")

the **atmosphere**, a gaseous envelope surrounding the earth; the **hydrosphere**, the waters filling the depressions and covering almost 71% of the earth; and the **lithosphere**, the solid part of the earth which underlies the atmosphere and hydrosphere; and the biosphere, composed of all living organisms.

In this section, we concentrate specifically on Nature's influence and/or impact related to the lithosphere. The atmosphere, hydrosphere, and biosphere divisions are individually addressed and discussed in the subsequent sections of the text.

Earth Geology

It is important to point out that geology is not only the study of earth as we see it today, but the history of the earth as it has evolved to its present condition. In attempting to pin down a definition of geology (to explain what it is and what it is about), consider the definition provided by Press and Siever (2001):

"Earth is a unique place, home to more than a million life forms, including ourselves. No other planet yet discovered has the same delicate balance necessary to sustain life. Geology is the science that studies the Earth—how it was born, how it evolved, how it works, and how we can help it."

Did You Know?

- The earth has evolved (changed) throughout its history, and will continue to evolve.
- The earth is about 4.6 billion years old, human beings have been around for only the past 2 million years.

The term geology is derived from the Greek geo, "earth," plus logos, "discourse or study of." Paraphrasing Press and Siever's (2001) definition of geology in simplistic terms we can define it as the science which deals with the study of the origin, structure, and history of the earth and its inhabitants as recorded in the rocks.

Geology consists of the sciences of mathematics, physics (geophysics and seismology), and chemistry (petrology and geochemistry); these are the sciences that set the general principles for all the other sciences. Additionally, geology is composed of or interrelated to the sciences that describe the great systems that make up the universe: astronomy (planetary geology) and biology (paleontology). Some geologists, the specialists, deal with very narrow and specialized parts of geology including mineralogy, meteorology, botany, zoology, and others. Geological science also includes exposure to the principles of sociology and psychology. In a nutshell and based on personal experience, it can be said with some certainty that the practicing geologist, whether a specialist or not, should and must be a generalist. That is, to be effective in practice, the geologist must have a well-rounded exposure to and some knowledge of just about everything related to the other scientific disciplines.

Structure of Earth

Earth is made up of three main compositional layers: crust, mantle, and core. The crust has variable thickness and composition: Continental crust is 10–50 km thick while the oceanic crust is 8–10 km thick. The **elements** silicon, oxygen, aluminum, and iron make up the earth's crust. Like the shell of an egg, the earth's crust is brittle and can break. Earth's continental crust is 35 kilometers thick. The oceanic crust is 7 kilometers thick.

Based on seismic (earthquake) waves that pass through the earth, we know that below the crust is the mantle, a dense, hot layer of semi-solid (plastic-like liquid) rock approximately 2,900 km thick. The mantle, which contains silicon, oxygen, aluminum, and more iron, magnesium, and calcium than the crust, is hotter and denser because temperature and pressure inside the earth increase with depth. According to USGS (1996), as a comparison, the mantle might be thought of as the white of a boiled egg. The 30 km thick transitional layer between the mantle and crust is called the Moho layer. The temperature at the top of the mantle is 870°C. The temperature at the bottom of the mantle is 2,200°C.

At the center of the earth lies the core, which is nearly twice as dense as the mantle because its composition is metallic [Iron (Fe)-Nickel (NI) alloy] rather than stony. Unlike the yolk of an egg however, the earth's core is actually made up of two distinct parts: a 2,200 km-thick liquid outer core and a 1,250 km-thick solid inner core. As the earth rotates, the liquid outer core spins, creating the earth's magnetic field.

Earth's Geological Processes

Earlier it was pointed out that geology is about materials—materials that make up the earth. The materials that make up the earth, of course, are mainly rocks (including dust, silt, sand, soil). Rocks in turn are composed of minerals and minerals are composed of atoms.

Earth processes are constantly acting upon and within the Earth to change it; to re-shape it. These processes are absolutely relentless forces of nature.

Examples of these on-going processes include formation of rocks; chemical cementation of sand grains together to form rock; construction of mountain ranges; and erosion of mountain ranges. These are internal processes that get their energy from the interior of the earth—most from radioactive decay (nuclear energy). Other examples of ongoing earth processes include those that are more apparent to us (external processes) because they occur relatively quickly and are visible. These include volcanic eruptions, dust storms, mudflows, and beach erosion. The energy source for these processes is solar and gravitational. It is important to point out that many of these processes are cyclical in nature. The two most important cyclical processes are the hydrologic (water) cycle (discussed in detail later in the text) and the rock cycle.

Rock Cycle (aka Geologic Cycle)

With time and changing conditions, the igneous, sedimentary, and metamorphic rocks of earth are subject to alteration by the processes of weathering (erosion), volcanism, and tectonism. Known as the **rock cycle** (see Figure 12), this series of events (or group of changes) represents a response to earth materials to various forms of energy. During the cycle, each rock type is altered or destroyed when it is forced out of its equilibrium conditions. As shown in Figure 12, most surface rocks started out as igneous rocks (rocks produced by crystallization from a liquid). When igneous rocks are exposed at the surface they are subject to weathering. Erosion moves particles into rivers and oceans where they are deposited to become sedimentary rocks. Sedimentary rocks can be buried or pushed to deeper levels in the earth, when changes in pressure and temperature cause them to become metamorphic rocks. At high temperatures metamorphic rocks may melt to become magmas. Magmas rise to the surface, crystallize to become igneous rocks, and the process starts over. Keep in mind that the cycles are not always completed, for there can be many short circuits along the way, as indicated in Figure 12.

Did You Know?

All rock that is on earth (except for meteorites) today is made of the same stuff as the rocks that dinosaurs and other ancient life forms walked, crawled, or swam over. While the stuff that rocks are made from stays the same, the rocks do not. Over millions of years, rocks are recycled into other rocks. Moving tectonic plates help to destroy and form many types of rocks (Jennifer Bergman, 2005).

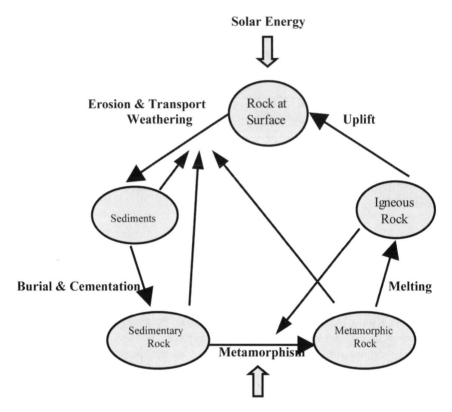

FIGURE 12. Rock cycle. If uninterrupted, it will continue completely around the outer margin of the diagram. However, as shown by the arrows, the cycle may be interrupted or "short-circuited."

The Lithosphere

The lithosphere is of prime importance to the geologist. This, the solid, inorganic, rocky crust portion of the earth, is composed of rocks and minerals which, in turn, comprise the continental masses and ocean basins. The rocks of the lithosphere are of three basic types, igneous, sedimentary, and metamorphic.

Soil

The soil is the sepulcher and the resurrection of all life in the past. The greater the sepulcher the greater the resurrection. The greater the resurrection the greater the growth. The life of yesterday seeks the earth today that that new life may come from it tomorrow. The soil is composed of stone flour and organic matter (humus) mixed; the greater the store of organic matter the greater the fertility.

—John Walton Spencer (1898)

Perhaps no term causes more confusion in communication between various groups of average persons, soil geologists, soil scientists, soil engineers, and earth scientists than the word soil. In simple terms, soil can be defined as the topmost layer of decomposed rock and organic matter which usually contains air, moisture, and nutrients, and can therefore support life. Most people would have little difficulty in understanding and accepting this simple definition. Then why are various groups confused on the exact meaning of the word soil? Quite simply, confusion reigns because soil is not simple—it is quite complex. In addition, the term soil has different meanings to different groups (like **pollution**, the exact definition of soil is a personal judgment call). Let's take a look at how some of these different groups view soil.

The average person: seldom gives soil a first or second thought. Why should they?—soil isn't that big a deal—that important—it doesn't impact their lives, pay their bills, or feed their bulldog, right?

Not exactly. Not directly.

The average person seldom thinks about soil as soil. He or she may think of soil in terms of dirt, but hardly ever as soil. Why is this? Having said the obvious about the confusion between soil and dirt, let's clear up this confusion.

First of all, soil is not dirt. Dirt is misplaced soil—soil where we don't want it, contaminating our hands and fingernails, clothes, automobiles, tracked in on the floor. Dirt we try to clean up, and to keep out of our living environments.

Secondly, soil is too special to be called dirt. Why? Because soil is mysterious, and whether we realize it or not, essential to our existence. Because we think of it as common, we relegate soil to an ignoble position. As our usual course of action, we degrade it, abuse it, throw it away, contaminate it, ignore it—we treat it like dirt, and only feces hold a more lowly status than it does. Soil deserves better.

Why?

Again, because soil is not dirt—how can it be? It is not filth, or grime, or squalor. Instead soil is clay, air, water, sand, loam, organic detritus of former life-forms (including humans), and most important, the amended fabric of Earth itself; if water is Earth's blood, and air is Earth's breath, then soil is its flesh and bone and marrow—simply put, soil is the substance that most life depends on.

Soil scientists: (or pedologists) are interested in soils as a medium for plant growth. Their focus is on the upper meter or so beneath the land surface (this is known as the weathering zone, which contains the organic-rich material that supports plant growth) directly above the unconsolidated parent material. Soil scientists have developed a classification system for soils based on the physical, chemical, and biological properties that can be observed and measured in the soil.

Soils engineers: are typically soil specialists who look at soil as a medium that can be excavated using tools. Soils engineers are not concerned with the plant-growing potential of a particular soil, but rather are concerned with a particular soil's ability to support a load. They attempt to determine (through examination and testing) a soil's particle size, particle-size distribution, and the plasticity of the soil.

Earth scientists: (or geologists) have a view that typically falls between pedologists and soils engineers—they are interested in soils and the weathering processes as past indicators of climatic conditions, and in relation to the geologic formation of useful materials ranging from clay deposits to metallic ores.

Would you like to gain new understanding of soil? Take yourself out to a plowed farm field somewhere, anywhere. Reach down and pick a handful of soil, and look at it—really look at it closely. What are you holding in your hand? Look at the two descriptions that follow, and you may gain a better understanding of what soil actually is and why it is critically important to us all.

1. A handful of soil is alive, a delicate living organism—as lively as an army of migrating caribou and as fascinating as a flock of egrets. Literally teeming with life of incomparable forms, soil deserves to be classified as an independent ecosystem, or more correctly stated as many ecosystems.
2. When we pick up a handful of soil, exposing Earth's stark bedrock surface, it should remind us (and maybe startle us) to the realization that without its thin living soil layer, Earth is a planet as lifeless as our own moon (Spellman 1998).

Soil Formation

We use soil for our daily needs (and very existence) but we do not sufficiently take account of its slow formation and fast loss. Simply, we take soil for granted. It's always been there—with the implied corollary that it will always be there—right? But where does soil come from?

In the classic 1941 text, *Factors of Soil Formation* by Jenny, he discussed what he called the factors of soil formation. Jenny pointed out that soil is formed as a result of the interaction of many factors, the most important of which are:

- **Climate**—key components of climate in soil formation are moisture and temperature.
- **Organisms**—vegetation, fauna, including microorganisms, and humans.
- **Relief**—major topographical features such as mountains, valleys, ridges, crests, sinks, plateaus, and floodplains.
- **Parent Material**—has a decisive effect on the properties of soils.
- **Time**—the value of a soil forming factor may change with time (e.g., climate change).

Did You Know?

Earthworms (annelids) are the most important of the soil forming fauna in temperate regions, being supported to a variable extent by the small arthropods and the larger burrowing animals (rabbits, moles). Earthworms are also important in tropical soils, but in general the activities of termites, ants, and beetles are of greater significance, particularly in the subhumid to semiarid savanna of Africa and Asia. Earthworms build up a stone-free layer at the soil surface, and organic matter that is accessible to microbial attack is then much greater (White, 1987).

Soil was formed, and in a never-ending process, it is still being formed. However, as mentioned, soil formation is a slow process—one at work over the course of millennia, as mountains are worn away to dust by employing the factors of soil formation listed above and affected by bare rock succession.

Any activity, human or natural, that exposes rock to air begins the process. Through the agents of physical and chemical weathering, through extremes of heat and cold, through storms and earthquake and entropy, bare rock is gradually worn away. As its exterior structures are exposed and weakened, plant life appears to speed the process along.

Lichens cover the bare rock first, growing on the rock's surface, etching it with mild acids and collecting a thin film of soil that is trapped against the rock and clings. This changes the conditions of growth so much that the lichens can no longer survive and are replaced by mosses.

The mosses establish themselves in the soil trapped and enriched by the lichens, and collect even more soil. They hold moisture to the surface of the rock, setting up another change in environmental conditions.

Well established mosses hold enough soil to allow herbaceous plant seeds to invade the rock. Grasses and small flowering plants move in, sending out fine root systems that hold more soil and moisture, and work their way into minute fissures in the rock's surface. More and more organisms join the increasingly complex community.

Weedy shrubs are the next invaders, with heavier root systems that find their way into every crevice. Each stage of succession affects the decay of the rock's surface and adds its own organic material to the mix. Over the course of time, mountains are worn away, eaten away to soil, as time, plants, weather, and extremes of weather work on them.

The parent material, the rock, becomes smaller and weaker as the years, decades, centuries, and millennia go by, creating the rich, varied, and valuable mineral resource we call soil.

References and Recommended Reading

Air Quality Criteria. 1968. Staff report, Subcommittee on Air and Water Pollution, Committee on Public Works, U.S. Senate, 94–411.

American Heritage Dictionary of the English Language, 4th ed. 2000. Boston: Houghton Mifflin.

Bergman, J. 2005. *Rocks and the Rock Cycle*. Accessed 05/23/08 @ http://www.windows.ucar.edu/ tour/ling=/earth/geology/rocks_intro.html.

Blue, J. 2007. Descriptor Terms. *Gazetteer of Planetary Nomenclature*. USGS. http://planetarynames, wr.usgs.gov/jsp/apped5.jsp.

Campbell, N.A. 2004. *Biology: Concepts & Connections*, 4th CD-Rom ed. Benjamin-Cummings Publishing Company.

Goshorn, D. 2006. Proceedings—DELMARVA Coastal Bays Conference III: Tri-State Approaches to Preserving Aquatic Resources. USEPA.

Huxley, T.H. 1876. *Science & Education, Volume III, Collected Essays*. New York: D. Appleton & Company.

Jenny, H. 1941. *Factors of Soil Formation*. New York: McGraw-Hill.

Jones, A.M. 1997. *Environmental Biology*. New York: Routledge.

Keeton, W.T. 1996. *Biological Science*. R.S. Means Company.

King, R.M. 2003. *Biology Made Simple*. New York: Broadway Books.

Koch, R. 1882. *Uber die Atiologie der Tuberkulose*. In *Verhandlungen des Knogresses fur Innere Medizin*. Wiesbaden: Erster Kongress.

Koch, R. 1884. *Mitt Kaiser Gesundh* 2: 1–88.

Koch R. 1893. *J. Hyg. Inf.* 14: 319–333.

Larsson, K.A. 1993. Prediction of the pollen season with a cumulated activity method. *Grana*, 32: 111–114.

Med Net. 2006. *Definition of Koch's Postulates*. Medicine Net.com.

NASA. 2000. *A Mystery of Earth's Wobble Solved: It's the Ocean*. Accessed 05/24/08 @ http://www.jpl.nasa.gov/releases/2000/chandlerwobble.html.

Press, R., and Siever, F. 2001. *Earth*, 3rd ed. New York: W.H. Freeman.

SGM. 2006. *The Scientific Method, Fish Health and Pfiesteria.* University of Maryland: NOAA.

Spellman, F.R. 1998. *Environmental Science and Technology: Concepts and Applications.* Lancaster, PA: Technomic Publishing Company.

Spellman, F.R. 2008. *Geology for Non-Geologists.* Lanham, MD: Government Institutes.

Spellman, F.R. 2008. *The Science of Air: Concepts and Applications.* Boca Raton, FL: CRC Press.

Spellman, F.R., and Whiting, N.E. 2006. *Environmental Science and Technology,* 2nd ed. Lanham, MD: Government Institutes.

Spieksma, F.T. 1991. Aerobiology in the Nineties: Aerobiology and pollinosis. *International Aerobiology Newsletter,* 34: 1–5.

USEPA. 2006. *What is the scientific method?* Accessed @ http://www.epa.gov /maia/ html/ scientific.html.

USGS. 1996. This dynamic earth: The Story of plate tectonics. pubs.usgs.gov/gip/ dynamic/

White, R.E. 1987. *Introduction to the Principles and Practices of Soil Science.* Palo Alto, CA: Blackwell Scientific Publications.

3

Reshaping Earth

One touch of nature makes the whole world kin.

—William Shakespeare (1564–1616), *Troilus and Cressida*

Introduction

BOTH ENDOGENIC AND EXOGENIC FORCES work to reshape the surface of the earth. *Endogenic* forces are internal forces that are powered by Earth's internal heat engine—for example, faulting, plate tectonics, earthquakes, and volcanoes. John Muir (1912) described the shocks and outbursts of endogenic forces as "love-beats of Nature's heart." Examples of endogenic forces are clearly and dramatically illustrated by the Teton Mountains which were created by faulting and the on-going eruptions of Mt. St. Helens in Washington State and Mt. Shishaldin in Alaska.

Exogenic (means derived or originating externally) refers to external processes and phenomena that occur on or above the Earth's surface. Examples of exogenic forces include weathering effects, erosion by water, wind, and glacial mechanics, comet and meteoroid impacts, and radiation from the sun.

Endogenic Forces

As mentioned, endogenic forces include volcanism, plate tectonics, and earthquakes. In this section we discuss these processes.

Volcanism

Volcanism is the phenomenon connected with volcanoes and volcanic activity. To gain understanding of this process one must have a basic understanding of the dynamics involved. In light of this we begin by explaining that endogenic volcanism brings magma from the mantle within the planet and rises to the surface as a volcanic eruption. Magmas are large bodies of molten rock deeply buried within the earth. These magmas are less dense than surrounding rocks, and will therefore move upward. In the upward movement, sometimes magmatic materials are poured out upon the surface of the earth as, for example, when lava flows from a volcano. These igneous rocks are volcanic or extrusive rocks; they form when the magma cools and crystallizes on the surface of the earth (Craters of the Moon, Idaho, is a good example). Under certain other conditions, magma does not make it to the earth's surface and cools and crystallizes within earth's crust. These intruding rock materials harden and form intrusive or plutonic rocks.

Magma is made up of molten silicate material and may include already formed crystals and dissolved gases. The term magma applies to silicate melts within the earth's crust. When magma reaches the surface it is referred to as lava. The chemical composition of magma is controlled by the abundance of elements in the Earth. The elements oxygen, silicon, aluminum, hydrogen, sodium, calcium, iron, potassium, and manganese comprise 99% of magma. Because oxygen is so abundant, chemical analyses are usually given in terms of oxides. Silicon dioxide (SiO_2; also known as silica) is the most abundant oxide. Because magma gas expands as pressure is reduced, magmas have an explosive character. The flow (or viscosity) of magma depends on temperature, composition, and gas content. Higher silicon dioxide content and lower temperature magmas have higher viscosity.

Magma consists of three types: basaltic, andesitic, and rhyolitic. Table 3.1 summarizes the characteristics of each type.

TABLE 3.1
Characteristics of Magma Types

Magma Type	Solidified Volcanic	Solidified Plutonic	Chemical Composition	Temperature
Basaltic	Basalt	Gabbro	45-55% silicon dioxide	1000-1200 °C
Andesitic	Andesite	Diorite	55-65% silicon dioxide	800-1000 °C
Rhyolitic	Rhyolite	Granite	65-75% silicon dioxide	650-800 °C

Intrusive Rocks

Intrusive (or plutonic) rocks are rocks that have solidified from molten mineral mixtures beneath the surface of the earth. Intrusive rocks that are deeply buried tend to cool slowly and develop a coarse texture. On the other hand, those intrusive rocks near the surface that cool more quickly are finer textured. The shape, size, and arrangement of the grains comprising it determine the texture of igneous rocks. Because of crowded conditions under which mineral particles are formed, they are usually angular and irregular in outline. Typical intrusive rocks include

- **gabbro**—a heavy, dark-colored igneous rock consisting of coarse grains of feldspar and augite
- **peridotite**—is a rock in which the dark minerals are predominant
- **granite**—is the most common and best-known of the coarse-textured intrusive rocks
- **syenite**—resembles granite, but is less common in its occurrence and contains little or no quartz

Extrusive Rocks

Extrusive (or volcanic) rocks are those that pour out of craters of volcanoes or from great fissures or cracks in the earth's crust, and make it to the surface of the earth in a molten state (liquid lava). Extrusive rocks tend to cool quickly, and typically have small crystals (because fast cooling does not allow large crystals to grow). Some cool so rapidly that no crystallization occurs, and volcanic glass is produced.

Some of the more common extrusive rocks include felsite, pumice, basalt, and obsidian.

- **Felsite**—very fine textured igneous rocks
- **Pumice**—frothy lava that solidifies while steam and other gases bubble out of it
- **Basalt**—world's most abundant fine-grained extrusive rock
- **Obsidian**—volcanic glass; cools so fast that there is no formation of separate mineral crystals

Bowen's Reaction Series

The geologist, Norman L. Bowen, back in the early 1900's, was able to explain why certain types of minerals tend to be found together, while others are almost never associated with one another. Bowen found that minerals tend to form in specific sequences in igneous rocks, and these sequences could be assembled into a composite sequence. The idealized progression which he determined is still accepted as the general model (see Figure 13) for the evolution of magmas during the cooling process.

In order to better understand Bowen's Reaction Series, it is important to define key terms:

- **Magma**—molten igneous rock
- **Felsic**—white pumice
- **Pumice**—textured form of volcanic rock; a solidified frothy lava
- **Extrusion**—magma intruded or emplaced beneath the surface of the earth
- **Feldspar**—the family of minerals including microcline, orthoclase, and plagioclase
- **Mafic**—a mineral containing iron and magnesium
- **Aphanitic**—mineral grains too small to be seen without a magnifying glass
- **Phaneritic**—mineral grains large enough to be seen without a magnifying glass
- **Reaction Series**—a series of minerals in which a mineral reacts to change to another

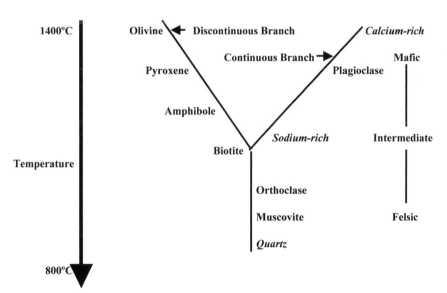

FIGURE 13. Bowen's Reaction Series

Source: Adapted from GeoMan (2008) accessed @ http://Jersey.uoregon.edu/~mstrick/AskGeoMan/geoQuerry32.html

- **Rock-forming Mineral**—the minerals commonly found in rocks. Bowen's Reaction Series lists all of the common ones in igneous rocks
- **Specific Gravity**—the relative mass or weight of a material compared to the mass or weight of an equal volume of water

Some igneous rocks are named according to textural criteria:

- **Scoria**—porous
- **Pumice**—vesicular
- **Obsidian**—glass
- **Tuff**—cemented ash
- **Breccia**—cemented fragments
- **Permatite**—extremely large crystals
- **Aplite**—sugary texture, quartz & feldspar
- **Porphyry**—fine matrix, large crystals

The Discontinuous Reaction Series

The left side of Figure 13 shows a group of mafic or iron-magnesium bearing minerals—olivine, pyroxene, amphibole, and biotite. If the chemistry of the melt is just right, these minerals react discontinuously to form the next mineral in the series. If there is enough silica in the igneous magma melt, each mineral will change to the next mineral lower in the series as the temperature drops. Descending down Bowen's Reaction Series, the minerals increase in the proportions of silica in their composition. In basaltic melt, as shown in Figure 13, olivine will be the first mafic mineral (silicate mineral rich in magnesium and iron) to form. When the temperature is low enough to form pyroxene, all of the olivine will react with the melt to form pyroxene, and pyroxene will crystallize out of the melt. At the crystallization temperature of amphibole, all the pyroxene will react with the melt to form amphibole, and amphibole will crystallize. At the crystallization temperature of biotite, all of the amphibole will react to form biotite, and biotite will crystallize. Thus, all igneous rocks should only have biotite; however, this is not the case. In crystallizing olivine, if there is not enough silica to form pyroxene, then the reaction will not occur and olivine will remain. Additionally, if the temperature drops too fast for the reaction to take place (volcanic magma eruption) the reaction will not have time to occur; thus the rock will solidify quickly and the mineral will remain olivine.

The Continuous Reaction Series

The right side of Figure 13 shows the plagioclases. Plagioclase minerals have the formula $(Ca, Na)(Al, Si)_3O_8$. The highest temperature plagioclase has only calcium (Ca). The lowest temperature plagioclase has only sodium (Na). In between, these ions mix in a continuous series from 100% Ca and 0% Na at the highest temperature to 50% Ca and 50% Na at the middle temperature to 0% Ca and 100% Na at the lowest temperature. In a basaltic melt, for example, the first plagioclase to form could be 100% Ca and 0% Na plagioclase. As the temperature drops, the crystal reacts with the melt to form 99% Ca and 1% Na plagioclase and 99% Ca and 1% Na plagioclase, which then crystallizes. These then react to form 98% Ca and 2% Na and the same composition would crystallize and so forth. All of this happens continuously, provided there is enough time for the reactions to take place, and enough sodium, aluminum, and silica in the melt to form each new mineral. The end result will be a rock with plagioclases with the same ratio of Ca to Na as the starting magma.

✓ **Important Point:** The magma temperature and the chemical composition of the Magma determine what minerals crystallize and thus what kind of igneous rock we get.

Eruption of Magma

The volcanic processes which lead to the deposition of extrusive igneous rocks can be studied in action today and help us to explain the textures of ancient rocks with respect to depositional processes. Some of the major features of volcanic processes and landforms are discussed in this section. The information below is from USGS (2008) *Principal Types of Volcanoes.*

Types of Volcanoes

Geologists generally group volcanoes into four main kinds—cinder cones, composite volcanoes, shield volcanoes, and lava domes.

- **Cinder Cones**—are the simplest type of volcano. They are built from particles and blobs of congealed lava ejected from a single bent. As the gas-charged lava is blown violently into the air, it breaks into small fragments that solidify and fall as cinders around the vent to form a circular or oval cone. Most cinder cones have a bowl-shaped crater at the summit and rarely rise more than a thousand feet or so above their surroundings. Cinder cones are numerous throughout western North America as well as throughout other volcanic terrains of the world.
- **Composite Volcanoes**—Some of earth's grandest mountains are composite volcanoes—sometimes called stratovolcanoes. They are typically

steep-sided, symmetrical cones of large dimension. They are built of alternating layers of **lava flows**, volcanic ash, cinders, blocks, and bombs,
and may rise as much as 8,000 feet above their bases. Most composite
volcanoes have a crater at the summit which contains a central vent or a
clustered group of vents. Lavas either flow through breaks in the crater
wall or issue from fissures on the flanks of the cone. Lava, solidified
within the fissures, forms dikes that act as ribs which greatly strengthen
the cone. The essential feature of a composite volcano is a conduit system
through which magma from a reservoir deep in the earth's crust rises to
the surface. The volcano is built up by the accumulation of material
erupted through the conduit and increases in size as lava, cinders, ash, etc.,
are added to its slopes.

• **Shield Volcanoes**—are built almost entirely of fluid lava flows. Flow after
  flow pours out in all directions from a central summit vent, or group of
  vents, building a broad, gently sloping cone of flat, domical shape, with a
  profile much like that of a warrior's shield. They are built up slowly by
  the accretion of thousands of highly fluid lava flows called basalt lava that
  spread widely over great distances and then cool as thin, gently dipping
  sheets. Lava also commonly erupts from vents along fractures (rift zones)
  that develop on the flanks of the cone. Some of the largest volcanoes in
  the world are shield volcanoes (see Figure 14).

• **Lava Domes**—are formed by relatively small bulbous masses of lava too
  viscous to flow any great distance; consequently, on extrusion, the lava
  piles over and around its vent. A dome grows largely by expansion from
  within. As it grows, its outer surface cools and hardens, then shatters,
  spilling loose fragments down its sides. Some domes form craggy knobs
  or spines over the volcanic vent, whereas others form short, steep-sided

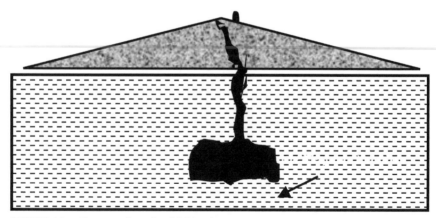

FIGURE 14.   Cross-section of a shield volcano

lava flows known as "coulees" (Fr. "to flow"). **Volcanic domes** commonly occur within the craters or on the flanks of large composite volcanoes.

Types of Volcanic Eruptions

The type of volcanic eruption is often labeled with the name of a well-known volcano where characteristic behavior is similar—hence the use of such terms as "Strombolian," "Vulcanian," "Vesuvian," "Pelean," "Hawaiian," and others.

- **Strombolian-type Eruption**—is in constant action with huge clots of molten lava bursting from the summit crater to form luminous arcs through the sky. Collecting on the flanks of the cone, lava clots combine to stream down the slopes in fiery rivulets.
- **Vulcanian-type Eruption**—is characterized by very viscous lavas; a dense cloud of ash-laden gas explodes from the crater and rises high above the peak. Steaming ash forms a whitish cloud near the upper level of the cone.
- **Vesuvian Eruption**—is characterized by great quantities of ash-laden gas which are violently discharged to form a cauliflower-shaped cloud high above the volcano.
- **Pelean-type Eruption (or "Nuee Ardente"—glowing cloud)**—is characterized by its explosiveness. It erupts from a central crater with violent explosions which eject great quantities of gas, volcanic ash, dust, incandescent lava fragments, and large rock fragments.
- **Hawaiian-type Eruption (or quiet)**—is characterized by less viscous lavas which permit the escape of gas with a minimum of explosive violence. In fissure-type eruptions, a fountain of fiery lava erupts to a height of several hundred feet or more. Such lava may collect in old pit craters to form lava lakes, or form cones, or feed radiating flows.
- **Phreatic (or steam-blast) Eruption**—driven by explosive expanding steam resulting from cold ground or **surface water** coming into contact with hot rock or magma. The distinguishing feature of phreatic explosions is that they only blast out fragments of preexisting solid rock from the volcanic conduit; no new magma is erupted.
- **Plinian Eruption**—is a large explosive event that forms enormous dark columns of tephra (solid material ejected) and gas high into the stratosphere. Such eruptions are named for Pliny the Younger, who carefully described the disastrous eruption of Vesuvius in 79 A.D. This eruption generated a huge column of tephra into the fall. Many thousands of people evacuated areas around the volcano, but about 2,000 were killed, including Pliny the Elder.

Did You Know?

The longest historical dome-building eruption is still occurring at Santiaguito Dome, which is erupting on the southeast flank of Santa Maria volcano in Guatemala; the dome began erupting in 1922.

Intrusions

Intrusive (or plutonic igneous) rocks have been intruded or injected into the surrounding rocks. Some of these intrusions are invisible because they are imbedded at great depth; consequently, igneous intrusive bodies may be seen only after the underlying rocks have been removed by erosion.

Intrusions are of two types: Concordant intrusions which are parallel to layers of rocks, and discordant intrusions which cut across layers. Some of the more common intrusive bodies (plutons) are discussed below.

Concordant Intrusions

- **Sills**—are tabular bodies of igneous rocks that spread out as essentially thin, horizontal sheets between beds or layers of rocks.
- **Laccoliths**—are lens-like, mushroom-shaped, or blister-like intrusive bodies, usually near the surface, that have relatively flat under surfaces and arched or domed upper surfaces. They differ from sills in that they are thicker in the center and become thinner near their margins.
- **Lopoliths**—are mega-sills, usually of gabbro or diorite that may cover hundreds of square kilometers and be kilometers thick. They often have a concave structure and are differentiated. That is, they take so long to harden that heavy minerals have a chance to sink and light minerals can rise.

Discordant Intrusions

- **Dikes**—are thin, wall-like sheets of magma intruded into fractures in the crust.
- **Stocks or plutons**—are small irregular intrusions.
- **Batholiths**—are the largest of igneous intrusions and are usually granitic and cover hundreds or thousands of square kilometers.

Did You Know?

Most obsidian is black, but red, green, and brown obsidian is known. Obsidian forms when magma is cooled so quickly that individual minerals cannot crystallize.

Volcanic Landforms

Volcanic landforms (or volcanic edifices) are controlled by the geological processes that form them and act on them after they have formed. Four principal types of volcanic landforms are formed: plateau basalts or lava plains, volcanic mountains, craters, and calderas.

- **Plateau Basalts and Lava Plains**—are formed when great floods of lava are released by fissure eruptions (instead of central vents) and spread in sheet-like layers over the earth's surface, forming broad plateaus. Some of these plateaus are quite extensive. For example, the Columbia River Plateau of Oregon, Washington, Nevada, and Idaho is covered by 200,000 square miles of basaltic lava.
- **Volcanic Mountains**—are mountains that are composed of the volcanic products of central eruptions and are classified as cinder cones (conical hills), composite cones (stratovolcanoes), and lava domes (shield volcanoes).
- **Volcanic Craters**—are circular, funnel-shaped depressions, usually less than 1 km in diameter, that form as a result of explosions that emit gases and tephra.
- **Calderas**—are much larger depressions, circular to elliptical in shape, with diameters ranging from 1 km to 50 km. Calderas form as a result of collapse of a volcanic structure. The collapse results from evacuation of the underlying magma chamber.

Thermal Areas

Thermal areas are locations where volcanic or other igneous activity takes place as is evidenced by the presence and/or action of volcanic gases, steam, or hot water escaping from the earth.

- **Fumaroles**—are vents where gases, either from a magma body at depth or steam from heated **groundwater**, emerge at the surface of the earth.
- **Hot Springs or Thermal Springs**—are areas where hot water comes to the surface of the earth. Cool groundwater moves downward and is heated by a body of magma or hot rock. A hot **spring** results if this hot water can find its way back to the surface, usually along fault zones.
- **Geysers**—result if the hot spring has a plumbing system that allows for the accumulation of steam from the boiling water. When the steam pressure builds so that it is higher than the pressure of the overlying water in the system, the steam will move rapidly toward the surface, causing the eruption of the overlying water. Some geysers, like Old

Faithful in Yellowstone Park, erupt at regular intervals, but most geysers are quite erratic in their performance. The time between eruptions is controlled by the time it takes for the steam pressure to build in the underlying plumbing system.

Earthquakes

Earthquakes are another significant endogenic process. One thing is certain; anyone who has witnessed (been exposed to) one or more of over a million or so earthquakes that occur each year on Earth is unlikely to forget such occurrences. Even though most earthquakes are insignificant, a few thousand of these produce noticeable effects such as tremors and/or ground shaking. The passage of time has shown that about 20 earthquakes each year cause major damage and destruction. It is estimated that about 10,000 people die each year because of earthquakes.

Over the millennia, the effect of damaging earthquakes has been obvious to those who witnessed the results. However, the cause of earthquakes has not been as obvious. For example, the cause of earthquakes has initially shifted from the blaming of super-incantations of mythical beasts, to the wrath of Gods, to unexplainable magical occurrences, and/or to just normal, natural phenomena occasionally required to retain Earth's structural integrity; that is, providing Earth with a periodic form of feedback to keep "things" in balance. More specifically, we can say with correctness, in our view, that an earthquake on Earth provides our planet with a sort of a geological homeostasis needed to maintain life as we know it.

Through the ages earthquakes have also come under the attention and eventually the pen of the world's greatest writers. Consider, for example, Voltaire's classic satirical novel, *Candide*, published in 1759, in which he mercilessly satirizes science and, in particular, earthquakes. Voltaire based the following comments on the 1755 Great Lisbon, Portugal, Earthquake which was blamed for causing the deaths of more than 60,000 people.

> Dr. Pangloss says to Candide (on viewing the total devastation of Lisbon):
> "The heirs of the dead will benefit financially; the building trade will enjoy a boom. Private misfortune must not be overrated. These poor people in their death agonies, and the worms about to devour them, are playing their proper and appointed part in God's master plan."

Although we still do not know what we do not know about earthquakes and their causes, we have evolved from using witchcraft, psychic gobbledygook, or magic to explain their origins, to the scientific methods employed today. In the first place, we do know that earthquakes are caused by the sudden release of energy along faults. Earthquakes are usually followed by a series of smaller earthquakes that we called aftershocks. Aftershocks represent further adjustments of rock along the fault. There are currently no reliable methods for predicting when earthquakes will occur.

In regard to the cause(s) or origin(s) of earthquakes, we have developed a couple of theories. One of these theories explains how earthquakes occur via *elastic rebound*. That is, according to elastic rebound theory, subsurface rock masses subjected to prolonged pressures from different directions will slowly bend and change shape. Continued pressure sets up strains so great that the rocks will eventually reach their elastic limit and rupture (break) and suddenly snap back into their original unstrained state. It is the snapping back (elastic rebound) that generates the seismic waves radiating outward from the break. The greater the stored energy (strain), the greater the release of energy.

The coincidence of many active volcanic belts with major belts of earthquake activity (*seismic* and *volcanic activity*) may indicate that volcanoes and earthquakes may have a common cause. Plate interactions commonly cause both earthquakes (tectonic earthquakes) and volcanoes.

Seismology

Even though *seismology* is the study of earthquakes, it is actually the study of how seismic waves behave in the Earth. The source of an earthquake is called the *hypocenter* or *focus* (i.e., the exact location within the Earth were seismic waves are generated). The *epicenter* is the point on the Earth's surface directly above the focus. Seismologists want to know where the focus and epicenter are located so a comparative study of the behavior of the earthquake event can be made with previous events—in an effort to further understanding.

Seismologists use instruments to detect, measure, and record seismic waves. Generally, the instrument used is the *seismograph*, which has been

Did You Know?

Any geometry textbook should tell you the math behind finding the exact epicenter location. For instance, to find the epicenter of an earthquake, a process called triangulation is used. Three seismographs are used and the time of detection for each allows the location of the epicenter. Another instance that we all may be more familiar with is used in cell phones so that we get reception from the cell tower that we are closest to and/or have the strongest signal from.

around for a long time. Modern updates have upgraded these instruments from the paper or magnetic tape strip to electronically recorded data that is input to a computer. A study of the relative arrival times of the various types of waves at a single location can be used to determine the distance to the epicenter. To determine the exact epicenter location, records from at least three widely separated seismograph stations are required.

Seismic Waves

As mentioned, some of the energy released by an earthquake travels through the Earth. The speed of a seismic wave depends on the density and elasticity of the materials through which they travel. Seismic waves come in several types as described below:

- **P-Waves**—Primary, Pressure, or Push-Pull Waves (arrive first—first detected by seismographs) are compressional waves (expand and contract) that travel through the earth (solids, liquids, or gases) at speeds of from 3.4 to 8.6 miles per second. P waves move faster at depth, depending on the elastic properties of the rock through which they travel. P waves are the same thing as sound waves.
- **S-Waves**—Secondary or Shear Waves travel with a velocity (between 2.2 and 4.5 miles per second) that depends only on the rigidity and density of the material through which they travel. They are the second set of waves to arrive at the seismograph and will not travel through gases or liquids; thus the velocity of S-waves through gas or liquids is zero.
- **Surface Waves**—Several types, travel along the Earth's outer layer or surface or on layer boundaries in the Earth. These are rolling, shaking waves that are the slowest waves but the ones that do the damage in large earthquakes.

Earthquake Magnitude and Intensity

The size of an earthquake is measured using two parameters—energy released (magnitude) and damage caused (intensity).

Earthquake Magnitude—The size of an earthquake is usually given in terms of its Richter Magnitude. Richter Magnitude is a scale devised by Charles Richter that measures the amplitude (height) of the largest recorded wave at a specific distance from the earthquake. A better measure is the Richter Scale which measures the total amount of energy released by an earthquake as recorded by seismographs. The amount of energy released is related to the Richter Scale by the equation:

$$\text{Log E} = 11.8 + 1.5 \text{ M} \qquad (3.1)$$

where

Log = the logarithm to the based 10
E = the energy released in ergs
M = is the Richter Magnitude

In using equation 3.1 to calculate Richter Magnitude, it quickly becomes apparent that we see that each increase in 1 in Richter Magnitude yields a 31 fold increase in the amount of energy released. Thus, a magnitude 6 earthquake releases 31 times more energy than a magnitude 5 earthquake. A magnitude 9 earthquake releases 31 × 31 or 961 times more energy than a magnitude 7 earthquake.

Did You Know?

While it is correct to say that for each increase of 1 in the Richter Magnitude, there is a tenfold increase in amplitude of the wave, it is incorrect to say that each increase of 1 in Richter Magnitude represents a tenfold increase in the size of the earthquake.

Earthquake Intensity—is a rough measure of an earthquake's destructive power (i.e., size and strength—how much the earth shook at a given place near the source of an earthquake). To measure earthquake intensity, Mercalli in 1902 devised an intensity scale of earthquakes based on the impressions of people involved; movement of furniture and other objects; and damage to buildings. The shock is most intense at the epicenter, which, as noted earlier, is located on the surface directly above the focus.

Mercalli's intensity scale uses a series of numbers (based on a scale of 1 to 12) to indicate different degrees of intensity (see Table 3.2). Keep in mind that this scale is somewhat subjective, but it provides a qualitative, but systematic, evaluation of earthquake damage.

TABLE 3.2
Modified Mercalli Intensity Scale

Intensity	Description
I	Not felt except under unusual conditions
II	Felt by only a few on upper floors
III	Felt by people lying down or seated
IV	Felt indoors by many, by few outside
V	Felt by everyone, people awakened
VI	Trees sway, bells ring, some objects fall
VII	Causes alarm, walls and plaster crack
VIII	Chimneys collapse, poorly constructed buildings seriously damaged
IX	Some houses collapse, pipes break
X	Ground cracks, most buildings collapse
XI	Few buildings survive, bridges collapse
XII	Total destruction

Plate Tectonics

Within the past 45 or 50 years geologists have developed the theory of plate tectonics (tectonics: Greek, "builder"). The theory of plate tectonics deals with the formation, destruction, and large scale motions of great segments of Earth's surface (crust), called *plates*. This theory relies heavily on the older concepts of continental drift (developed during the first half of the twentieth century) and seafloor spreading (understood during the 1960s) which help to explain the cause of earthquakes and volcanic eruptions, and the origin of fold mountain systems.

Crustal Plates

Earth's crustal plates are composed of great slabs of rock (lithosphere), about 100 km thick that cover many thousands of square miles (they are thin in comparison to their length and width); they float on the ductile asthenosphere, carrying both continents and oceans. Many geologists recognize at least eight main plates and numerous smaller ones. These *main* plates include

- African Plate covering Africa—Continental plate
- Antarctic Plate covering Antarctica—Continental plate
- Australian Plate covering Australia—Continental plate
- Eurasian Plate covering Asia and Europe—Continental plate
- Indian Plate covering Indian subcontinent and a part of Indian Ocean—Continental plate
- Pacific Plate covering the Pacific Ocean—Oceanic plate

- North American Plate covering North America and northeast Siberia—Continental plate
- South American Plate covering South America—Continental plate

The *minor* plates include

- Arabian Plate
- Caribbean Plate
- Juan de Fuca Plate
- Cocos Plate
- Nazea Plate
- Philippine Plate
- Scotia Plate

Plate Boundaries

As mentioned, the asthenosphere is the ductile, soft, plastic-like zone in upper mantle on which the crustal plates ride. Crustal plates move in relation to one another at one of three types of plate boundaries: convergent (collision boundaries), divergent (spreading boundaries), and transform boundaries. These boundaries between plates are typically associated with deep-sea trenches, large faults, fold mountain ranges, and mid-oceanic ridges.

- **Convergent Boundaries**—(or active margins) develop where two plates slide towards each other commonly forming either a subduction zone (if one plate subducts or moves underneath the other) or a continental collision (if the two plates contain continental crust). In continental collisions, neither plate is subducted. Instead, the crust tends to buckle and be pushed upward or sideways.
- **Divergent Boundaries**—occur where two plates slide apart from each other. Oceanic ridges, which are examples of these divergent boundaries, are where new oceanic, melted lithosphere materials well-up, resulting in basaltic magmas which intrude and erupt at the oceanic ridge; in turn creating new oceanic lithosphere and crust (new ocean floor). Along with volcanic activity, the mid-oceanic ridges are also areas of seismic activity.
- **Transform Plate Boundaries**—or shear/constructive boundaries, do not separate or collide; rather, they slide past each other in a horizontal manner with a shearing motion. Most transform boundaries occur where oceanic ridges are offset on the sea floor. The San Andreas Fault in California is an example of a transform fault.

Exogenic Forces

As mentioned, exogenic refers to external processes and phenomena that occur on or above the Earth's surface. Many of the results of these processes, because they are surface phenomena, are visible with time. Some might think these surface processes that alter the face of Earth are destructive. We take a different view. Actually, we share John Muir's view of nature at work sculpting Earth's surface to her liking. Muir said, "we see that everything in Nature called destruction must be creation—a change from beauty to beauty (1869)." Another view of nature as master sculptor: "In regard to surface rock formations, weathering and erosion are both creators and executioners! As soon as rock is lifted above sea level, weather starts to break it up. Water, ice, and chemicals split, dissolve, or rot the rocky surface until it crumbles. Mixed with water and air and plant and animal remains, crumbled rock forms soil . . . when holding part of Nature's work in one's hand, what could possibly be more significant than a handful of soil" (Spellman 2010)? In this section we discuss those tools that change beauty to beauty: weathering, glaciation, wind erosion, mass wasting, desertification, and stream (water flow) erosion.

Weathering

One can't obtain the slimmest slice of understanding of the geology of Earth without understanding Earth processes. For example, the weathering process is the first step in the erosion process that causes the breakdown of rocks, either to form new minerals that are stable on the surface of Earth, or to break the rocks down to smaller particles. Simply, weathering (which projects itself on all surface material above the **water table**) is the general term used for all the ways in which a rock may be broken down; therefore, to understand the causes of the sculpting of rocks to other forms (e.g., to soil) we first must understand the process of weathering.

Factors That Influence Weathering

The factors that influence weathering include:

- **Rock Type and Structure**—Each mineral contained in rocks has a different susceptibility to weathering. A rock with bedding planes, joints, and fractures provides pathways for the entry of water, leading to more rapid weathering. Differential weathering (rocks erode at differing rates) can occur when rock combinations consist of rocks that weather faster than more resistant rocks.

- **Slope**—On steep slopes weathering products may be quickly washed away by rains. Wherever the force of gravity is greater than the force of friction holding particles upon a slope, these tend to slide downhill.
- **Climate**—Higher temperatures and high amounts of water generally cause chemical reactions to run faster. Rates of weathering are higher in warmer than in colder dry climates.
- **Animals**—Rodents, earthworms, and ants burrow into soil and bring material to the surface where it can be exposed to the agents of weathering.
- **Time**—depends on slope, animals, and climate.

Categories of Weathering Processes

Although weathering processes are separated, it is important to recognize that these processes work in tandem to break down rocks and minerals to smaller fragments. Geologists recognize two categories of weathering processes:

1. **Physical (or Mechanical) Weathering**—disintegration of rocks and minerals by a physical or mechanical process.
2. **Chemical Weathering**—involves the decomposition of rock by chemical changes or solution.

Physical Weathering

Physical weathering involves the disintegration of a rock by physical processes. These include freezing and thawing of water in rock crevices, disruption by plant roots or burrowing animals, and the changes in volume that result from chemical weathering with the rock. These and other physical weathering processes are discussed below.

- **Development of Joints**—Joints are another way that rocks yield to stress. Joints are fractures or cracks in which the rocks on either side of the fracture have not undergone relative movement. Joints form as a result of expansion due to cooling, or relief of pressure as overlying rocks are removed by erosion. They form free space in rock by which other agents of chemical or physical weathering can enter (unlike faults that show offset across the fracture). They play an important part in rock weathering as zones of weakness and water movement.
- **Crystal Growth**—As water percolates through fractures and pore spaces it may contain ions that precipitate to form crystals. When crystals grow they can cause the necessary stresses needed for mechanical rupturing of rocks and minerals.

- **Heat**—it was once thought that daily heating and cooling of rocks was a major contributor to the weathering process. This view is no longer shared by most practicing geologists. However, it should be pointed out that sudden heating of rocks from forest fires may cause expansion and eventual breakage of rock.
- **Biological Activities**—Plant and animal activities are important contributors to rock weathering. Plants contribute to the weathering process by extending their root systems into fractures and growing, causing expansion of the fracture. Growth of plants and their effects are evident in many places where they are planted near cement work (streets, brickwork, and sidewalks). Animal burrowing in rock cracks can break rock.
- **Frost Wedging**—is often produced by alternate freezing and thawing of water in rock pores and fissures. Expansion of water during freezing causes the rock to fracture. Frost wedging is more prevalent at high altitudes where there may be many freeze-thaw cycles. One classic and striking example of weathering of Earth's surface rocks by frost wedging is illustrated by the formation of Hoodoos in Bryce Canyon National Park, Utah (see Figure 15). "Although Bryce Canyon receives a meager 18

FIGURE 15. Hoodoos in Bryce Canyon, Utah
Photo by Frank R. Spellman

inches of **precipitation** annually, it's amazing what this little bit of water can do under the right circumstances (NPS 2008)!" Approximately 200 freeze-thaw cycles occur annually in Bryce. During these periods, snow and ice melt in the afternoon and water seeps into the joints of the Bryce or Claron Formation. When the sun sets, temperatures plummet and the water refreezes, expanding up to 9% as it becomes ice. This frost wedging process exerts tremendous pressure or force on the adjacent rock and shatters and pries the weak rock apart. The assault from frost wedging is a powerful force but, at the same time, rain water (the universal solvent), which is naturally acidic, slowly dissolves away the limestone, rounding off the edges of these fractured rocks, and washing away the debris. Small rivulets of water round down Bryce's rime, forming gullies. As gullies are cut deeper, narrow walls of rock known as fins begin to emerge. Fins eventually develop holes known as windows. Windows grow larger until their roofs collapse, creating hoodoos (see Figure 15). As old hoodoos age and collapse, new ones are born (NPS 2008).

Did You Know?

Bryce Canyon National Park lies along the high eastern escarpment of the Paunsaugunt Plateau in the Colorado Plateau region of southern Utah. Its extraordinary geological character is expressed by thousands of rock chimneys (hoodoos) that occupy amphitheater-like alcoves in the Pink Cliffs, whose bedrock host is Claron Formation of the Eocene age (Davis and Pollock 2003).

Hoodoo Pronunciation: 'hu-du—*Noun* Etymology: West African; from voodoo.

A natural column of rock in western North America often in fantastic form.

Source: Merriam-Webster Online (www.m-w.com)

Chemical Weathering

Chemical weathering involves the decomposition of rock by chemical changes or solution. Rocks that are formed under conditions present deep within the Earth are exposed to conditions quite different (i.e., surface temperatures and pressures are lower on the surface and copious amounts of free water and oxygen are available) when uplifted onto the surface. The chief processes are oxidation, carbonation and hydration, and solution in water above and below the surface.

The Persistent Hand of Water

Because of its unprecedented impact on shaping and reshaping Earth, at this point in the text, it is important to point out that given time, nothing, absolutely nothing on Earth is safe from the heavy hand of water. The effects of water sculpting by virtue of movement and accompanying friction will be covered later in the text. For now, in regards to water exposure and chemical weathering, the main agent responsible for chemical weathering reactions is not water movement, but instead is water and weak acids formed in water.

The acids formed in water are solutions that have abundant free Hydrogen$^+$ ions. The most common weak acid that occurs in surface waters is carbonic acid. Carbonic acid (H_2CO_3) is produced when atmospheric carbon dioxide dissolves in water; it exists only in solution. Hydrogen ions are quite small and can easily enter crystal structures, releasing other ions into the water.

$$H_2O \; + \quad CO_2 \quad \rightarrow \quad H_2CO_3 \quad \rightarrow \quad H^+ \quad + \quad HCO_3^-$$

water   carbon dioxide   carbonic acid   hydrogen ion   bicarbonate ion

Types of Chemical Weathering Reactions

As mentioned, chemical weathering breaks rocks down chemically by adding or removing chemical elements, and changes them into other materials. Again, as stated, chemical weathering consists of chemical reactions, most of which involve water. Types of chemical weathering involve:

- **Hydrolysis**—is a water-rock reaction that occurs when an ion in the mineral is replaced by H$^+$ or OH$^-$.
- **Leaching**—Ions are removed by dissolution into water.
- **Oxidation**—Oxygen is plentiful near Earth's surface; thus, it may react with minerals to change the oxidation state of an ion.
- **Dehydration**—occurs when water or a hydroxide ion is removed from a mineral.
- **Complete Dissolution**—all of the mineral is completely dissolved in water.

Glaciation

Geologically, not that long ago, many parts of Earth were covered with massive sheets of ice. Moreover, the geologic record shows that this most re-cent ice-sheet-covering of large portions of Earth's surface is not a one-time phenomenon; instead, Earth has experienced several glaciation periods as well as interglacials like the one we are presently experiencing. Although the ice from the last ice age has now retreated from most of Europe, Asia, and North

America, it has left traces of its influence across the whole face of the landscape in jagged mountain peaks, gouged-out upland valleys, swamps, changed river courses, and boulder-strewn, table-flat prairies in the lowlands.

Ice covers about 10% of all land and ~12% of the oceans. Most of this ice is contained in the polar ice sea, polar sheets and ice caps, valley glaciers, and piedmont glaciers formed by valley glaciers merging on a plain. In the geology of the present time, the glaciers of today are not that significant in the grand scheme of things. However, it is the glaciation of the past with its accompanying geologic evidence left behind by ancient glaciers that is important. This geologic record indicates that the Earth's climate has undergone fluctuations in the past, and that the amount of the Earth's surface covered by glaciers has been much larger in the past than in the present. In regards to the effects of past glaciation, one need only look at the topography of the western mountain ranges in the northern part of North America to view the significant depositional processes of glaciers.

Glaciation is a geological process that modifies land surface by the action of glaciers. For those who study glaciation and glaciers, the fact that glaciations have occurred so recently in North America and Europe accounts for extant evidence, allowing the opportunity to study the undeniable results of glacial erosion and deposition. This is the case, of course, because the forces involved with erosion—weathering, mass wasting, and stream erosion—have not had enough time to remove the traces of glaciation from Earth's surface. Glaciated landscapes are the result of both glacial erosion (glaciers transport rocks and erode surfaces) and glacial deposition (glaciers transport material that melts and deposits material).

Glaciers

A *glacier* is a thick mass of slowly moving ice, consisting largely of recrystallized snow that shows evidence of downslope or outward movement due to the pull of gravity. Glaciers can only form at latitudes or elevations above the snowline (the elevation at which snow forms and remains present year round). Glaciers form in these areas if the snow becomes compacted, forcing out the air between the snowflakes. The weight of the overlying snow causes the snow to recrystallize and increase its grain-size, until it increases its density and becomes a solid block of ice.

Types of Glaciers

The various types of glaciers include:

- **Mountain Glaciers**—are relatively small glaciers which occur at higher elevations in mountainous regions. A good example of mountain glaciers can be seen in the remaining glaciers of Glacier National Park, Montana.

- **Valley Glaciers**—are tongues of ice that as snow and ice accumulate and spill down a valley filling it with ice, perhaps for scores of miles (see Figures 16a–d). When a valley glacier extends down to sea level, it may carve a narrow valley into the coastline. These are called *fjord glaciers*. The narrow valleys they carve that later become filled with seawater after the ice melts are called *fjords*. When a valley glacier extends down a valley and then covers a gentle slope beyond the mountain range, it is called a *piedmont glacier*. If valley glaciers cover a mountain range, they are called *ice caps*.
- **Continental Glaciers (Ice Sheets)**—are the largest glaciers. They cover Greenland and Antarctica and contain about 95% of all glacial ice on Earth.
- **Ice Shelves**—are sheets of ice floating on water and attached to land. They may extend hundreds of miles from land and reach thicknesses of several thousand feet.
- **Polar Glaciers**—are always below the melting point at the surface and do not produce any melt water.
- **Temperate Glaciers**—are at a temperature and pressure level near the melting point throughout the body except for a few feet of ice. This layer is subjected to annual temperature fluctuations.

FIGURE 16a.   Matanuska Glacier, Alaska
Photo by JoAnn Chapman, with permission

FIGURE 16b.   Inside Matanuska Glacier, Alaska, a valley glacier
Photo by JoAnn Chapman, with permission

FIGURE 16c.   Matanuska Glacier, Alaska
Photo by JoAnn Chapman, with permission

FIGURE 16d. Matanuska Glacier and Lake, Alaska
Photo by JoAnn Chapman, with permission

Glacier Characteristics

The primary characteristics displayed by glaciers are changes in size and movement. A glacier changes in size by the addition of snowfall, compaction, and recrystallization. This process is known as *accumulation*. Glaciers also shrink in size (due to temperature increases). This process is known as *ablation*.

Earth's gravity, pushing, pulling, and tugging almost everything toward Earth's surface is involved with the movement of glaciers. Gravity moves glaciers to lower elevations by two different processes:

- **Basal Sliding**—This type of glacier movement occurs when a film of water at the base of the glacier reduces friction by lubricating the surface and allowing the whole glacier to slide across its underlying bed.
- **Internal Flow**—also called creep, forms fold structures and results from deformation of the ice crystal structure. The crystals slide over each other like a deck of cards. This type of flow is conducive to the formation of crevasses in the upper portions of the glacier. Generally, crevasses form when the lower portion of a glacier flows over sudden changes in topography.

Did You Know?

Within a glacier the velocity constantly changes. The velocity is slow next to the base of the glacier and where it is in contact with valley walls. The velocity increases toward the center and upper parts of the glacier.

Glacial Erosion

Glacial erosion has a powerful effect on land that has been buried by ice and has done much to shape our present world. Both valley and continental glaciers acquire tens of thousands of boulders and rock fragments, which, frozen into the sole of the glacier, act like thousands of files, gouging and rasping the rocks (and everything else) over which the glaciers pass. The rock surfaces display fluting, striation, and polishing effects of glacial erosion. The form and direction of these grooves can be used to show the direction in which the glaciers move.

Glacial erosion manifests itself in small scale erosional features, landform production by mountain glaciers, and ice cap and ice sheet produced landforms. These are described in the following.

- **Small-scale Erosional Features**—include glacial striations and polish. *Glacial striations* are long parallel scratches and glacial grooves that are produced at the bottom of temperate glaciers by rocks embedded in the ice scraping against the rock underlying the glacier. *Glacial polish* is characteristic of rock that has a smooth surface produced as a result of fine-grained material embedded in the glacier acting like sandpaper on the underlying surface.
- **Landforms Produced by Mountain Glaciers**—These erosion-produced features include:

  - **Cirques**—are bowl-shaped valleys formed at the heads of glaciers and below arêtes and horned mountains; they often contain a small lake called a **tarn**.
  - **Glacial Valleys**—are valleys that once contained glacial ice and become eroded into a "U" shape in cross section. "V"-shaped valleys are the result of stream erosion.
  - **Arêtes**—are sharp ridges formed by headward glacial erosion.
  - **Horns**—are sharp, pyramidal mountain peaks formed when headward erosion of several glaciers intersect.

- **Hanging Valleys**—Yosemite's Bridalveil Falls is a waterfall that plunges over a hanging valley. Generally, hanging valleys result in tributary streams that are not able to erode to the base level of the main stream; therefore, the tributary stream is left at higher elevation than the main stream, creating a hanging valley and sometimes spectacular waterfalls.
- **Fjords**—are submerged, glacially deepened, narrow inlets with sheer, high sides, a U-shaped cross profile, and a submerged seaward sill largely formed of end moraine.

- *Landforms produced by Ice Caps and Ice Sheets*

- **Abrasional Features**—are small-scale abrasional features in the form of glacial polish and striations that occur in temperate environments beneath ice caps and ice sheets.
- **Streamlined Forms**—Sometimes called "basket of eggs" topography, the land beneath a moving continental ice sheet is molded into smooth elongated forms called *drumlins*. Drumlins are aligned in the direction of ice flow, their steeper, blunter ends point toward the direction from which the ice came.

Glacial Deposits

All sediment deposited as a result of glacial erosion is called *Glacial Drift*. The sediment deposited, glacial drift, consists of rock fragments that are carried by the glacier on its surface, within the ice and at its base.

- **Ice Land Deposits**

  - **Till (or Rock Flour)**—is nonsorted glacial drift deposited directly from ice. Consisting of a random mixture of different sized fragments of angular rocks in a matrix of fine grained, sand- to clay-sized fragments, till was produced by abrasion within the glacier. After undergoing diagenesis and turning to rock, till is called *tillite*.
  - **Erratics**—is a glacially deposited rock, fragment, or boulder that rests on a surface made of different rock. Erratics are often found miles from their source and by mapping the distribution pattern of erratics geologists can often determine the flow directions of the ice that carried them to their present locations. No one has described a glacial erratic better than William Wordsworth (1807) in his classic poem *The Leech-Gatherer (or, Resolution and Independence)*.

As a huge stone is sometimes seen to lie
Couched on the bald top of an eminence;
Wonder to all who do the same espy,
By what means it could thither come, and whence;
So that it seems a thing endued with sense:
Like a Sea-beast crawled forth, that on a shelf
Of rock or sand reposeth, there to sun itself.

- **Moraines**—are mounds, ridges, or ground coverings of unsorted debris, deposited by the melting away of a glacier. Depending on where it formed in relation to the glacier, moraines can be:
  - **Ground Moraines**—are till-covered areas deposited beneath the glacier and result in a hummocky topography with lots of enclosed small basins.
  - **End Moraines and Terminal Moraines**—are ridges of unconsolidated debris deposited at the low elevation end of a glacier as the ice retreats due to ablation (melting). They usually reflect the shape of the glacier's terminus.
  - **Lateral Moraines**—are till deposits that were deposited along the sides of mountain glaciers.
  - **Medial Moraines**—When two valleys glaciers meet to form a larger glacier, the rock debris along the sides of both glaciers merge to form a medial moraine (runs down the center of a valley floor).
- **Glacial Marine Drift (Icebergs)**—are glaciers that reach lake shores or oceans and calve off into large icebergs which then float on the water surface until they melt. The rock debris that the icebergs contain is deposited on the lakebed or ocean floor when the iceberg melts.
- **Stratified Drift**—is glacial drift that can be picked up and moved by meltwater streams which can then deposit that material as stratified drift.
- **Outwash Plains**—melt runoff at the end of a glacier is usually choked with sediment and forms braided streams, which deposit poorly sorted stratified sediment in an outwash plain. They usually are flat, interlocking alluvial fans.
- **Outwash Terraces**—form if the outwash streams cut down into their outwash deposits, forming river terraces called outwash terraces.
- **Kettle Holes**—are depressions (sometimes filled by lakes; e.g., Minnesota, the land of a thousand lakes) due to melting of large blocks of stagnant ice, found in any typical glacial deposit.
- **Kames**—are isolated hills of stratified material formed from debris that fell into openings in retreating or stagnant ice.

- **Eskers**—are long, narrow, and often branching sinuous ridges of poorly sorted gravel and sand formed by deposition from former glacier streams.

Wind Erosion

Wind alone is not a significant agent of erosion; it is very limited in extent and effect. It is largely confined to desert regions, but even there it is limited to a height of about 18 inches above ground level. Wind does have the power, however, to transport, to deposit, and to erode sediment.

Did You Know?

Wind is common in deserts because the air near the surface is heated and rises, cooler air comes in to replace hot rising air and this movement of air results in winds. Also, arid desert regions have little or no soil moisture to hold rock and mineral fragments.

Wind Sediment Transport

Sediment near the ground surface is transported by wind in a process called *saltation* (Latin, saltus, "leap"). Similar to what occurs in the bed load of streams, wind saltation refers to short jumps (leaps) of grains dislodged from the surface and leaping a short distance. As the grains fall back to the surface they dislodge other grains that then get carried by wind until they collide with ground to dislodge other particles. Above ground level, wind can swoop down to the surface and lift smaller particles, suspending them (making them airborne) in the wind and possibly propelling them long distances.

Wind-Driven Erosion

As mentioned, wind by itself has little if any effect on solid rock. But, in arid and semiarid regions, wind can be an effective geologic agent anywhere that it is strong enough (possesses high enough velocity) to pick up a load of rock fragments (which may become effective tools of erosion in the land-forming process). Wind can erode by *deflation* and *abrasion*.

Deflation—(or blowing away) is the lowering of the land surface due to removal of fine-grained particles by the wind. Deflation concentrates the coarser grained particles at the surface, eventually resulting in a relatively smooth surface composed only of the coarser-grained fragments that cannot be transported

---

Did You Know?

Sand ripples occur as a result of large grains accumulating as smaller grains are transported away. Ripples form in lines perpendicular to wind direction. Wind blown dust is sand-sized particles which generally do not travel very far in the wind, but smaller-sized fragments can be suspended in the wind for much longer distances.

---

by the wind. Such a coarse-grained surface is called *desert pavement* (see Figure 17). Some of these coarser-grained fragments may exhibit a dark, enamel-like coat of iron or manganese called *desert varnish*.

Deflation may create several types of distinctive features. For example, *lag gravels* are formed when the wind blows away finer rock particles, leaving behind a residue of coarse gravel and stones. *Blowouts* may be developed where wind has scooped out soft unconsolidated rocks and soil.

Abrasion—Wind abrades (sand blasts) by picking up sand and dust particles which are transported as part of its load. Abrasion is restricted to a distance of about a meter or two above the surface because sand grains are lifted a short distance. The destructive action of these wind-blown abrasives may wear away wooden telephone poles and fence posts, and abrade, scour, or groove solid rock surfaces.

Wind abrasion also plays a part in the development of such landforms as isolated rocks (pedestals and table rocks) which have had their bases undercut by wind-blown sand and grit. *Ventifacts* are another interesting and relatively

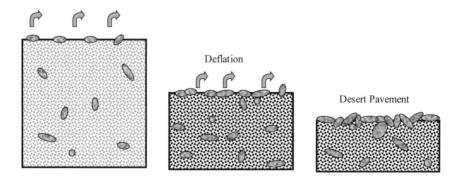

FIGURE 17.   Wind-driven deflation processes

common product of wind erosion. These are any bedrock surface or stone or pebble that has been abraded or shaped by wind-blown sediment in a process similar to sand blasting. Ventifacts are formed when the wind blows sand against the side of the stone, shaping it into a flat, polished surface. At a much larger scale, elongate ridges called *yardangs* form by the abrasion and streamlining rock structures oriented parallel to the prevailing wind direction.

Wind Deposition

The velocity of the wind and the size, shape, and weight of the rock particles determines the manner in which wind carries its load. Wind-transported materials are most commonly derived from flood plains, beach sands, glacial deposits, volcanic explosions, and dried lake bottoms—places containing light ash and loose, weathered rock fragments.

The wind is capable of transporting large quantities of material for very great distances. The wind deposits sediment when its velocity decreases to the point where the particles can no longer be transported. Initially (in a strong wind), part of the sediment load rolls or slides along the ground (bed load). Some sand particles move by a series of leaping or bounding movements (saltation). And lighter dust may be transported upward (suspension) into higher, faster moving wind currents, traveling many thousands of miles.

As mentioned, the wind will begin to deposit its load when its velocity is decreased or when the air is washed clean by falling rain or snow. A decrease in wind velocity may also be brought about when the wind strikes some barrier-type obstacle (fences, trees, rocks, human-made structures) in its path. As the air moves over the top of the obstacle, streamlines converge and the velocity increases. After passing over the obstacle, the streamlines diverge and the velocity decreases. As the velocity decreases, some of the load in suspension can no longer be held in suspension, and thus drops out to form a deposit. The major types of wind-blown or eolian deposits are dunes and loess.

Dunes

Sand dunes are asymmetrical mounds with a gentle slope in the upwind direction and a steep slope on the downwind side. Dunes vary greatly in size and shape and form when there is a ready supply of sand, a steady wind, and some kind of obstacle or barrier such as rocks, fences, or vegetation to trap some of the sand. Sand dunes form when moving air slows down on the downwind side of an obstacle. Dunes may reach heights up to 500 m and cover large areas. Types of sand dunes include barchan, transverse, longitudinal, and parabolic.

- **Barchan Dunes**—are crescent-shaped dunes characterized by two long, curved extensions pointing in the direction of the wind, and a curved slip face on the downwind side of the dune. These dunes are formed in areas where winds blow steadily and from a single direction.
- **Transverse Dunes**—form along sea coasts and lake shores and may be fifteen feet high and half a mile in length. Transverse dunes develop with their long axis at right angles to the wind.
- **Longitudinal Dunes**—are long ridge-like dunes that develop parallel to the wind.
- **Parabolic Dunes**—are U-shaped dunes with convex noses trailed by elongated arms. They are usually stabilized by vegetation, and occur where there is abundant vegetation, a constant wind direction, and an abundant sand supply.

Loess

Loess is a yellowish, fine-grained, non-stratified material carried by the wind and accumulated in deposits of dust. The materials forming loess are derived from surface dust originating primarily in deserts, river floodplains, deltas, and glacial outwash deposits. Loess is cohesive and possesses the property of forming steep bluffs with vertical faces such as the deposits found in the pampas of Argentina and the lower Mississippi River valley.

Mass Wasting

Mass wasting, or mass movement, takes place as earth materials (loose uncemented mixture of soil and rock particles known as *regolith*) move downslope in response to gravity without the aid of a transporting medium such as water, ice, or wind—though these factors play a role in regolith movement. This type of erosion is apt to occur in any area with slopes steep enough to allow downward movement of rock debris. Some of the factors which help gravity overcome this resistance are discussed below.

- **Gravity**—The heavy hand of gravity constantly pulls everything, everywhere toward Earth's surface. On a flat surface, parallel to Earth's surface, the constant force of gravity acts downward. This downward force prevents gravitational movement of any material that remains on or parallel to a flat surface.

  On a slope, the force of gravity can be resolved into two components: a component acting perpendicular to the slope, and a component acting tangential to the slope. Thus, material on a slope is pulled inward in a direction that is perpendicular (the glue) to the slope. This helps prevent material from sliding downward. However, as stated previously, on a slope, another component of gravity exerts a force (a constant tug) that acts to pull material down a slope, parallel to the surface of the slope. Known as *shear stress*, this force of gravity exerts stress in direct relationship to the steepness of the slope. That is, shear stress increases as the slope steepens. In response to increased shear stress, the perpendicular force (the glue) of gravity decreases.

• **Water**—Even though mass wasting may occur in either wet or dry materials, water greatly facilitates downslope movements; it is an important agent in the process of mass wasting. Water will either help hold material together (act as glue—demonstrated in building beach sandcastles with slightly dampened sand), increasing its angle of repose, or cause it to slide downward like a liquid (acting like a lubricant). Water may soften clays, making them slippery, add weight to the rock mass, and, in large amounts, may actually force rock particles apart, thus reducing soil cohesion.

• **Freezing and Thawing**—Earlier the erosive power of frost wedging (water contained in rock and soil expands when frozen) was pointed out. Mass-wasting in cold climates is governed by the fact that water is frozen as ice during long periods of the year, especially in high altitude regions. Ice, although it is solid, does have the ability to flow (glacial-movement effect), and alternate periods of freezing and thawing can also contribute to movement, and in some instances, ice expansion may be great enough to force rocks downhill.

• **Undercutting**—occurs when streams erode their banks or surf action along a coast undercuts a slope making it unstable. Undercutting can also occur when human-made excavations remove support and allow overlying material to fall.

• **Organic Activities**—Whenever animals burrow into the ground, they disturb soil materials, casting rocks out of their holes as they dig; these are commonly piled up downslope. Eventually, weather conditions and the constant force applied by gravity can put these piles into motion. Animals also contribute to mass wasting whenever they walk on soil surfaces; their motions can knock materials downhill.

• **Shock Waves or Vibrations**—A sudden strong shock or vibration, such as an earthquake, faulting, blasting, and heavy traffic, can trigger slope instability. Minor shocks like heavy vehicles rambling down the road, wind blowing the trees, or human-made explosions can also trigger mass-wasting events such as landslides.

Did You Know?

When shear on a slope decreases, material may still be stuck to the slope and prevented from moving downward by the force of friction. It may be held in place by the frictional contact between the particles making up that material. Contact between the surfaces of the particles creates a certain amount of tension that holds the particles in place at an angle. The steepest angle at which loose material on a slope remains motionless is called the angle of repose (generally about 35 degrees). Particles with angle edges that catch on each other also tend to have a higher angle of repose than those that have become rounded through weathering and that simply roll over each other.

Kinds of Mass Movements

A landslide is a mass movement that occurs suddenly and violently. In contrast, soil creep is mass movement that is almost imperceptible. These processes can be divided into two broad categories: rapid and slow movements. *Rapid Movements* include landslides, slumps, mudflows, and earthflows. *Slow movements* include soil creep and solifluction.

Rapid Movements

- **Landslides**—are by far the most spectacular and violent of all mass movements. Landslides are characterized by the sudden movement of great quantities of rock and soil downslope. Such movements typically occur on steep slopes that have large accumulations of weathered material. Precipitation in the form of rain or snow may seep into the mass of steeply sloping rock debris, adding sufficient weight to start the entire mass sliding.
- **Slumps**—special landslides occur along curved surfaces. The upper surface of each slump block remains relatively undisturbed, as do the individual blocks. Slumps leave arcuate (*L.* curved like a bow) scars or depressions on the hill slope. Heavy rains or earthquakes usually trigger slumps. Slump is a common occurrence along the banks of streams or the walls of steep valleys.
- **Mudflows**—are highly fluid, high velocity mixtures of sediment and water that have a consistency of wet concrete. Mass wasting of this type typically occurs as certain arid or semi-arid mountainous regions are subjected to unusually heavy rains.

- **Earthflows**—are usually associated with heavy rains and move at velocities between several centimeters and hundreds of meters per year. They
usually remain active for long periods of time. They generally tend to be
narrow tongue-like features that begin at a scarp or small cliff.

Slow Movements

- **Soil Creep**—This continuous movement, usually so slow as to be imperceptible, normally occurs on almost all slopes that are moist but not steep
enough for landslides. Soil creep is usually accelerated by frost wedging,
alternate thawing and freezing, and by certain plant and animal activities.
Evidence for creep is often seen in bent trees, offsets in roads and fences,
and inclined utility poles.
- **Solifluction**—This downslope movement is typical of areas where the
ground is normally frozen to considerable depth—arctic, subarctic, and
high mountain regions. The actual soil flowage occurs when the upper portion of the mantle rock thaws and becomes water-saturated. The underlying, still frozen subsoil acts as a slide for the sodden mantle rock which will
move down even the gentlest slope.

Did You Know?

Landslides constitute a major geologic hazard because they are widespread, occur in all 50 states and U.S. territories, and caused $1–2
billion in damages and more than 25 fatalities on average each year.
Expansion of urban and recreational developments into hillside
areas leads to more people that are threatened by landslides each
year. Landslides commonly occur in connection with other major
natural disasters such as earthquakes, volcanoes, wildfires, and floods
(USGS 2008).

Desertification

Deserts are areas where the amount of precipitation received is less than
the potential evaporation (<10 in./year); they cover roughly 30% of the
Earth's land surface—areas we think of as arid. *Desertification* occurs in hot
areas far from sources of moisture, areas isolated from moisture by high
mountains, in coastal areas along which there are onshore winds and coldwater currents, and/or high pressure areas where descending air masses
produce warm, dry air.

According to USGS (1997), the world's great deserts were formed by natural processes interacting over long intervals of time. During most of these times, deserts have grown and shrunk independently of human activities. Desertification does not occur in linear, easily mappable patterns. Deserts advance erratically, forming patches on their borders. Scientists question whether desertification, as a process of global change, is permanent or how and when it can be halted or reversed.

Water Flow (Streams): Transport of Material (Load)

Water flow in streams generally is turbulent. Turbulence exerts a shearing force that causes particles to move along the stream bed by pushing, rolling, and skipping, referred to as bed load. This same shear causes turbulent eddies that entrain particles in suspension (called the suspended load—particles size under 0.06 mm). Entrainment is the incorporation of particles when stream velocity exceeds the entraining velocity for a particular particle size.

Did You Know?

Entrainment is a natural extension of erosion and is vital to the movement of stationary particles in changing flow conditions. Remember, all sediments ultimately derive from erosion of basin slopes, but the immediate supply usually derives from the stream channel and banks, while the bed load comes from the streambed itself and is replaced by erosion of bank regions.

The entrained particles in suspension (suspended load) also include fine sediment, primarily clays, silts, and fine sands that require only low velocities and minor turbulence to remain in suspension. These are referred to as wash load (under 0.002 mm), because this load is "washed" into the stream from banks and upland areas (Gordon et al. 1992; Spellman 1996).

Thus the suspended load includes the wash load and coarser materials (at lower flows). Together, the suspended load and bed load constitute the solid load. It is important to note that in bedrock streams, the bed load will be a lower fraction than in alluvial streams where channels are composed of easily transported material.

A substantial amount of material is also transported as the dissolved load. Solutes (ions) are generally derived from chemical weathering of bedrock and soils, and their contribution is greatest in sub-surface flows and in regions of limestone geology.

The relative amount of material transported as solute rather than solid load depends on basin characteristics, lithology (i.e., the physical character of rock) and hydrologic pathways. In areas of very high runoff, the contribution of solutes approaches or exceeds sediment load, whereas in dry regions, sediments make up as much as 90% of the total load.

Deposition occurs when stream competence (i.e., the largest particle that can be moved as bed load and the critical erosion—competent—velocity is the lowest velocity at which a particle resting on the streambed will move) falls below a given velocity. Simply stated: the size of the particle that can be eroded and transported is a function of current velocity.

Sand particles are the most easily eroded. The greater the mass of larger particles (e.g., coarse gravel) the higher the initial current velocities must be for movement. However, smaller particles (silts and clays) require even greater initial velocities because of their cohesiveness and because they present smaller, streamlined surfaces to the flow. Once in transport, particles will continue in motion at somewhat slower velocities than initially required to initiate movement, and will settle at still lower velocities.

Particle movement is determined by size, flow conditions, and mode of entrainment. Particles over 0.02 mm (medium-coarse sand size) tend to move by rolling or sliding along the channel bed as traction load. When sand particles fall out of the flow, they move by saltation or repeated bouncing. Particles under 0.06 mm (silt) move as suspended load and particles under 0.002 (clay), indefinitely, as wash load. A considerable amount of particle sorting takes place because of the different styles of particle flow in different sections of the stream (Richards 1982; Likens 1984).

Unless the supply of sediments becomes depleted, the concentration and amount of transported solids increases. However, discharge is usually too low, throughout most of the year, to scrape or scour, shape channels, or move significant quantities of sediment in all but sand-bed streams, which can experience change more rapidly. During extreme events, the greatest scour occurs and the amount of material removed increases dramatically.

Mt. St. Helens: Lahars[1]

Speaking of an extreme event, we have used the May 18, 1980, eruption of Mt. St. Helens in this text and again in this section to illustrate the impact and influence of a major natural occurrence on reshaping Earth's surface

1. Information in this section is taken from Various USGS cited sources. Anecdotal information is provided by co-author Frank R. Spellman based on his 30 years of observation and onsite research.

features. In regard to the erosional impact of water flow, few natural occurrences or events have a more profound impact on Earth's surface than debris flows, mudflows, and lahars generated by volcanic eruption. The Mt. St. Helens lahars resulted when hot ash from the May 18 eruption fell on and melted snow and glacial ice on the upper slopes of Mt. St. Helens.

In discussing the impact of the Mt. St. Helens eruption and the mudflow, debris flow, and lahars it is important to define what these terms mean. According to Brantley and Power (1985), *lahar* is an Indonesian word describing mudflows and debris flows that originate from the slopes of a volcano. Both types of flows contain a high concentration of rock debris to give them the internal strength necessary to transport huge boulders as well as trees, buildings, and bridges and to exert extremely high impact forces against objects in their paths. Debris flows are coarser and less cohesive than mudflows. As lahars become diluted in a downstream direction they become hyperconcentrated stream flows.

In addition, inflow of debris and sediment into streams caused by natural events can be increased and/or decreased as a result of human activities. For example, poor agricultural practices and deforestation greatly increase erosion; thus increasing inflow of debris and sediment into streams. Human-made structures such as dams and channel diversions can, on the other hand, greatly reduce sediment inflow. Keep in mind that we pointed out earlier that a human-made dam was installed in the lower reaches of the Toutle River to impede the flow of Mt. St. Helens mudflow into the Columbia River.

References and Recommended Reading

Abbott, P.L. 1996. *Natural Disasters*. New York: Wm. C. Brown Publishing Co.

Anderson, J.G., Bodin, P., Brune, H.N., Prince, J., Singh, S.K., Quaas, R., and Onate, R. 1986. Strong ground motion from the Michoacan, Mexico earthquake. *Science*. V. 233, 1043–49.

Associated Press, in the *Lancaster New Era* (Lancaster, PA). "Tougher air pollution standards too costly, Midwestern states say." September 25, 1998.

Associated Press, in the *Lancaster New Era* (Lancaster, PA). "Ozone hole over Antarctica at record size." September 28, 1998.

Associated Press, in the *Virginian-Pilot* (Norfolk, VA). "Does warming feed El Niño?" p. A-15, December 7, 1997.

ASTM. 1969. *Manual on water*. Philadelphia: American Society for Testing and Materials.

Atkinson, L., and Sancetta, C. 1993. Hail and farewell. *Oceanography* 6 (34).

Barnes, F.A. 1987. *Canyon Country Arches and Bridges*. Self published.

Blatt, H., Berry, W.B.N., and Brande, S. 1991. *Principles of Stratigraphic Analysis.* Boston: Blackwell Scientific Publications.

Brady, N. C., and Weil, R. R. 1996. *The Nature and Properties of Soils* (11th ed.). Upper Saddle River, NJ: Prentice-Hall.

Browning, J.M. 1973. Catastrophic rock slides. Mount Huascaran, north-central, Peru, May 32, 1970. *Bulletin American Association of Petroleum Geologists* 57: 1335–1341.

Carson, R. 1962. *Silent Spring.* Boston: Houghton Mifflin.

Chernicoff, S. 1999. *Geology.* Boston: Houghton Mifflin.

Ciardi, J. 1997. From Stoneworks, in *The Collected Poems of John Ciardi,* Cifelli, E. M. (ed.). Fayettville: University of Arkansas Press.

Coch, N.K. 1995. *Geohazards, Natural and Human.* New York: Prentice-Hall.

Davis, G.H., and Pollock, G.L. 2003. Geology of Bryce Canyon National Park, Utah. In *Geology of Utah's Parks and Monuments,* 2nd ed. Eds. Sprinkel, D.A., et al. Salt Lake City: Utah Geological Association.

Dolan, E. F. 1991. *Our Poisoned Sky.* New York: Cobblehill Books.

Eagleman, J. 1983. *Severe and Unusual Weather.* New York: Van Nostrand Reinhold.

Eswaran, H. 1993. "Assessment of global resources: Current status and future needs." *Pedologie,* XL111, 19–39.

Foth, H. D. 1978. *Fundamentals of Soil Science,* 6th ed. New York: John Wiley & Sons, 1978.

Franck, I., and Brownstone, D. 1992. *The Green Encyclopedia.* New York: Prentice-Hall.

Francis, P. 1993. *Volcanoes, A Planetary Perspective.* New York: Oxford University Press.

GeoMan. 2008. *Bowen's Reaction Series.* Accessed @ http://jersey.uorgeon.edu/~mstrick/ Ask GeoMan/ geo Querry32.html.

Giller, P.S., and Jalmqvist, B. 1998. *The Biology of Streams and Rivers.* Oxford: Oxford University Press.

Goodwin, P.H. 1998. *Landslides, Slumps, and Creep.* New York: Franklin Watts.

Gordon, N.D., McMahon, T.A., and Finlayson, B.L. 1992. *Stream Hydrology: An Introduction for Ecologists.* Chichester: John Wiley & Sons.

Gould, S.J. 1989. *Wonderful Life.* New York: W.W. Norton.

Hansen, J.E. et al. 1986. "Climate Sensitivity to Increasing Greenhouse Gases," Greenhouse Effect and Sea Level Rise: A Challenge for this Generation, ed., M. C. Barth and J. G. Titus, New York: Van Nostrand Reinhold.

Hansen, J.E. et al. 1989. Greenhouse Effect of Chlorofluorocarbons and Other Trace Gases. *Journal of Geophysical Research* 94 (November): 16,417–16,421.

Harris, E.C. 1979. *Principles of Archaeological Stratigraphy.* New York: Academic Press.

Holmes, A. 1937. *The Age of Earth.* London: Nelson.

Holmes, A. 1978. *Principles of Physical Geology,* 3rd ed., New York: John Wiley & Sons.

Jennings, T. 1999. *Landslides and Avalanches.* North Mankato, MN: Thame-side Press.

Keller, E.A. 1985. *Environmental Geology,* 4th ed. New York: Merrill Publishing Co.

Kemmer, F. N. *Water: The Universal Solvent.* Oak Ridge, IL: NALCO Chemical Company.

Kiersh, G.A. 1964. Vaiont reservoir disaster. *Civil Engineering* 34.

Konigsburg, E. M. 1996. *The View From Saturday.* New York: Scholastic.

Lane, Megan. 2000. It's raining fish! BBC, August 7.

Leopold, L.B. 1994. A View of the River. Cambridge, MA: Harvard University Press.

Levin, H.L. 2003. *The Earth Through Time,* 7th ed. Hoboken, NJ: John Wiley & Sons.

Levin, H.L. 2005. *The Earth Through Time,* 8th ed. Hoboken, NJ: John Wiley & Sons.

Likens, W.M. 1984. Beyond the Shoreline: A Watershed Ecosystem Approach. *Vert. Int. Ver. Theor. Aug Liminol.* 22: 1–22.

Lyman, J., and Fleming, R.H. 1940. Composition of Seawater. *J Mar Res* **3**: 134–146.

Macmillan Publishing Company. 1992. *The Way Nature Works.* New York: Macmillan Publishing Company.

McKnight, T. 2004. *Geographica: The Complete Illustrated Atlas of the World.* New York: Barnes and Noble Books.

Morisawa, M. 1968. *Stream: Their Dynamics and Morphology.* New York: McGraw-Hill.

Mowat, F. 1957. *The Dog Who Wouldn't Be.* New York: Willow Books.

Murck, B.W., Skinner, B.J., and Porter, S.C. 1997. *Dangerous Earth, An Introduction to Geologic Hazards.*

NOAA. 2008. Global Warming: Frequently asked questions. Accessed 11/21/08 @http://lwf.ncdc.noaa.gov/oa/climate/globalwarming.html.

NPS. 2008. *The Hoodoo.* Washington, DC: National Park Service.

Oreskes, N. (ed). 2003. *Plate Tectonics: An Insiders History of the Modern Theory of the Earth.* New York: Westview.

Palmer, D. 2005. *Earth Time: Exploring the Deep Past from Victorian England to the Grand Canyon.* New York: John Wiley & Sons.

Richards, K. 1982. *Rivers: Form and Processes in Alluvial Channels.* London: Mehuen.

Skinner, B.J., and Porter, S.C. 1995. *The Dynamic Earth, An Introduction to Physical Geology,* 3rd ed. New York: John Wiley & Sons.

Soil Survey Staff. 1994. *Keys to Soil Taxonomy.* Washington, DC: USDA Natural Resources Conservation Service.

Spellman, F.R. 1996. *Stream Ecology and Self-Purification.* Lancaster, PA: Technomic Publishing Company.

Spellman, F.R. 1998. *The Science of Environmental Pollution.* Boca Raton: CRC Press.

Spellman, F.R. 2010. *Geography for Nongeographers.* Lanham, MD: Government Institutes.

Spellman, F.R., and Whiting, N.E. 2006. *Environmental Science and Technology,* 2nd edition. Lanham, MD: Government Institutes.

Stanley, S.M. 1999. *Earth System History.* New York: W.H. Freeman.

Stephens, J.C. et al. 1984. Organic soils subsidence. *Geological Society of American Reviews in Engineering Geology,* v. VI, p.3.

Sverdrup, H.U., Johnson, M.W., and Fleming, R.H. 1942. *The Oceans: Their Physics, Chemistry and General Biology.* New York: Prentice-Hall.

Swanson, D.A., Wright, T.H., and Helz, R.T. 1975. Linear vent systems and estimated rates of magma production and eruption of the Yakima basalt on the Columbia Plateau. *American Journal of Science* 275: 877–905.

Tarbuck, E.J., and Lutgens, F.K. 2000. *Earth Science*. Upper Saddle River, NJ: Prentice-Hall.

Tarbuck, E.J., Lutgens F.K., and Tasa, D. 2007. *Earth: An Introduction to Physical Geology*, 9th ed. Upper Saddle River, NJ: Prentice-Hall.

Tilling, R.I. 1984. *Eruptions of Mount St. Helens: Past Present and Future*. Department of the Interior, U.S. Geological Survey.

Time Magazine. Global Warming: It's Here . . . and Almost Certain to Get Worse. August 24, 1998.

Tomera, A. N. 1989. *Understanding Basic Ecological Concepts*. Portland, ME: J. Weston Walch, Publisher.

Turcotte, D.L., and Schubert, G. 2002. *Geodynamics: Second Edition*. New York: John Wiley & Sons.

USA TODAY. "Global warming: Politics and economics further complicate the issue," p. A-1, 2, December 1, 1997.

USDA. 1975. *Soil Survey Staff, Soil Classification, A Comprehensive System*. Washington, DC: USDA Natural Resources Conservation Service.

USDA. 1975. *Soil Taxonomy: A Basic System of Soil Classification for Making and Interpreting Soil Surveys*. Washington, DC.: USDA Natural Resources Conservation Service.

USGS. 1989. Lessons learned from the Loma Prieta, California, earthquake of October 17, 1989. Circular 1045.

USGS. 1997. Desertification. Accessed 7/08/08 @ http://pubs.usgs.gov/gip/deserts/desertificaiton/.

USGS. 2000. *Photo Glossary of Volcano Terms*. Accessed 06/01/08 @ http:volcanoes.usgs.gov/Products/Pglossary/LavaCascade.html.

USGS. 2001. *Radiometric Time Scale*. Washington, DC: Accessed 10/20/08 @ http://pubs.usgs.

USGS. 2006. *Sinkholes*. Accessed 7/06/08 @ http://ga.wwater.usgs.gov/edu/earthqw sinkholes.htm.

USGS. 2007. *U.S. Geological Survey Fact sheet 2007-3015: U.S. Geological Survey Geologic Names committee*, 2007. Washington, DC: USGS.

USGS. 2008. *Earthquake Hazards Program*. Accessed 09/14/08 @ http://earthquakes.usgs.gov/ learning/topics/megaqk_facts_fantasy.php.

USGS. 2008. *Landslide Hazards Program*. Accessed 7/22/08 @ http://landslides.usgs.gov/. gov/gip/geotime/radiometric.html.

USGS. 2008. *Principal Types of Volcanoes*. Accessed 05/31/08 @ http://pubs.usgs.gov/ gip/volc/ types.html.

USGS. 2008. *The Geologic Time Scale*. Accessed 10/10/08 @ http://vulcan.wr.usgs.gov/ Glossary/geo_time_scale.html.

Voltaire. 1991. *Candide*. New York: Dover Publishing Inc.

Vreeland, R.H. 1994. *Nature's Bridges and Arches, Volume 1—General Information*, 2nd ed. Self published.

Walker, J. 1992. *Avalanches and Landslides*. New York: Gloucester Press.

Williams, H., and McKinney, A.R. 1979. *Volcanology*. New York: Freeman & Copper Co.

III

THE ATMOSPHERE: EARTH'S BREATH

This most excellent canopy, the air, look you, this brave o'erhanging firmament, this majestical roof fretted with golden fire.

—Shakespeare, *Hamlet*

4

The Atmosphere

The sky was delicious—sweet enough for the breath of angels. Every draught of it gave a separate and distinct piece of pleasure. I do not believe that Adam and Eve ever tasted better in their balmiest nook.

—John Muir, *The Craftsman*, 1905

Introduction

SEVERAL THEORIES of cosmogony attempt to explain the origin of the universe. Without speculating on the validity of any one theory, the following is simply the authors' view.

Earth's Atmosphere Is Born

The time: 4.5 billion years ago.

Before the universe there was time. Only time; otherwise, the vast void held only darkness—everywhere.

Overwhelming darkness.

Not dim . . . not murky . . . not shadowy, or not unlit. Simple nothingness—nothing but darkness, a shade of black so intense we cannot fathom or imagine it today—Light had no existence—this was black of blindness, of burial in the bowels of the Earth, the blackness of no other choice.

With time—eons of time—darkness came to a sudden, smashing, shattering, annihilating, scintillating, cataclysmic end (a beginning)—and there was light . . . light everywhere. This new force replaced darkness and lit up the expanse without end, creating a brightness fed by billions of glowing round masses so powerful as to renounce and overcome the darkness that had come before.

With the light was heat-energy which shone and warmed and transformed into mega-mega-mega trillions of super-excited ions, molecules, and atoms—heat of unimaginable proportions, forming gases—gases we don't even know how to describe, how to quantify, let alone how to name. These were gases that we do not know what we do not know about them. But gases they were—and they were everywhere.

With light, energy, heat, and gases present, the stage was set for the greatest show of all time, anywhere—ever: the formation of the Universe.

Over time—time in stretches we can't imagine, so vast we can't contemplate them meaningfully—the heat, the light, the energy and gases all came together and grew, like an expanding balloon, into one solid glowing mass. But it continued to grow, with the pangs, sweating, and moans accompanying any birthing, until it had reached the point of no return—explosion level. And it did; it exploded with the biggest bang of all time (with the biggest bang hopefully of all time—one none of us wants to ever see or feel or not feel).

The Big Bang sent masses of hot gases in all directions—to the farthest reaches of anything, everything—into the vast, wide, measureless void. Clinging together as they rocketed, soared, and swirled, forming galaxies that gradually settled into their arcs through the void, constantly propelled away from the force of their origin, these masses began their eternal evolution.

Two masses concern us . . . the Sun and Earth.

Forces well beyond the power of the Sun (beyond anything imaginable) stationed this massive gaseous orb approximately 93,000,000 miles from the dense molten core enveloped in cosmic gases and the dust of time that eventually became the insignificant mass we now call Earth.

Distant from the Sun, Earth's mass began to cool, slowly; the progress was slower than we can imagine, but cool it did. While the dust and gases cooled, Earth's inner core, mantle, and crust began to form—no more a quiet or calm evolution than the revolution that cast it into the void had been.

Downright violent was this transformation—the cooling surface only a facade for the internal machinations going on inside, out-gassing from huge, deep destructive vents (we would call them volcanoes today) erupting continuously—never stopping, blasting away, delivering two main ingredients: magma and gas.

The magma worked to form the primitive features of Earth's early crust. The gases worked to form Earth's initial atmosphere—our point of interest: the atmosphere. Without atmosphere, what is there?

About 4 billion years before the present, Earth's early atmosphere was chemically reducing, consisting primarily of methane, ammonia, water vapor, and hydrogen—for life as we know it today, an inhospitable brew.

Earth's initial atmosphere was not a calm, quiet, quiescent environment—to the contrary, it was an environment best characterized as dynamic—ever changing—where bombardment after bombardment by intense, bond-breaking ultra-violet light, along with intense lightning and radiation from radionuclides, provided energy to bring about chemical reactions that resulted in the production of relatively complicated molecules, including amino acids and sugars (building blocks of life).

About 3.5 billion years before the present, primitive life formed in two radically different theaters: on Earth and below the primordial seas near hydrothermal vents that spotted the wavering, water-covered floor.

Initially, on Earth's unstable surface, very primitive life forms derived their energy from fermentation of organic matter formed by chemical and photochemical processes, then gained the ability to produce organic matter (CH_2O) by photosynthesis.

Thus the stage was set for the massive biochemical transformation that resulted in the production of almost all the atmosphere's O_2.

The O_2 initially produced was quite toxic to primitive life forms. However, much of this oxygen was converted to iron oxides by reaction with soluble iron. This process formed enormous deposits of iron oxides—the existence of which provides convincing evidence for the liberation of O_2 in the primitive atmosphere.

Eventually, enzyme systems developed that enabled organisms to mediate the reaction of waste-product oxygen with oxidizable organic matter in the sea. Later, the mode of waste gradient disposal was utilized by organisms to produce energy by respiration, which is now the mechanism by which non-photosynthetic organisms obtain energy. In time, O_2 accumulated in the atmosphere. In addition to providing an abundant source of oxygen for respiration, the accumulated atmospheric oxygen formed an ozone (O_3) shield—the O_3 shield absorbs bond-rupturing ultra-violet radiation.

With the O_3 shield protecting tissue from destruction by high energy ultra-violet radiation, the Earth, although still hostile to life forms we are familiar with, became a much more hospitable environment for life (self-replacing molecules), and life forms were enabled to move from the sea (where they flourished next to the hydrothermal gas vents) to the land. And from that point on to the present, Earth's atmosphere became more life-form friendly.

Earth's Thin Skin

Shakespeare likened it to a majestic overhanging roof (constituting the transition between its surface and the vacuum of space); others have likened it to the skin of an apple. Both these descriptions of our atmosphere are fitting, as is being described as the Earth's envelope, veil, or gaseous shroud. The atmosphere is more like the apple skin, however. This thin skin, or layer, contains the life-sustaining oxygen (21%) required by all humans and many other life forms; the carbon dioxide (0.03%) so essential for plant growth; the **nitrogen** (78%) needed for chemical conversion to plant nutrients; the trace gases such as methane, argon, helium, krypton, neon, xenon, ozone, and hydrogen; and varying amounts of water vapor and airborne particulate matter. Life on earth is supported by this atmosphere, solar energy, and other planet's magnetic fields.

Gravity holds about half the weight of a fairly uniform mixture of these gases in the lower 18,000 feet of the atmosphere; approximately 98% of the material in the atmosphere is below 100,000 feet.

Atmospheric pressure (or air pressure—measured by a barometer) varies from 1000 millibars (mb) at sea level to 10 mb at 100,000 feet. From 100,000 to 200,000 feet the pressure drops from 9.9 mb to 0.1 mb and so on.

The atmosphere is considered to have a thickness of 40–50 miles. The atmosphere is composed of five layers, from bottom to top: Troposphere, stratosphere, mesosphere, thermosphere, and exosphere. However, here we are primarily concerned with the troposphere, the part of the earth's atmosphere that extends from the surface to a height of about 27,000 ft above the poles, about 36,000 ft in midlatitudes, and about 53,000 ft over the equator. As mentioned, above the troposphere is the stratosphere, a region that increases in temperature with altitude (the warming is caused by absorption of the sun's radiation by ozone) until it reaches its upper limit of 260,000 ft.

Troposphere

Extending above earth approximately 27,000 feet, the troposphere is the focus of this text because people, plants, animals, and insects live here and depend on this thin layer of gases. Moreover, all of the Earth's weather takes place within the troposphere. The troposphere has the highest density of gas molecules. The troposphere begins at ground level and extends 7.5 miles up into the sky where it meets with the second layer called the stratosphere.

Did You Know?

It was pointed out earlier that the gases that are so important to life on Earth are primarily contained in the troposphere. Note that another important substance contained in the troposphere is water vapor. Along with being the most remarkable of the trace gases contained in the troposphere, water vapor is also the most variable. Considerable attention is paid to water moisture of the air surrounding Earth. Unlike the other trace gases in the atmosphere, water vapor alone exists in gas, solid, and liquid forms. It also functions to add and remove heat from the air when it changes from one form to another. Infrared light, a frequency of electromagnetic radiation, is absorbed by gases in the troposphere. Water vapor (in conjunction with air-borne particles) is essential for the stability of earth's **ecosystem**. This water vapor-particle combination interacts with the global circulation of the atmosphere and produces the world's weather, including clouds and precipitation.

Stratosphere

The stratosphere begins at the 7.5 mile point and reaches 21.1 miles into the sky. In the rarified air of the stratosphere, the significant gas is ozone (life-protecting ozone—not to be confused with pollutant ozone), which is produced by the intense electromagnetic ultraviolet radiation from the sun. In quantity, the total amount of ozone in the atmosphere is so small that if it were compressed to a liquid layer over the globe at sea level, it would have a thickness of less than 3/16 in.

Ozone contained in the stratosphere can also impact (add to) ozone in the troposphere. Normally, the troposphere contains about 20 parts per billion parts of ozone. On occasion, however, via the jet stream, this concentration can increase to 5–10 times higher than average.

Did You Know?

The troposphere, stratosphere, mesosphere, and thermosphere act together as a giant safety blanket. They keep the temperature on the earth's surface from dipping to extreme icy cold that would freeze everything solid, or from soaring to blazing heat that would burn up all life.

A Jekyll and Hyde View of the Atmosphere

When non-city dwellers look up into that great natural canopy above our heads, they see many features provided by our world's atmosphere that we know and enjoy: the blueness and clarity of the sky, the color of a rainbow, the spattering of stars reaching every corner of blackness, the magical colors of a sunset. The air they breathe carries the smell of ocean air, the refreshing breath of clean air after a thunderstorm, the light touch of warmth or cold, and the beauty contained in a snowflake.

But the atmosphere sometimes personifies another face—Mr. Hyde's face. The terrible destructiveness of a hurricane, a tornado, a monsoon, a typhoon, or a hailstorm, the wearying monotony of winds carrying dust and rampaging windstorms carrying fire up a hillside—these are some of the terrifying aspects of the other face of the disturbed atmosphere.

The atmosphere can also personify a Hyde-like face whenever man is allowed to pour his filth (pollution) into it. That is the view of it afforded from patches here and there that are not blocked by his buildings. Pollution rising from humankind's enterprises can mask the stars and make the visible sky a dirty yellow-brown or at best a sickly pale blue.

Fortunately, Earth's atmosphere is self-healing. Air-cleansing is provided by clouds and the global circulation system; they constantly work to purge the air of pollutants. Only when air pollutants overload Nature's Way of rejuvenating its systems to their natural state are we faced with the repercussions that can be serious, even life-threatening.

Did You Know?

The sun emits a frequency of electromagnetic radiation known as visible light. Earth emits a frequency of electromagnetic radiation known as infrared light.

Atmospheric Particulate Matter

Along with gases and water vapor Earth's atmosphere is literally a boundless arena for particulate matter of many sizes and types. Atmospheric particulates vary in size from 0.0001 to 10,000 microns. Particulate size and shape has a direct bearing on visibility. For example, a spherical particle in the 0.6-micron range can effectively scatter light in all directions, reducing visibility.

The types of airborne particulates in the atmosphere vary widely, with the largest sizes derived from volcanoes, tornadoes, waterspouts, burning embers from forest fires, and seed parachutes, spider webs, pollen, soil particles, and living microbes.

The smaller particles (the ones that scatter light) include fragments of rock, salt and spray, smoke, and particles from forested areas. The largest portion of airborne particulates is invisible. They are formed by the condensation of vapors, chemical reactions, photochemical effects produced by ultraviolet radiation, and ionizing forces that come from radioactivity, cosmic rays, and thunderstorms.

Airborne particulate matter is produced either by mechanical weathering, breakage, and solution or by the vapor-to-condensation-to-crystallization process (typical of particulates from a furnace of a coal-burning power plant).

As you might guess, anything that goes up must eventually come down. This is typical of airborne particulates also. Fallout of particulate matter depends, obviously, mostly on their size—less obviously, on their shape, density, weight, airflow, and injection altitude. The residence time of particulate matter also is dependent on the atmosphere's cleanup mechanisms (formation of clouds and precipitation) that work to remove them from their suspended state.

Some large particulates may only be airborne for a matter of seconds or minutes with intermediate sizes able to stay afloat for hours or days. The finer particulates may stay airborne for a much longer duration: for days, weeks, months, and even years.

Particles play an important role in atmospheric phenomena. For example, particulates provide the nuclei upon which ice particles are formed, cloud condensation forms, and for condensation to take place. Obviously, the most important role airborne particulates play is in cloud formation. Simply put, without clouds life as we know it would be much more difficult and cloudbursts that eventually erupt would cause such devastation that it is hard to imagine or contemplate.

The situation just described could also result whenever massive forest fires and volcanic action take place. These events would release a superabundance of cloud condensation nuclei which would overseed the clouds, causing massive precipitation to occur. If natural phenomena such as forest fires and volcanic eruptions can overseed clouds and cause massive precipitation, then what effect would result from man-made pollutants entering the atmosphere at unprecedented levels? This question and other pollution-related questions will be answered in subsequent chapters.

Moisture in the Atmosphere

Hath the rain a father? or who hath begotten the drops of dew? Out of
whose womb came the ice? and the hoary frost of heaven, who hath gen-
dered it? . . . Can't thou lift up thy voice to the clouds, that abundance of
water may cover thee?

—Job 38:28–29, 34

On a hot day when clouds build up signifying that a storm is imminent, we
do not always appreciate what is happening.

What is happening?

This cloud buildup actually signals that one of the most vital processes in
the atmosphere is occurring: the condensation of water as it is raised to higher
levels and cooled within strong updrafts of air created either by convection
currents, turbulence, or physical obstacles like mountains. The water originated from the surface—evaporated from the seas, from the soil, or transpired by vegetation. Once within the atmosphere, however, a variety of events
combine to convert the water vapor (produced by evaporation) to water
droplets. The air must rise and cool to its dew point, of course. At dew point,
water condenses around minute airborne particulate matter to make tiny
cloud droplets forming clouds—clouds from which precipitation occurs.

Whether created by the Sun heating up a hillside, by jet aircraft exhausts,
or factory chimneys, there are actually only ten major cloud types. The deliv-
erers of countless millions of tons of moisture from the Earth's atmosphere,
they form even from the driest desert air containing as little as 0.1% water
vapor. They not only provide a visible sign of motion but also indicate change
in the atmosphere portending weather conditions that may be expected up to
48 hours ahead. In this chapter we take a brief look at the nature and consequences of these cloud-forming processes.

Cloud Formation

The atmosphere is a highly complex system; the effect of change in any other
single property tends to be transmitted to many other properties. The most
profound affect on the atmosphere is the result of alternate heating and cooling of the air, which causes adjustments in relative humidity and buoyancy;
they cause condensation, evaporation, and cloud formation.

The temperature structure of the atmosphere (along with other forces that
propel the moist air upward) is the main force behind the form and size of
clouds. Exactly how does temperature affect atmospheric conditions? For one
thing, temperature (that is, heating and cooling of the surface atmosphere)

causes vertical air movements. Let's take a look at what happens when air is heated.

Let's start with a simple parcel of air in contact with the ground. As the ground is heated, the air in contact with it will warm also. This warm air increases in temperature and expands. Remember, gases expand on heating much more than liquids or solids, so this expansion is quite marked. In addition, as the air expands, its density falls (meaning that the same mass of air now occupies a larger volume). You've heard that warm air rises? Because of its lessened density, this parcel of air is now lighter than the surrounding air and tends to rise. Conversely, if the air cools, the opposite occurs—it contracts, its density increases, and it sinks. Actually, alternate heating and cooling are intimately linked with the process of evaporation, condensation, and precipitation.

But how does a cloud actually form? Let's look at another example.

On a sunny day, some patches of ground warm up more quickly than others because of differences in topography (soil and vegetation, etc.). As the surface temperature increases, heat passes to the overlying air. Later, by mid-morning, a bulbous mass of warm, moisture-laden air rises from the ground. This mass of air cools as it meets lower atmospheric pressure at higher altitudes. If cooled to its dew point temperature, condensation follows and a small cloud forms. This cloud breaks free from the heated patch of ground and drifts with the wind. If it passes over other rising air masses, it may grow in height. The cloud may encounter a mountain and be forced higher still into the air. Condensation continues as the cloud cools; if the droplets it holds become too heavy, they fall as rain.

Major Cloud Types

As you review the list of cloud types below, keep in mind that clouds whose names incorporate the word "nimbus" or the prefix "nimbo-" are clouds from which precipitation is falling. Earlier, it was mentioned that there are ten major cloud types.

Stratiform genera contain the following species:

Cirrus
Cirrostratus
Cirrocumulus
Altostratus
Altocumulus
Stratus
Stratocumulus
Nimbostratus

Cumuliform genera contain the following species:

Cumulus
Cumulonimbus

From the list above it is apparent that the cloud groups are classified into a system that uses Latin words to describe the appearance of clouds as seen by an observer on the ground. Table 4.1 summarizes the four principal components of this classification system (Ahrens 1994).

TABLE 4.1
Summary of Components of Cloud Classification System

Latin Root	Translation	Example
cumulus	heaped/puffy	fair weather cumulus
stratus	layered	altostratus
cirrus	curl of hair/wispy	cirrus
nimbus	rain (thunder & lightning)	cumulonimbus

Further classification identifies clouds by height of cloud base. For example, cloud names containing the prefix "cir-" as in cirrus clouds, are located at high levels while cloud names with the prefix "alto-" as in altostratus, are found at middle levels. This module introduces several cloud groups. The first three groups are identified based upon their height above the ground. The fourth group consists of vertically developed clouds, while the final group consists of a collection of miscellaneous cloud types.

Let's take a closer look at each of these cloud types.

A *stratus* cloud is a featureless, gray, low-level cloud. Its base may obscure hilltops or occasionally extend right down to the ground, and because of its low altitude, it appears to move very rapidly on breezy days. Stratus can produce drizzle or snow, particularly over hills, and may occur in huge sheets covering several thousand miles.

Cumulus clouds also seem to scurry across the sky, reflecting their low altitude. These small, dense, white, fluffy, flat-based clouds are typically short-lived, lasting no more than 10–15 minutes before dispersing. They are typically formed on sunny days, when localized convection currents are set up: these currents can form over factories or even brush fires, which may produce their own clouds.

Cumulus may expand into low-lying horizontally layered, massive *stratocumulus*, or into extremely dense, vertically developed, giant *cumulonimbus* with a relatively hazy outline and a glaciated top which is up to 7 miles in diameter. These clouds typically form on summer afternoons; their high, flattened tops contain ice, which may fall to the ground in the form of heavy showers of rain or hail.

Rising to middle altitudes, the bluish-gray layered *altostratus* and rounded, fleecy, whitish-gray *altocumulus* appear to move slowly because of their greater distance from the observer.

Cirrus (meaning tuft of hair) clouds are made up of white narrow bands of thin, fleecy parts and are relatively common over northern Europe, and generally ride the jet stream rapidly across the sky.

Cirrocumulus are high altitude clouds composed of a series of small, regularly arranged cloudlets in the form of ripples or grains; they are often present with cirrus clouds in small amounts. *Cirrostratus* are high altitude, thin, hazy clouds, usually covering the sky and giving a halo effect surrounding the sun or moon.

Did You Know?

Clouds play an important role in boundary layer meteorology and air quality. Convective clouds transport pollutants vertically, allowing an exchange of air between the boundary layer and the free troposphere. Cloud droplets formed by heterogeneous nucleation on aerosols grow into rain droplets through condensation, collision, and coalescence. Clouds and precipitation scavenge pollutants from the air. Once inside the cloud or rain water, some compounds dissociate into ions and/or react with one another through aqueous chemistry. Another important role for clouds is the removal of pollutants trapped in rain water and its deposition onto the ground.

Source: EPA/600/R-99/030

Summary

The process of evaporation (converting moisture into vapor) supplies moisture to the lower atmosphere. The prevailing winds then circulate the moisture and mix it with drier air elsewhere.

Water vapor is only the first stage of the precipitation cycle; the vapor must be converted into liquid form. This is usually achieved by cooling, either rapidly, as in convection, or slowly, as in cyclonic storms. Mountains also cause uplift, but the rate will depend upon their height and shape and the direction of the wind.

To actually produce precipitation, the cloud droplets must become large enough to reach the ground without evaporating. The cloud must possess the right physical properties to enable the droplets to grow. If the cloud lasts long enough for growth to take place, then precipitation will usually occur. Precipitation results from a delicate balance of counteracting forces, some leading to droplet growth and others to droplet destruction.

Did You Know?

Contrails are clouds formed around the small particles (aerosols) which are in aircraft exhaust. When these persist after the passage of the plane they are indeed clouds, and are of great interest to researchers. Under the right conditions, clouds initiated by passing aircraft can spread with time to cover the whole sky.

References and Recommended Reading

Ahrens, D. 1994. *Meteorology Today: an Introduction to Weather, Climate and the Environment*, 5th ed. West Publishing Company.

EPA. 2007. *Air Pollution Control—Atmosphere*. Accessed 12/28/09 @ http://www .epa.gov/air/oaqpseog /course422/apl.html.

EPA. 2007. *Air Pollution Control Orientation Course: Air Pollution*. Accessed 01/05/10 @ http://www.epa.gov/air/oaqps/eog/course422/ap.1.html.

NASA. 2008. *Observing Cloud Type*. Washington, DC: National Aeronautics and Space Administration.

NOAA. 2007. Cloud Types. Accessed 10/29/10 @ http://www.gfdl.NOAA.gov/~01/ weather /clouds.html.

Spellman, F.R. 2009. *The Science of Air*, 2nd ed. Boca Raton, FL: CRC Press.

Spellman, F.R., and Whiting, N. 2006. *Environmental Science & Technology: Concepts and Applications*, 2nd ed. Rockville, MD: Government Institutes.

5

Atmospheric Water Phenomena

The Rainy Day
The day is cold, and dark, and dreary;
It rains, and the wind is never weary;
The vine still clings to the moldering wall,
But at every gust the dead leaves fall,
And the day is dark and dreary.

My life is cold, and dark, and dreary;
It rains, and the wind is never weary;
My thoughts still cling to the moldering Past,
But the hopes of youth fall thick in the blast
And the days are dark and dreary.

Be still, sad heart! And cease repining;
Behind the clouds is the sun still shining;
Thy fate is the common fate of all,
Into each life some rain must fall,
Some days must be dark and dreary.

—Henry Wadsworth Longfellow

Introduction

THE PRINCIPAL ACTIONS brought on by weather systems that affect land and sea and the humans, animals, and vegetation thereon are winds and precipitation. The latter comes in a variety of forms as discussed below. Most weather of consequence to people occurs in storms. These may be local in origin but more commonly are carried to locations in wide areas along pathways followed by active air masses consisting of Highs and Lows. The key ingredient in storms is water, either as a liquid or as a vapor. The vapor acts like a gas and thus contributes to the total pressure of the atmosphere, making up a small but vital fraction of the total (NASA 2007). The change of state from solid to a vapor (gas) is referred to as sublimation.

Precipitation is found in a variety of forms. Which form actually reaches the ground depends upon many factors: for example, atmospheric moisture content, surface temperature, intensity of updrafts, and method and rate of cooling.

Water vapor in the air will vary in amount depending on sources, quantities, processes involved, and air temperature. Heat, mainly as solar irradiation but with some contributed by the earth and human activity and some from change of state processes, will cause some water molecules either in water bodies (oceans lakes, rivers) or in soils to be excited thermally and escape from their sources. This is called evaporation; if water is released from trees and other vegetation, the process is known as **evapotranspiration**. The evaporated water, or moisture, that enters the air is responsible for a state called humidity. Absolute humidity is the weight of water vapor contained in a given volume of air. The Mixing Ratio refers to the mass of the water vapor within a given mass of dry air. At any particular temperature, the maximum amount of water vapor that can be contained is limited to some amount; when that amount is reached the air is said to be saturated for that temperature. If less than the maximum amount is present, then the property of air that indicates this is its Relative Humidity (RH), defined as the actual water vapor amount compared to the saturation amount at the given temperature; this is usually expressed as a percentage. RH also indicates how much moisture the air can hold above its stated level which, after attaining, could lead to rain. The lowest RH during a day usually occurs after sunrise.

When a parcel of air attains or exceeds RH = 100% condensation will occur and water in some state will begin to organize as some type of precipitation. One familiar form is dew, which occurs when the saturation temperature or some quantity of moisture reaches a temperature at the surface at which condensation sets in, leaving the moisture to coat the ground (especially obvious on lawns).

The term *dew point* has a more general use, being that temperature at which an air parcel must be cooled to become saturated; that is, the dew point is the temperature at which vapor condenses to water. Dew frequently forms when the current air mass contains excessive moisture after a period of rain but the air is now clear; the dew precipitates out to coat the surface (noticeable on vegetation). Ground fog is a variant in which lowered temperatures bring on condensation within the near surface air as well as the ground. When warm moist air moves over a cold surface advection is formed.

The other types of precipitation are listed in Table 5.1 along with descriptive characteristics related to each type.

TABLE 5.1
Types of Precipitation

Type	Approximate Size	State of Water	Description
Mist	0.005 to 0.05 mm	Liquid	Droplets large enough to be felt on face when air is moving 1 meter/second. Associated with stratus clouds.
Drizzle	Less than 0.5 mm	Liquid	Small uniform drops that fall from stratus clouds, generally for several hours.
Rain	0.5 to 5 mm	Liquid	Generally produced by nimbostratus or cumulonimbus clouds. When heavy, size can be highly variable from one place to another.
Sleet	0.5 to 5 mm	Solid	Small, spherical to lumpy ice particles that form when raindrops freeze while falling through a layer of sub-freezing air. Because the ice particles are small, any damage is generally minor. Sleet can make travel hazardous.
Glaze	Layers 1 mm to 2 cm thick	Solid	Produced when supercooled raindrops freeze on contact with solid objects. Glaze can form a thick covering of ice having sufficient weight to seriously damage trees and power lines.
Rime	Variable accumulation	Solid	Deposits usually consisting of ice feathers that point into the wind. These delicate frostlike accumulations form as supercooled cloud or fog droplets encounter objects and freeze on contact.
Snow	1 mm to 2 cm	Solid	The crystalline nature of snow allows it to assume many shapes, including six-sided crystals, plates, and needles. Produced in supercooled clouds where water vapor is deposited as crystals that remain frozen during their descent.
Hail	5 mm to cm or larger	Solid	Precipitation in the form of hard, rounded pellets or irregular lumps of ice. Produced in large convective, cumulonimbus clouds, where frozen ice particles and supercooled water coexist.
Graupel	2 mm to 5 mm	Solid	Sometimes called "soft hail," graupel forms on rime and collects on snow crystals to produce irregular masses of "soft" ice. Because these particles are softer than hailstones, they normally flatten out upon impact.

Source: NASA 2007.

Evaporation and transpiration are complex processes which return moisture to the atmosphere. The rate of evapotranspiration depends largely on two
factors: (1) how saturated (moist) the ground is and (2) the capacity of the
atmosphere to absorb the moisture. In this chapter we discuss the factors responsible for both precipitation and evapotranspiration.

Precipitation

As mentioned, if all the essentials are present, precipitation occurs when the
dew point is reached. However, it was also pointed out that it is quite possible
for an air mass or cloud containing water vapor to be cooled below the dew
point without precipitation occurring. In this state the air mass is said to be
supercooled.

How then are droplets of water formed? Water droplets form around microscopic foreign particles already present in the air. These particles on which
the droplets form are called *hygroscopic nuclei.* They are present in the air
primarily in the form of dust, salt from sea water evaporation, and from combustion residue. These foreign particles initiate the formation of droplets that
eventually fall as precipitation. To have precipitation, larger droplets or drops
must form. This may be brought about by two processes: (1) coalescence (collision) or (2) the Bergeron process.

Coalescence

Simply put, *coalescence* is the fusing together of smaller droplets into larger
ones. The variation in the size of the droplets has a direct bearing on the efficiency of this process. Raindrops come in different sizes and can reach diameters up to 7 mm. Having larger droplets greatly enhances the coalescence
process.

But what actually goes on inside a cloud to cause rain to fall? To answer this
question, we must take a look inside a cloud to see exactly what processes
occur to make rain—rain that actually falls as rain. Rainmaking is based on
the essentials of the Bergeron process.

Bergeron Process

Named after the Swedish meteorologist who suggested it, the *Bergeron process*
is probably the most important process for the initiation of precipitation. To
gain understanding on how the Bergeron process works, let's look at what
actually goes on inside a cloud to cause rain.

Within a cloud made up entirely of water droplets there will be a variety of droplet sizes. The air will be rising within the cloud anywhere from 10–20 cm per second (depending on type of cloud). As the air rises, the drops become larger through collision and coalescence; many will reach drizzle size. Then the updraft intensifies up to 50 cm per second (and more), which reduces the downward movement of the drops, allowing them more time to become even larger. When the cloud becomes approximately 1 km deep, small raindrops of 700 m diameter are formed.

The droplets, because of their small size, do not freeze immediately, even when the temperatures fall below 0°C. Instead, the droplets remain unfrozen in a supercooled state. However, when the temperature drops as low as −10°C, ice crystals may start to develop among the water droplets. This mixture of water and ice would not be particularly important but for a peculiar characteristic or property of water. Therefore at −10°C, air saturated with respect to liquid water is super-saturated relative to ice by 10% and at −20°C by 21%. Thus, ice crystals in the cloud tend to grow and become heavier at the expense of the water droplets.

Eventually, the ice crystals sink to the lower levels of the cloud where temperatures are only just below freezing. When this occurs, they tend to combine (the supercooled droplets of water act as an adhesive) and form snowflakes. When the snowflakes melt, the resulting water drops may grow further by collision with cloud droplets before they reach the ground as rain. The actual rate at which water vapor is converted to raindrops depends on three main factors: (1) the rate of ice crystal growth, (2) supercooled vapor, and (3) the strength of the updrafts (mixing) in the cloud.

Did You Know?

Have you ever thought about that raindrop? You know the one; the one that struck you in the head? Or did you just reach up and whisk it aside and move on not giving it a first or second thought? If you did happen to think about that raindrop, did you recognize it as a type of meteor, more correctly as a hydrometeor? If not, you should have, because a raindrop is a product of condensation or precipitation of moisture within the earth's atmosphere and is known technically as a hydrometeor. The fact is all precipitation types are hydrometeors by definition, including virga, which is precipitation which evaporates before reaching the ground.

— *Glossary of Meteorology* (2009)

Types of Precipitation

We stated that in order for precipitation to occur, water vapor must condense, which occurs when water vapor ascends and cools. Three mechanisms by which air rises allowing for precipitation to occur are convectional, orographic, and frontal, which are discussed in this section. Moreover, to broaden the reader's knowledge of rain or hydrometeors, we have also added anecdotal information and unusual phenomena such as colored and cloudless rain to the discussion.

Convectional Precipitation

Convectional precipitation is the spontaneous rising of moist air due to instability. This type of precipitation is usually associated with thunderstorms and occurs in the summer because localized heating is required to initiate the convection cycle. We have discussed that upward-growing clouds are associated with convection. Since the updrafts (commonly called a "thermal") are usually strong, cooling of the air is rapid and lots of water can be condensed quickly, usually confined to a local area; and a sudden summer downpour may occur as a result.

Convectional thunderstorm clouds are also described as supercells.

Convective thunderstorms are the most common type of atmospheric instability that produces lightning followed by thunder. Lightning is one of the most spectacular phenomena witnessed in storms.

Did You Know?

A lightning bolt can attain an electric potential up to 30 million volts and current as much as 10,000 amps. It can cause air temperatures to reach 10,000°C. But a bolt's duration is extremely short (fractions of a second). Although a bolt can kill people it hits, most can survive.

Orographic Precipitation

Orographic precipitation is a straightforward process, characteristic of mountainous regions; almost all mountain areas are wetter than the surrounding lowlands. This type of precipitation arises when air is forced to rise over a mountain or mountain range. The wind, blowing along the surface of the Earth, ascends along topographic variations. Where air meets this extensive barrier, it is forced to rise. This ascending wind usually gives rise to cooling and encourages condensation and thus orographic precipitation on the windward side of the mountain range.

Frontal Precipitation

Frontal precipitation results when two different fronts (or the boundary between two air masses characterized by varying degrees of precipitation), at different temperatures, meet. The warm air mass (since it is lighter) moves up and over the colder air mass. The cooling is usually less rapid than in the vertical convection process because the warm air mass moves up at an angle, more of a horizontal motion.

Colored Rain

History records many instances of unusual objects falling with the rain—in 2000, in an example of raining animals, a small water spout in the North Sea sucked up a school of fish a mile offshore, depositing them shortly afterwards in the United Kingdom (Lane 2000). Colored rain is less rare than raining animals, and can often be explained by the airborne transport of various particulates and/or pollutants. All in all, colored rain is not mysterious; instead, it's simply unusual. Short excerpts describing a few unusual colored rain incidents follow.

- Australia, January 1897—"There was a heavy fall of rain, which was full of red dust, and the next morning the whole landscape was red. The 'rain of blood,' it was called, and really it looked like it. The funny thing was that it occurred on a public holiday and on a hot day, so that all the holiday-makers who were caught in it had their clothes stained a deep red; and as many people were dressed in white, you may guess what they looked like" (*Symons's Monthly Meteorological Magazine*, 32: 27, 1897).
- England, April 1884—"Black Rain.—Yesterday afternoon (April 28) a violent thunderstorm raged over the district between Church Stretton and Much Wenlock. Torrents of rain fell, seemingly a mixture of ink and water in equal proportions. One old man here says he never saw anything like it but once. I certainly never saw such a colored rain, and I intend to have a bottle of it analyzed. Even this afternoon the little brooks are quite black, and the ruts in the roads look as if ink and water had been poured into them.—Rev. R.I. Buddicombe, Ticklerton, Church Stretton" (*Nature*, 30: 32, 1884). Keep in mind that in the 1880s coal was the energy source of choice and was burned in immense quantities in Europe.

- Kentucky, March 1867—"It seems that the days of miracles have not yet passed. On the night of the 12th inst. [March 12] we in this section had a copious fall of rain of about two and a half inches, and such vessels as were left standing out were found to contain water impregnated with a yellow substance such as is contained in the inclosed vial. We learned today from Bowling Green, fourteen miles distant, that it was the same there, and the inhabitants, believing it to be sulphur, are somewhat alarmed, not knowing but what it is the beginning of a preparation of the great fire in which sinners expect to find themselves ensconced in a coming day! Whatever it is, we are not chemists enough to make out. Clothes that were left lying out were made yellow with the substance. It seems to be odorless—has the resemblance of farina contained in the anthers of plants" (*Scientific American*, 16: 233, 1867).

Cloudless Rain

On April 23, 1800, between 9 and 10 P.M., "Philadelphia . . . was visited by a very curious phenomenon. A shower of rain of at least twenty minutes' continuance, and sufficiently plentiful to wet the clothes of those exposed to it, fell when the heavens immediately overhead were in a state of the most perfect serenity. Throughout the whole of it, the stars shone with undiminished luster. Not a cloud appeared, except one to the east and another to the west of the city, each about fifteen degrees distant from the zenith" (*American Journal of Science*, 2: 1: 178, 1839). Instances such as this one are usually explained away as being a result of high-altitude winds wafting rain horizontally over long distances.

Evapotranspiration

Another important part or process of the hydrological cycle (though it is often neglected because it can rarely be seen) is *evapotranspiration*. More complex than precipitation, evaporation and transpiration is a land-atmosphere interface process whereby a major flow of moisture is transferred from ground level to the atmosphere. It returns moisture to the air, replenishing that lost by precipitation, and it also takes part in the global transfer of energy. The rate of evapotranspiration depends largely on two factors: (1) how moist the ground is and (2) the capacity of the atmosphere to absorb the moisture. Therefore, the greatest rates are over the tropical oceans where moisture is always available, and the long hours of sunshine and steady trade winds evaporate vast quantities of water.

Just how much moisture is returned to the atmosphere via transpiration? In answering this question, Table 5.2 makes clear, for example, that in the United States alone, about two-thirds of the average rainfall over the U.S. mainland is returned via evaporation and transpiration.

TABLE 5.2
Water Balance in the United States (in bgd*)

Precipitation	4,200
Evaporation and transpiration	3,000
Runoff	1,250
Withdrawal	310
Irrigation	142
Industry (utility cooling water)	142
Municipal	26
Consumed (irrigation loss)	90
Returned to streams	220

*bgd = billion gallons per day
Source: National Academy of Sciences, 1962

Evaporation

Evaporation is the process by which a liquid is converted into a gaseous state. Evaporation takes place (except when air reaches saturation at 100% humidity) almost on a continuous basis. It involves the movement of individual water molecules from the surface of Earth into the atmosphere, a process occurring whenever a vapor pressure gradient exists from the surface to the air (i.e., whenever the humidity of the atmosphere is less than that of the ground). Evaporation also requires energy (derived from the sun or from sensible heat from the atmosphere or ground): 2.48 x 10⁶ Joules to evaporate each kilogram of water at 10°C.

Transpiration

A related process, *transpiration* is the loss of water from a plant by evaporation. Just as we release water vapor when we breathe, plants do, too—although the term "transpire" is more appropriate than "breathe." Most water is lost from the leaves through pores known as **stomata**, whose primary function is to allow gas exchange between the plant's internal tissues and the atmosphere. Transpiration from the leaf surfaces causes a continuous upward flow of water from the roots via the **xylem**, which is known as the transpiration stream.

Transpiration occurs mainly by day, when the stomata open up under the influence of sunlight. Acting as evaporators they expose the pure moisture (the plant's equivalent of perspiration) in the leaves to the atmosphere. If the vapor pressure of the air is less than that in the leaf cells, the water is transpired.

The amount of water that plants transpire varies greatly geographically and over time. There are a number of factors that determine transpiration rates:

- **Temperature**—Transpiration rates go up as the temperature goes up, especially during the growing season, when the air is warmer due to stronger sunlight and warmer air masses. Higher temperatures cause the plant cells which control the openings (stoma) where water is released to the atmosphere to open, whereas colder temperatures cause the openings to close.
- **Relative Humidity**—As the relative humidity of the air surrounding the plant rises, the transpiration rate falls. It is easier for water to evaporate into dryer air than into more saturated air.
- **Wind and Air Movement**—Increased movement of the air around a plant will result in a higher transpiration rate. This is somewhat related to the relative humidity of the air, in that as water transpires from a leaf, the water saturates the air surrounding the leaf. If there is no wind, the air around the leaf may not move very much, raising the humidity of the air around the leaf. Wind will move the air around, with the result that the more saturated air close to the leaf is replaced by drier air.
- **Soil-moisture Availability**—When moisture is lacking, plants can begin to senesce (premature aging, which can result in leaf loss) and transpire less water.
- **Type of Plant**—Plants transpire water at different rates. Some plants which grow in arid regions, such as cacti and succulents, conserve precious water by transpiring less water than other plants (USGS 2010).

As you might guess, because of transpiration, far more water passes through a plant than is needed for growth. In fact, only about 1% or so is actually used in plant growth. For example, during a growing season, a leaf will transpire many more times more water than its own weight. An acre of corn gives off about 2,000–4,000 gallons (11,400–15,100 liters) of water each day, and a large oak tree can transpire 40,000 gallons (151,000 liters) per year. The excess movement of moisture through the plant is important to the plant because the water acts as a solvent, transporting vital nutrients from the soil into the roots and carrying them through cells of the plant. Obviously, without this vital process plants would die.

Evapotranspiration: The Process

Although evapotranspiration plays a vital role in cycling water over Earth's land masses, it seldom is appreciated. In the first place, distinguishing between evaporation and transpiration is often difficult. Both processes tend to be operating together, so the two are normally combined to give the composite term *evapotranspiration*.

Governed primarily by atmospheric conditions, energy is needed to power the process. Wind also plays an important role, which acts to mix the water molecules with the air and transport them away from the surface. The primary **limiting factor** in the process is lack of moisture at the surface (soil is dry). Evaporation can continue only so long as there is a vapor pressure gradient between the ground and the air.

Did You Know?

A new study reported by LiveScience (2010) points out that "the soils in large areas of the Southern Hemisphere, including large parts of Australia, Africa and South America, have been drying up in the past decade. Most climate models have suggested that evapotranspiration would increase with global warming, because of increased evaporation of water from the ocean and more precipitation overall (water that can evaporate). The new research, published online this week in the journal Nature, found that's exactly what was happening from 1982 to the late 1990s. But in 1998, this significant increase in evapotranspiration—about 0.3 inches (7 millimeters) of water per year—slowed dramatically or stopped. In large portions of the world, soils are not becoming drier than they used to be, releasing less water and offsetting some moisture increases elsewhere. Because the data only goes back for a few decades, the researchers say they can't be certain whether the change is part of the natural variability of climate or part of a longer-lasting global change."

Albino Rainbow

Have you ever looked up into the sky and seen eleven Suns? Have you been at sea and witnessed the towering, spectacular Fata Morgana? (Do you even know what Fata Morgana is?—if not, hold on, we'll get to it shortly.) How about a "glory"—have you ever seen one? Or how about the albino rainbow—have you seen one lately? Do you know what these are? They will be described shortly.

Normally, when we look up into the sky, we see what we expect to see: an ever-changing backdrop of color, with dynamic vistas of blue sky, white, puffy clouds, gray storms, and gold and red sunsets. On some occasions, however, when atmospheric conditions are just right, we can look up at the sky or out upon the horizon and see the strange phenomena (or lights in the sky) mentioned above. What causes these momentary wonders?

Because Earth's atmosphere is composed of gases (air)—it is actually a sea of molecules. These molecules of air scatter the blue, indigo, and violet shorter wavelengths of light more than the longer orange and red wavelengths, which is why the sky appears blue.

What are wavelengths of light? Simply put, a wavelength of light actually refers to the electromagnetic spectrum. The portion of the spectrum visible to the human eye falls between the infrared and ultraviolet wavelengths. The colors that make up the visible portion of the electromagnetic spectrum are commonly abbreviated by the acronym ROY G BIV (Red, Orange, Yellow, Green, Blue, Indigo, and Violet).

The word *light* is commonly given to visible electromagnetic radiation. However, only the frequency (or wavelength) distinguishes visible electromagnetic radiation from the other portions of the spectrum.

Let's get back to the sky; that is, to looking up into the sky. Have you ever noticed that right after a rain shower, how dark a shade of blue the sky appears? Have you looked out upon the horizon at night or in the morning and noticed that the Sun's light gives off a red sky? This phenomenon is caused by sunlight passing through large dust particles, which scatter the longer wavelengths. Have you ever noticed that fog and cloud droplets, with diameters larger than the wavelength of light scatter all colors equally and make the sky look white? Maybe you have noticed that fleeting greenish light that appears just as the Sun sets? It occurs because different wavelengths of light, are *refracted* (bent) in the atmosphere by differing amounts. Because green light is refracted more than red light by the atmosphere, green is the last to disappear.

What causes rainbows? A rainbow is really nothing more than an airborne prism. When sunlight enters a raindrop, refraction and reflection take place, splitting white light into the spectrum of colors from red to blue and making a rainbow.

Earlier a "glory" was mentioned. Interactions of light waves can produce a glory, an optical phenomenon. A glory often appears as an iconic Saint's halo about the head of the observer, which is produced by a combination of diffraction, reflection and refraction (backscattered light) towards its source by a cloud of uniformly sized water droplets. For example, if you were standing on a mountain, with the sun to your back, you may cast a shadow on the fog in the valley. Your shadow may appear to be surrounded by colored halos—a glory. The glory is caused by light entering the edges of tiny droplets and being returned in the same direction from which it arrived. These light waves interfere with each other, sometimes canceling out and sometimes adding to each other.

A real-life example of a group in 1893 at Gausta Mountain, Norway, that saw well-defined glories was reported in *Nature*, 48: 391, 1893: "We mounted to the flagstaff in order to obtain a better view of the scenery, and there we at once observed in the fog, in an easterly direction, a double rainbow forming a complete circle and seeming to be 20 to 30 feet distance from us. In the middle of this we all appeared as black, erect, and nearly life-size silhouettes. The outlines of the silhouettes were so sharp that we could easily recognize the figures of each other, and every movement was reproduced. The head of each individual appeared to occupy the centre of the circle, and each of us seemed be standing on the inner periphery of the rainbow."

Why do we sometimes see multiple suns? Reflection and refraction of light by ice crystals can create bright halos in the form of arcs, rings, spots, and pillars. Mock suns (sun dogs) may appear as bright spots 22° or 46° to the left or right of the sun. A one-sided mock sun was observed by several passengers aboard the ship *Fairstar* in the Sea of Timor on June 15, 1965. "About 10 minutes before sunset, Mrs. N.S. noticed to her surprise a second sun, much less bright than the real one, somewhat to the left, but at the same height above the horizon. There was nothing at the symmetrical point to the right" (Weather, 21: 250, 1966). Sun pillars occur when ice crystals act as mirrors, creating a bright column of light extending above the Sun. Such a pillar of bright light may be visible even when the Sun has set.

What is the Fata Morgana? It is an illusion, a mirage, which often fools sailors into seeing mountain ranges floating over the surface of the ocean. Henry Wadsworth Longfellow had his own take on Fata Morgana, which basically explains the essence of the phenomenon:

Fata Morgana
O sweet illusions of song
That tempt me everywhere,
In the lonely fields, and the throng
Of the crowded thoroughfare!

I approach and ye vanish away,
I grasp you, and ye are gone;
But over by night and by day,
The melody soundeth on.

As the weary traveler sees
In desert or prairie vast,
Blue lakes, overhung with trees
That a pleasant shadow cast;

Fair towns with turrets high,
And shining roofs of gold,
That vanish as he draws nigh,
Like mists together rolled—

So I wander and wander along,
And forever before me gleams
The shining city of song,
In the beautiful land of dreams.

But when I would enter the gate
Of that golden atmosphere,
It is gone, and I wonder and wait
For the vision to reappear.

Then there is

Socrates: Didst thou never espy a Cloud in the sky,
                         Which a centaur or leopard might be?
                 Or a wolf or a cow?
Strepsiades: Very often, I vow:
                 And show me the cause, I entreat.
Socrates: Why, I tell you that these become just what they please . . .
                         —Aristophanes, *The Clouds*, ca. 420 B.C.E.
                 Trans. Benjamin Bickley Rogers (Cambridge, Mass, and London,
                 1960)

The flagship account of the Fata Morgana comes to us from friar Antonio Minasi who described a Fata Morgana seen across the Straits of Messina in 1773. "In Italian legend, Morgan Le Fay, or, in Italian, Fata Morgana, falls in love with a mortal youth and gives him the gift of eternal life in return for her love; when he becomes restless and bored with captivity, she summons up fairy spectacles for his entertainment" (In Antonia Minasi, *Dissertazioni*, Rome, 1773).

An albino rainbow (white rainbow or fogbow) is an eerie phenomenon that can only be seen on rare occasions in foggy conditions. They form when the sun or moon shine on minute droplets of water suspended in the air. The fog droplets are so small that the usual prismatic colors of the rainbow merge together to form a white arc opposite the sun or moon (Corliss, 1983; *Marine Observer*, 6: 11, 1919).

Atmospheric phenomena (lights in the sky) are real, apparent, and sometimes visible. Awe inspiring as they are, their significance—their actual existence—is based on physical conditions that occur in our atmosphere.

References and Recommended Reading

American Meteors Society. 2001. *Definition of terms by the IAU Commission 22, 1961.* Accessed 10/10/10 @ http://www.amsmeteors.org/define.html.

Corliss, W.R. 1983. *Handbook of Unusual Natural Phenomena.* New York: Anchor Press.

EPA. 2007. Basic Concepts in Environmental Sciences: Module 1 & 2. Accessed 12/30/07 @ http://www.epa.gov/eogapti1/module 1 and 2.

Heinlein, R.A. 1973. *Time Enough for Love.* New York: G.P. Putnam's Sons.

Hesketh, H.E. 1991. *Air Pollution Control: Traditional and Hazardous Pollutants.* Lancaster, PA: Technomic Publishing Company.

LiveScience. 2010. *Large Swaths of Earth Drying Up, Study Suggests.* Accessed 10/13/10 @ http://news.yahoo.com/s/livescienc/20101012/sc_livescience/largeswathofof fearthdryingupstudysuggests. . .

NASA. 2007. *Pascal's Principle and Hydraulics.* Accessed 12/29/07 @ http://www.grc .nasa.gov/WWW.

National Academy of Sciences. 1962. *Water Balance in the U.S.* National Research Council Publication 100-B.

Shipman, J.T., Adams, J.L., and Wilson, J.D. 1987. *An Introduction to Physical Science.* Lexington, MA: D.C. Heath and Company.

Spellman, F.R. 2007. *The Science of Water,* 2nd ed. Boca Raton, FL: CRC Press.

Spellman, F.R. 2009. *The Science of Air,* 2nd ed. Boca Raton, FL: CRC Press.

Spellman, F.R., and Whiting, N. 2006. *Environmental Science & Technology: Concepts and Applications.* Lanham, MD: Government Institutes.

USGS. 2010. *The Water Cycle: Evapotranspiration.* Accessed 10/07/10 @ http://ga.water .usgs.gov/edu/ watercycleevaportranspiration.html.

6

Atmosphere in Motion

Nature provides exceptions to every rule.

—Margaret Fuller, *The Dial*, 1843

Introduction

THERE ARE SCIENTISTS and engineers out there in the real world who will tell us that perpetual motion in or for any mechanized machine is a pipedream—it's simply wishful thinking and impossible. Have you ever pondered the most dynamic perpetual motion machine of them all—Earth's atmosphere? Probably not—but it is. It must be in a state of perpetual motion because it perpetually strives to eliminate the constant differences in temperature and pressure between different parts of the globe. How are these differences eliminated or compensated for? Answer: By its motion, which produces winds and storms.

Global Air Movement

Basically, winds are the movement of the Earth's atmosphere, which by its weight exerts a pressure on the Earth that we can measure using a barometer. Winds are often confused with air currents, but they are different. Wind is the horizontal movement of air or motion along the Earth's surface. Air currents, on the other hand, are vertical air motions collectively referred to as updrafts and downdrafts.

Global air movements explain how air and storm systems travel over the Earth's surface. The global air movements would be simple if the Earth did not rotate; the rotation was not tilted relative to the sun, and, the Earth had no water. Because of these conditions, the sun heats the entire surface, but where the sun is more directly overhead it heats the ground and atmosphere more. The result would be the equator becomes very hot with the hot air rising into the upper atmosphere. That air would then move toward the poles where it would become very cold and sink, then return to the equator. One large area of high pressure would be at each of the poles with a large belt of low pressure around the equator.

However, because the earth rotates, the axis is tilted (23.5), and there is more land mass in the Northern Hemisphere that in the Southern Hemisphere, the actual global pattern is much more complicated. Instead of one large circulation between the poles and the equator, there are three circulations:

1. **Hadley Cell**—Low latitude air movement toward the equator that with heating, rises vertically, with poleward movement in the upper atmosphere. This forms a convection cell that dominates tropical-subtropical climates.
2. **Ferrel Cell**—A mid-latitude mean atmosphere circulation cell for weather named by Ferrel in the nineteenth century. In this cell the air flows poleward and eastward near the surface and equatorward and westward at higher levels.
3. **Polar Cell**—Air rises, diverges, and travels toward the poles. Once over the poles, the air sinks, forming the polar highs. At the surface air diverges outward from the polar highs. Surface winds in the polar cell are easterly (polar easterlies).

Between each of these circulation cells are bands of high and low pressure at the surface. The high pressure band is located about 30°N/S latitude and at each pole. Low pressure bands are found at the equator and 50–60°N/S.

Usually, fair and dry/hot weather is associated with high pressure, with rainy and stormy weather associated with low pressure. You can see the result of these circulations on a globe. Look at the number of deserts located along the 30°N/S latitude around the world. Now, look at the region between 50–60°N/S latitude. These areas, especially the west coast of continents, tend to have more precipitation due to more storms moving around the earth at these latitudes (NWS 2010).

Global Wind Patterns

Throughout history, man has been both fascinated by and frustrated by winds. Man has written about winds almost from the time of the first written word. For example, Herodotus (and later Homer and many others) wrote about winds in his *The Histories*. Wind has had such an impact upon human existence that Man has given winds names that describe a particular wind, specific to a particular geographical area. Table 6.1 lists some of these winds, their colorful names, and the region where they occur. Some of these names are more than just colorful—the winds are actually colored. For example, the *harmattan* blows across the Sahara filled with red dust; mariners called this red wind the "sea of darkness."

TABLE 6.1
Assorted Winds of the World

Wind Name	Location
aajej	Morocco
alm	Yugoslavia
besharbar	Caucasus
biz roz	Afghanistan
brickfielder	Australia
chinook	America
datoo	Gibraltar
haboob	Sudan
imbat	North Africa
nafhat	Arabia
Samiel	Turkey
tsumuji	Japan
williwaw	Alaska

In regard to global wind patterns, the region of Earth receiving the Sun's direct rays is the equator. Here, air is heated and rises, leaving low pressure areas behind. Moving to about thirty degrees north and south of the equator, the warm air from the equator begins to cool and sink. Between thirty degrees latitude and the equator, most of the cooling sinking air moves back to the equator. The rest of the air flows toward the poles. The air movements toward the equator are called trade winds—warm, steady breezes that blow almost continuously. The Coriolis Effect makes the *trade winds* appear to be curving to the west, whether they are traveling to the equator from the south or north.

Did You Know?

The Coriolis Effect is an apparent deflection of moving objects when they are viewed from a rotating reference frame. In a reference frame with clockwise rotation, the deflection is to the left of the motion of the object; in one with anti-clockwise rotation, the deflection is to the right.

The trade winds coming from the south and the north meet near the equator. These converging trade winds produce general upward winds as they are heated, so there are not steady surface winds. This area of calm is called the *doldrums*, also called the "equatorial calms." In the day of the sailing ships, the doldrums' light, shifting, and sometimes complete absence were (are) notorious for trapping sailing ships for days (or even weeks) without enough wind to power their sails.

Between thirty and sixty degrees latitude, the winds that move toward the poles appear to curve to the east. Because winds are named for the direction in which they originate, these winds are called *prevailing westerlies*, or *westerlies*. Prevailing westerlies in the Northern Hemisphere are responsible for many of the weather movements across the United States and Canada.

Did You Know?

The westerlies are strongest in the winter hemisphere and times when the pressure is lower over the poles, while they are weakest in the summer hemisphere and when pressures are higher over the poles.

At about sixty degrees latitude in both hemispheres, the prevailing westerlies join with *polar easterlies* to reduce upward motion. The polar easterlies form when the atmosphere over the poles cools. This cool air then sinks and spreads over the surface. As the air flows away from the poles, it is turned to the west by the Coriolis Effect. Again, because these winds begin in the east, they are called easterlies. Many of these changes in wind direction are hard to visualize (NASA 2010).

Earth's Atmosphere in Motion

To state that Earth's atmosphere is constantly in motion is to restate the obvious. Anyone observing the constant weather changes around them is well aware of this phenomenon. Although constant weather changes are obvious, the importance of the dynamic state of our atmosphere is much less obvious.

As mentioned, the constant motion of Earth's atmosphere (air movement) consists of both horizontal (wind) and vertical (air currents) dimensions. The atmosphere's motion is the result of thermal energy produced from the heating of the earth's surface and the air molecules above. Because of differential heating of the Earth's surface, energy flows from the equator poleward.

Even though air movement plays the critical role in transporting the energy of the lower atmosphere, bringing the warming influences of spring and summer and the cold chill of winter, the effects of air movements on our environment are often overlooked, even though wind and air currents are fundamental to how nature functions. All life on earth has evolved with mechanisms dependent on air movement: pollen is carried by winds for plant reproduction; animals sniff the wind for essential information; wind power was the motive force that began the earliest stages of the industrial revolution. Now we see the effects of winds in other ways, too: Wind causes weathering (erosion) of the earth's surface; wind influences ocean currents; air pollutants and contaminants such as radioactive particles transported by the wind impact our environment.

Causes of Air Motion

In all dynamic situations, forces are necessary to produce motion and changes in motion—winds and air currents. The air (made up of various gases) of the atmosphere is subject to two primary forces: (1) gravity and (2) pressure differences from temperature variations.

Gravity (gravitational forces) holds the atmosphere close to the Earth's surface. Newton's law of universal gravitation states that every body in the universe attracts another body with a force equal to:

$$F = G \, m_1 m_2 \, r2 \qquad (6.1)$$

where:

F = Force
m_1 and m_2 = the masses of the two bodies
G = universal constant of 6.67×10^{-11} N x m^2/kg^2
r = distance between the two bodies

✓ **Important Point:** The force of gravity decreases as an inverse square of the distance between them.

Thermal conditions affect density, which in turn cause gravity to affect vertical air motion and planetary air circulation. This affects how air pollution is naturally removed from the atmosphere.

Although forces in other directions often overrule gravitational force, the ever-present force of gravity is vertically downward and acts on each gas molecule, accounting for the greater density of air near the Earth.

Atmospheric air is a mixture of gases, so the gas laws and other physical principles govern its behavior. The pressure of a gas is directly proportional to its temperature. Pressure is force per unit area ($P = F/A$), so a temperature variation in air generally gives rise to a difference in pressure of force. This difference in pressure resulting from temperature differences in the atmosphere creates air movement—on both large and local scales. This pressure difference corresponds to an unbalanced force, and when a pressure difference occurs, the air moves from a high- to a low-pressure region.

In other words, horizontal air movements (called advective winds) result from temperature gradients, which give rise to density gradients and subsequently, pressure gradients. The force associated with these pressure variations (pressure gradient force) is directed at right angles to (perpendicular to) lines of equal pressure (called isobars) and is directed from high to low pressure.

Look at Figure 18. The pressures over a region are mapped by taking barometric readings at different locations. Lines drawn through the points (locations) of equal pressure are called isobars. All points on an isobar are of equal pressure, which means no air movement along the isobar. The wind direction is at right angles to the isobar in the direction of the lower pressure. In Figure 18, notice that air moves down a pressure gradient toward a lower isobar like a ball rolls down a hill. If the isobars are close together, the pressure gradient force is large, and such areas are characterized by high wind speeds. If isobars are widely spaced (see Figure 18), the winds are light because the pressure gradient is small.

Localized air circulation gives rise to thermal circulation (a result of the relationship based on a law of physics whereby the pressure and volume of a gas is directly related to its temperature). A change in temperature causes a change in the pressure and/or volume of a gas. With a change in volume comes a change in density, since $P = m/V$, so regions of the atmosphere with different temperatures may have different air pressures and densities. As a

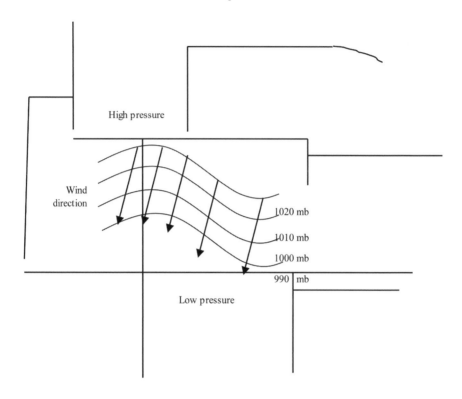

FIGURE 18. Isobars drawn through locations having equal atmospheric pressures. The air motion, or wind direction, is at right angles to the isobars and moves from a region of high pressure to a region of low pressure.

Source: Spellman and Whiting 2006

Did You Know?

Air pressure at any location whether it is on the earth's surface or up in the atmosphere depends on the weight of the air above. Imagine a column of air. At sea level, a column of air extending hundreds of kilometers above sea level exerts a pressure of 1013 millibars (mb). But, if you travel up the column to an altitude of 4.4 km (18,000 feet), the air pressure would be roughly half, or approximately 506 mb.

result, localized heating sets up air motion and gives rise to thermal circulation. To gain understanding of this phenomenon, consider Figure 19.

Once the air has been set into motion, secondary forces (velocity-dependent forces) act. These secondary forces are (1) Earth's rotation (Coriolis force) and (2) contact with the rotating earth (friction). The *Coriolis force*, named after its discoverer, French mathematician Gaspard Coriolis (1772–1843), is the effect of rotation on the atmosphere and on all objects on the Earth's surface. In the Northern Hemisphere, it causes moving objects and currents

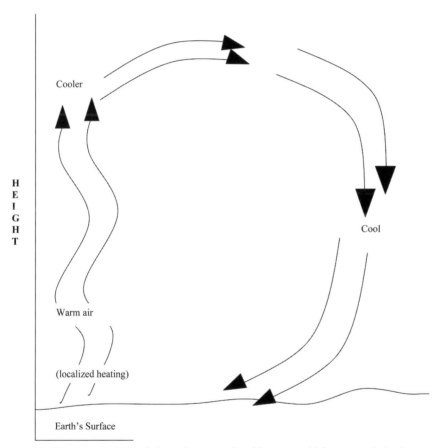

FIGURE 19.   Thermal circulation of air. Localized heating, which causes air in the region to rise, initiates the circulation. As the warm air rises and cools, cool air near the surface moves horizontally into the region vacated by the rising air. The upper, still cooler, air then descends to occupy the region vacated by the cool air.

Source: Spellman and Whiting 2006

to be deflected to the right; in the Southern Hemisphere, it causes deflection to the left, because of the Earth's rotation. Air, in large-scale north or south movements, appears to be deflected from its expected path. That is, air moving poleward in the Northern Hemisphere appears to be deflected toward the east; air moving southward appears to be deflected toward the west.

Figure 20 illustrates the Coriolis Effect on a propelled particle (analogous to the apparent effect of an air mass flowing from Point A to Point B). From Figure 20, the action of the Earth's rotation on the air particle as it travels north over the Earth's surface as Earth rotates beneath it from east to west, can

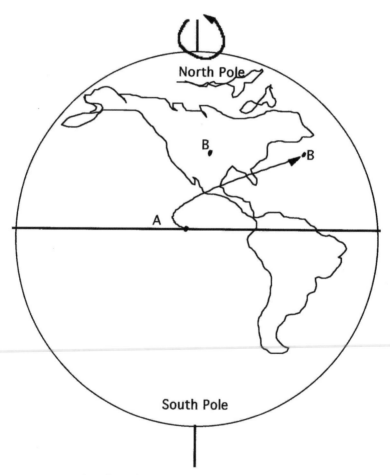

FIGURE 20.   The effect of the earth's rotation on the trajectory of a propelled particle.

Source: Spellman and Whiting 2006

be seen. Projected from point A to Point B, the particle will actually reach Point B because as it is moving in a straight line (deflected) the Earth rotates east to west beneath it.

Friction (drag) can also cause the deflection of air movements. This friction (resistance) is both internal and external. The friction of its molecules generates internal friction. Friction is also generated when air molecules run into each other. External friction is caused by contact with terrestrial surfaces. The magnitude of the frictional force along a surface is dependent on the air's magnitude and speed, and the opposing frictional force is in the opposite direction of the air motion.

Did You Know?

Friction, one of the major forces affecting the wind, comes into play near the earth's surface and continues to be a factor up to altitudes of about 500 to 1000 m. This section of the atmosphere is referred to as the Planetary or Atmospheric Boundary Layer. Above this layer, friction no longer influences the wind.

Local and World Air Circulation

Air moves in all directions, and these movements are essential for those of us on Earth: Vertical air motion is essential in cloud formation and precipitation. Horizontal air movement near the Earth's surface produces winds.

Wind is an important factor in human comfort, especially affecting how cold we feel. A brisk wind at moderately low temperatures can quickly make us uncomfortably cold. Wind promotes the loss of body heat, which aggravates the chilling effect, expressed through wind chill factors in the winter (see Table 6.2) and the heat index in the summer (see Table 6.3). These two scales describe the cooling effects of wind on exposed flesh at various temperatures.

Local winds are the result of atmospheric pressure differences involved with thermal circulations because of geographic features. Land areas heat up more quickly than do water areas, giving rise to a convection cycle. As a result, during the day, when land is warmer than the water, we experience a lake or sea breeze.

At night, the cycle reverses. Land loses its heat more quickly than water, so the air over the water is warmer. The convection cycle sets to work in the opposite direction and a land breeze blows.

In the upper troposphere (above 11 to 14 km, west to east flows) are very narrow fast-moving bands of air called jet streams. *Jet streams* have significant

TABLE 6.2
Wind Chill Chart

Wind MPH	Temperature (degrees Fahrenheit)											
	30	25	20	15	10	5	0	−5	−10	−15	−20	−25
5	25	19	13	7	1	−5	−11	−16	−22	−28	−34	−40
10	21	15	9	3	−4	−10	−16	−22	−28	−35	−41	−47
15	19	13	3	0	−7	−13	−19	−26	−32	−39	−45	−51
20	17	11	4	−2	−9	−15	−22	−29	−35	−42	−48	−55
25	16	9	3	−4	−11	−17	−24	−31	−37	−44	−51	−58
30	15	8	1	−5	−12	−19	−26	−33	−39	−46	−53	−60
35	14	7	0	−7	−14	−21	−27	−34	−41	−48	−55	−62
40	13	6	−1	−8	−15	−22	−29	−36	−43	−50	−57	−64
45	12	5	−2	−9	−16	−23	−30	−37	−44	−51	−58	−65
50	12	4	−3	−19	−17	−24	−31	−38	−45	−52	−60	−67
55	11	4	−3	−11	−18	−25	−32	−39	−46	−54	−61	−68
60	10	3	−4	−11	−19	−26	−33	−40	−48	−55	−62	−69

Note: Grey cells indicate frostbite occurs in 15 minutes or less.
Source: USA Today: http://www.usatoday.com/weather/resources/basics/windchill/wind-chill-chart.htm

effects on surface airflows. When jet streams accelerate, divergence of air occurs at that altitude. This promotes convergence near the surface and the formation of cyclonic motion. Deceleration causes convergency aloft and subsidence near the surface, causing an intensification of high-pressure systems.

Jet streams are thought to result from the general circulation structure in the regions where great high- and low-pressure areas meet.

Did You Know?

Because jet streams follow the boundaries between hot and cold air, and because these hot and cold air boundaries are most pronounced in winter, jet streams are strongest for both the Northern and Southern Hemisphere winters. Jet streams also "follow the sun" in that as the sun's elevation increases each day in the spring, the jet streams shift north moving into Canada by summer. As autumn approaches and the sun's elevation decreases, the jet stream move south in the U.S., helping bring cooler air to the country (NOAA 2010).

TABLE 6.3
Heat Index Chart (Temperature and Relative Humidity)

RH (%)	Temperature (°F)															
	90	91	92	93	94	95	96	97	98	99	100	101	102	103	104	105
90	119	123	128	132	137	141	146	152	157	163	168	174	180	186	193	199
85	115	119	123	127	132	136	141	145	150	155	161	166	172	178	184	190
80	112	115	119	123	127	131	135	140	144	149	154	159	164	169	175	180
75	109	112	115	119	122	126	130	134	138	143	147	152	156	161	166	171
70	106	109	112	115	118	122	125	129	133	137	141	145	149	154	158	163
65	103	106	108	111	114	117	121	124	127	131	135	139	143	147	151	155
60	100	103	105	108	111	114	116	120	123	126	129	133	136	140	144	148
55	98	100	103	105	107	110	113	115	118	121	124	127	131	134	137	141
50	96	98	100	102	104	107	109	112	114	117	119	122	125	128	131	135
45	94	96	98	100	102	104	106	108	110	113	115	118	120	123	126	129
40	92	94	96	97	99	101	103	105	107	109	111	113	116	118	121	123
35	91	92	94	95	97	98	100	102	104	106	107	109	112	114	116	118
30	89	90	92	93	95	96	98	99	101	102	104	106	108	110	112	114

Note: Exposure to full sunshine can increase HI values by up to 15°F.
Source: Weather Images: http://www.weatherimages.org/data/heatindex.html

References and Recommended Reading

Anthes, R.A. 1996. *Meteorology*, 7th ed. Upper Saddle River, NJ: Prentice-Hall.

Anthes, R.A., Cahir, J. J., Fraizer, A.B., and Panofsky, H.A. 1984. *The Atmosphere*, 3rd ed. Columbus, OH: Charles E. Merrill Publishing Company.

Ingersoll, A.P. 1983. The Atmosphere. *Scientific American* 249(33): 162–174.

Lutgens, F.K. and Tarbuck, E.J. 1982. *The Atmosphere, An Introduction to Meteorology.* Englewood Cliffs, NJ: Prentice-Hall.

Miller, G.R., Jr. 2004. *Environmental Science*, 10th ed. Australia: Thompson-Brooks/ Cole.

Moron, J.M., Morgan, M.D., and Wiersma, J. H. 1986. *Introduction to Environmental Science*, 2nd ed., New York: W. H. Freeman & Company.

NOAA. 2010. *The Jet Stream.* Accessed 10/15/10 @ http://www.srh.noaa.gov/netsteam/ glabal/jet.htm.

NWS. 2010. *Global Circulations.* Fort Worth, TX: National Weather Service.

Shipman, J.T., Adams, J.L., and Wilson, J.D. 1987. *An Introduction to Physical Science*, 5th ed. Lexington, MA: D.C. Heath & Company.

Spellman, F.R., and Whiting, N.E. 2006. *Environmental Science and Technology: Concepts and Applications*, 2nd ed. Lanham, MD: Government Institutes.

7

Weather and Climate

The Pharisees also with the Sadducees came and tempting desired him that he would show them a sign from heaven. He answered and said unto them, When it is evening ye say, it will be fair weather today for the sky is red and lowering. Oh ye hypocrites, ye can discern the face of the sky, but can ye not discern the signs of the times (Matthew 16:1–4).

> *Mean Weather*
> Intermittent rain, I've learned,
> Which forecasts tell about,
> Is rain that stops when I go in
> And starts when I come out.
> —Elizabeth Dolan
> In *The Breeze* 2, no. 8 (September 10, 1945): p. 6

Until man duplicates a blade of grass, nature can laugh at his so-called scientific knowledge.

> —Thomas Edison

Classical Weather Proverbs based on Scientific Facts:

> Clouds flying against the wind indicate rain.
> When walls are unusually damp, rain is to be expected.
> Rainbow in the morning, sailors take warning.
> Rainbow at night, sailor's delight.

Mackerel scales and mare's tails
Make lofty ships to carry low sails.

A Mackerel sky,
Not twenty-four hours dry.

Evening red and morning gray,
Set the traveler on his way;
Evening gray and morning red,
Bring down rain upon his head.

Introduction

A N EMINENT METEOROLOGIST once said, "A butterfly flapping its wings in
Brazil can cause a tornado in Texas." What the meteorologist was imply-
ing is true to a point (and in line with what some critics might say): because
of tiny nuances in Earth's weather patterns, making accurate, long-range
weather predictions is extremely difficult.

What is the difference between weather and climate? Some people get these
two confused, believing they mean the same thing, but they do not. In this
chapter you will gain a clear understanding of the meaning of and difference
between the two and also gain basic understanding of the role weather plays
in air pollution.

Did You Know?

Meteorology is the science concerned with the atmosphere and its
phenomena. The atmosphere is the media into which all air pollution
is emitted. The meteorologist observes atmospheric processes such as
temperature, density, (air) winds, clouds, precipitation, and other characteristics and endeavors to account for its observed structure and
evaluation (weather, in part) in terms of external influence and the
basic laws of physics (EPA 2005).

What's the Difference between Weather and Climate?

Weather is the state of the atmosphere, mainly with respect to its effect upon life
and human activities, over a short period of time; weather is defined primarily
in terms of heat, pressure, wind, and moisture. On the other hand, *climate* is the

long-term manifestations of weather or how the atmosphere behaves over relatively long periods of time.

In regard to weather, at high levels above the Earth, where the atmosphere thins to near vacuum, there is no weather; instead, weather is a near-surface phenomenon. This is evidenced clearly on a day-by-day basis where you see the ever-changing, sometimes dramatic, and often violent weather display.

In the study of air science and, in particular, of air quality, the following determining factors are directly related to the dynamics of the atmosphere, resulting in local weather. These factors include strength of winds, the direction they are blowing, temperature, available sunlight (needed to trigger photochemical reactions, which produce smog), and the length of time since the last weather event (strong winds and heavy precipitation) cleared the air.

Weather events (such as strong winds and heavy precipitation) that work to clean the air we breathe and replenish our water supplies are beneficial, obviously. However, few people would categorize the weather events such as tornadoes, hurricanes, and typhoons as beneficial. On the other hand, in terms of personification, if we can say that these events are simply a result of Mother Nature sneezing, scratching her back, or yawning, then our view may be different. The point is these events have the potential to kill people and destroy property and change landscapes, and so forth; how can anything natural be considered bad? The problem is we do not know what we do not know about these weather events. However, what we know and how we perceive certain weather events is different; that is, we sometimes look at some weather events as having both a positive and negative effect (again, that Jekyll and Hyde characterization). One such event is El Nino-Southern Oscillation, discussed below.

El Niño–Southern Oscillation

El Niño–Southern Oscillation, or ENSO, is a natural phenomenon that occurs every 2 to 9 years on an irregular and unpredictable basis. El Niño is a warming of the surface waters in the tropical eastern Pacific, which causes fish to disperse to cooler waters and, in turn, causes the adult birds to fly off in search of new food sources elsewhere.

Through a complex web of events, El Niño (which means "the child" in Spanish because it usually occurs during the Christmas season off the coasts of Peru and Ecuador) can have a devastating impact on all forms of marine life.

During a normal year, equatorial trade winds pile up warm surface waters in the western Pacific. Thunderheads unleash heat and torrents of rain. This heightens the east-west temperature difference, sustaining the cycle. The jet stream blows from north Asia to California. During an El Niño–Southern

Oscillation year, trade winds weaken, allowing warm waters to move east. This decreases the east-west temperature difference. The jet stream is pulled farther south than normal, picks up storms it would usually miss, and carries them to Canada or California. Warm waters eventually reach South America.

One of the first signs of its appearance is a shifting of winds along the equator in the Pacific Ocean. The normal easterly winds reverse direction and drag a large mass of warm water eastward toward the South American coastline. The large mass of warm water basically forms a barrier that prevents the upwelling of nutrient-rich cold water from the ocean bottom to the surface. As a result, the growth of microscopic algae that normally flourish in the nutrient-rich upwelling areas diminishes sharply, and that decrease has further repercussions. For example, El Niño–Southern Oscillation has been linked to patterns of subsequent droughts, floods, typhoons, and other costly weather extremes around the globe. Take a look at El Niño–Southern Oscillation's affect on the West Coast of the United States where ENSO has been blamed for West Coast hurricanes, floods, and early snowstorms. On the positive side, ENSO typically brings good news to those who live on the East Coast of the United States: a reduction in the number and severity of hurricanes.

Note that, in addition to reducing the number and severity of hurricanes, in October 1997 the Associated Press reported that a new study has shown that ENSO also deserves credit for invigorating plants and helping to control the pollutant linked to global warming. Researchers have found that El Niño causes a burst of plant growth throughout the world and this removes carbon dioxide from the atmosphere.

Atmospheric carbon dioxide (CO_2) has been increasing steadily for decades. The culprits are increased use of fossil fuels and the clearing of tropical rainforests. However, during an ENSO phenomenon, global weather is warmer, there is an increase in new plant growth and CO_2 levels decrease.

Not only does ENSO have a major regional impact in the Pacific, its influence extends to other parts of the world through the interaction of pressure, air flow, and temperature effects.

El Niño–Southern Oscillation is a phenomenon that, although not quite yet completely understood by scientists, causes both positive and negative results, depending upon where you live.

The Sun: The Weather Generator

The sun is the driving force behind weather. Without the distribution and reradiation to space of solar energy, we would experience no weather (as we know it) on Earth. The sun is the source of most of the Earth's heat. Of the gigantic amount of solar energy generated by the sun, only a small portion bombards earth. Most of the sun's solar energy is lost in space. A little over 40% of the sun's radiation reaching earth hits the surface and is changed to heat. The rest stays in the atmosphere or is reflected back into space.

Like a greenhouse, the earth's atmosphere admits most of the solar radiation. When solar radiation is absorbed by the earth's surface, it is reradiated as heat waves, most of which are trapped by carbon dioxide and water vapor in the atmosphere, which work to keep the Earth warm in the same way a greenhouse traps heat.

By now you are aware of the many functions performed by the Earth's atmosphere. You should also know that the atmosphere plays an important role in regulating the earth's heating supply. The atmosphere protects the earth from too much solar radiation during the day and prevents most of the heat from escaping at night. Without the filtering and insulating properties of the atmosphere, the Earth would experience severe temperatures similar to other planets.

On bright clear nights the Earth cools more rapidly than on cloudy nights because cloud cover reflects a large amount of heat back to Earth, where it is reabsorbed.

The Earth's air is heated primarily by contact with the warm Earth. When air is warmed, it expands and becomes lighter. Air warmed by contact with Earth rises and is replaced by cold air, which flows in and under it. When this cold air is warmed, it too rises and is replaced by cold air. This cycle continues and generates a circulation of warm and cold air, which is called *convection*.

At the earth's equator, the air receives much more heat than the air at the poles. This warm air at the equator is replaced by colder air flowing in from north and south. The warm, light air rises and moves poleward high above the Earth. As it cools, it sinks, replacing the cool surface air which has moved toward the equator.

The circulating movement of warm and cold air (convection) and the differences in heating cause local winds and breezes. Different amounts of heat are absorbed by different land and water surfaces. Soil that is dark and freshly plowed absorbs much more than grassy fields. Land warms faster that does water during the day and cools faster at night. Consequently, the air above such surfaces is warmed and cooled, resulting in production of local winds.

Winds should not be confused with air currents. Wind is primarily oriented toward horizontal flow. Air currents, on the other hand, are created by air moving upward and downward. Wind and air currents have direct impact on air pollution. Air pollutants are carried and dispersed by wind. An important factor in determining the areas most affected by an air pollution source is wind direction. Since air pollution is a global problem, wind direction on a global scale is important.

Along with wind, another constituent associated with earth's atmosphere is water. Water is always present in the air. It evaporates from the earth, two thirds of which is covered by water. In the air, water exists in three states: solid, liquid, and invisible vapor.

The amount of water in the air is called humidity. The *relative humidity* is the ratio of the actual amount of moisture in the air to the amount needed for saturation at the same temperature. Warm air can hold more water than cold. When air with a given amount of water vapor cools, its relative humidity increases; when the air is warmed, its relative humidity decreases.

Air Masses

An air mass is a vast body of air (a macroscale phenomena that can have global implications) in which the condition of temperature and moisture are much the same at all points in a horizontal direction. An air mass takes on the temperature and moisture characteristics of the surface over which it forms and travels, though its original characteristics tend to persist. The processes of radiation, convection, condensation, and evaporation condition the air in an air mass as it travels. Also, pollutants released into an air mass travel and disperse within the air mass. Air masses develop more commonly in some regions than in others. Table 7.1 summarizes air masses and their properties.

TABLE 7.1
Classification of Air Masses

Name	Origin	Properties	Symbol
Arctic	Polar regions	Low temperatures, low specific but high summer relative humidity, the coldest of the winter air masses	A
Polar continental*	Subpolar continental areas	Low temperatures (increasing with southward movement), low-humidity, remaining constant	cP
Polar maritime	Subpolar area and arctic region	Low temperatures increasing with movement, higher humidity	mP
Tropical continental	Subtropical high-pressure land areas	High temperatures, low moisture content	cT
Tropical maritime	Southern borders of oceanic subtropical, high-pressure areas	Moderate high temperatures, high relative and specific humidity	mT

* The name of an air mass, such as Polar continental, can be reversed to continental Polar, but the symbol, cP, is the same for either version.
Source: EPA 2005.

When two different air masses collide, a *front* is formed. A front is not a sharp wall but a zone of transition which is often several miles wide. Four frontal patterns—warm, cold, occluded, and stationary—can be formed by air of different temperatures. A *cold front* marks the line of advance of a cold air mass from below, as it displaces a warm air mass. A *warm front* marks the advance of a warm air mass as it rises up over a cold one.

When cold and warm fronts merge (the cold front overtaking the warm front) *occluded fronts* form. Occluded fronts can be called cold front or warm front occlusions. But, in either case, a colder air mass takes over an air mass that is not as cold.

The last type of front is the stationary front. As the name implies, the air masses around this front are not in motion. A stationary front can cause bad weather conditions that persist for several days.

Thermal Inversions and Air Pollution

Earlier, it was pointed out that during the day the sun warms the air near the Earth's surface. Normally, this heated air expands and rises during the day, diluting low-lying pollutants and carrying them higher into the atmosphere. Air from surrounding high-pressure areas then moves down into the low-pressure area created when the hot air rises (see Figure 21a). This continual

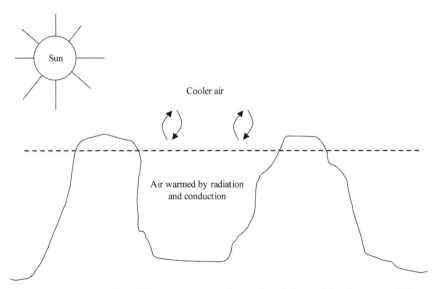

FIGURE 21a. Normal conditions. Air at Earth's surface is heated by the sun and rises to mix with the cooler air above it.

mixing of the air helps keep pollutants from reaching dangerous levels in the air near the ground.

Sometimes, however, a layer of dense, cool air is trapped beneath a layer of less dense, warm air in a valley or urban basin. This is called a *thermal inversion*. In effect, a warm-air lid covers the region and prevents pollutants from escaping in upward-flowing air currents (see Figure 21b). Usually these inversions trap air pollutants (i.e., plume dispersion is inhibited) at ground level for a short period of time. However, sometimes they last for several days when a high-pressure air mass stalls over an area, trapping air pollutants at ground level where they accumulate to dangerous levels.

The best known location in the United States where thermal inversions occur almost on a daily basis is in the Los Angeles basin. The Los Angeles basin is a valley with a warm climate, light winds, surrounded by mountains located near the Pacific Coast. Los Angeles is a large city with a large population of people and automobiles and possesses the ideal conditions for smog, which is worsened by frequent thermal inversions.

Climate

The global weather patterns we observe are directly affected by the earth's tilt, rotation, and land/sea distribution. While the weather varies from day-to-day at any particular location, over the years, the same type of weather will reoccur. The reoccurring "average weather" found in any particular place throughout

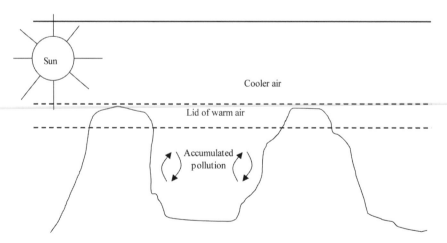

FIGURE 21b.   Thermal inversion. A layer of warm air forms a lid about the Earth, and the cooler air at the surface is unable to mix with the warm air above. Pollutants are trapped.

several years over is called climate. An apt and easy to remember expression of the difference between climate and weather was offered by Mark Twain:

Climate is what you expect, but weather is what you get.

Climate Divisions

German climatologist and amateur botanist Wladimir Köppen (1846–1940) divided the world's climates into several major categories (A, B, C, D, E, H) based upon general temperature profile related to latitude. These categories are as follows (NOAA 2010):

A—Tropical Climates
Tropical moist climates extend north and south from the equator to about 15° to 25° latitude. In these climates all months have average temperatures greater than 64°F (18°C) and annual precipitation greater than 59 inches.

B—Dry Climates
The most obvious climatic feature of this climate is that potential evaporation and transpiration exceed precipitation. These climates extend from 20–35° north and south of the equator and in large continental regions of the mid-latitudes often surrounded by mountains.

C—Moist Subtropical Mid-Latitude Climates
The climate generally has warm and humid summers with mild winters. Its extent is from 30–50° of latitude mainly on the eastern and western borders of most continents. During the winter, the main weather feature is the mid-latitude cyclone. Convective thunderstorms dominate summer months.

D—Moist Continental Mid-latitude Climates
Moist continental mid-latitude climates have warm to cool summers and cold winters. The location of these climates is poleward of the C climates. The average temperature of the warmest month is greater than 50°F (10°C), while the coldest month is less than –22°F (–30°C). Winters are severe with snowstorms, strong winds, and bitter cold form Continental Polar or Arctic air masses.

E—Polar Climates
Polar climates have year-round cold temperatures with the warmest month less than 50°F (10°C). Polar climates are found on the northern coastal areas of North America, Europe, Asia, and on the land masses of Greenland and Antarctica.

H—Highlands (This category was added after Köppen created this system.) Unique climates based on their elevation. Highland climates occur in mountainous terrain where rapid elevation changes cause rapid climatic changes over short distances.

Did You Know?

The classical length of record to determine the climate or rank any particular place is 30 years, as defined by the World Meteorological Organization (WMO). The quantities most often observed are temperature, precipitation, and wind. The "normals" are computed once every 10 years which helps to smooth out year-to-year variations. For example, the current 30-year normals were calculated from actual weather data that occurred during the 30 years of 1971–2000. In 2011, a new set of 30-year normals will be calculated during the day from the period of 1981–2010. So, when you hear what the normal high and low temperature for your location are, for example, they come from these 30-year averages (NOAA 2010).

Climate Subdivisions

Koppen further divided each major division (A, B, C, D, E, H) into smaller groups based upon precipitation and temperature patterns. These subcategories are as follows:

The **second letter** (lowercase) designates subdivisions based on precipitation pattern.

> **f**—wet year-round
> **s**—dry summer season
> **w**—dry winter season
> **m**—monsoon

The **third letter** (lowercase) designates subdivisions based in temperature pattern

> **a**—hot summer
> **b**—warm summer
> **c**—cool summer
> **d**—very cold winters

TABLE 7.2
The Complete Koppen Climates

Af	Tropical rainforest	No dry season. The driest month has at least 2.40" (61 mm) of rain. Rainfall is generally evenly distributed throughout the year. All average monthly temperatures are greater than 64° (18°C).
Am	Tropical monsoonal	Pronounced wet season. Short dry season. There are one or more months with less than 2.40" (61 mm). All average monthly temperatures are greater than 64°F (18°C). Highest annual temperature occurs just prior to the rainy season.
Aw	Tropical Savanna	Winter dry season. There are more than two months with less than 2.40" (61 mm). All average monthly temperatures are greater than 64°F (18°C).
BSh	Subtropical Steppe	Low-latitude dry. Evaporation exceeds precipitation on average but is less than potential evaporation. Average temperature is *more* than 64°F (18°C).
BSk	Mid-latitude Steppe	Mid-latitude dry. Evaporation exceeds precipitation on average but is less than potential evaporation. Average temperature is *less* than 64°F (18°C).
BWh	Subtropical Desert	Low-latitude desert. Evaporation exceeds precipitation on average but is *less than half* potential evaporation. Average temperature is more than 64°F (18°C). Frost is absent or infrequent.
BWk	Mid-latitude	Mid-latitude desert. Evaporation exceeds precipitation on average but is *less than half* potential evaporation. Average temperature is less than 64°F (18°C). Winter has below freezing temperatures.
Cfa	Humid	Mild with no dry season, hot summer. Average temperature of Subtropical warmest months are over 72°F (22°C). Average temperature of coldest month is under 64°F (18°C). Year round rainfall but highly variable.
Cfb	Marine West Coast	Mild with no dry season, warm summer. Average temperature of all months is *lower* than 72°F (22°C). At least four months with average temperatures over 50°F (10°C). Year round equally spread rainfall.
Cfc	Marine West Coast	Mild with no dry season, cool summer. Average temperature of all months is *lower* than 72°F (22°C). There are one to three months with average temperatures over 50°F (10°C). Year round equally spread rainfall.
Csa	Mediterranean	Mild with dry, hot summer. Warmest month has average temperature *more* than 72° (22°C). At least four months with average temperatures over 50°F (10°C). Frost danger in winter. At least three times as much precipitation during wettest winter months as in the driest summer month.

TABLE 7.2
The Complete Koppen Climates *(continued)*

Csb	Mediterranean	Mild with cool, dry summer. No month with average temperature of warmest months are over 72°F (22°C). At least four months with average temperatures over 50°F (10°C). Frost danger in winter. At least three times as much precipitation during wettest winter months as in the driest summer month.
Cwa	Humid Subtropical	Mild with dry winter, hot summer
Dfa	Humid Continental	Humid with hot summer
Dfb	Humid Continental	Humid with severe winter, no dry season, warm summer
Dfc	Subarctic	Severe winter, no dry season, cool summer
Dfd	Subarctic	Severe, very cold winter, no dry season, cool summer
Dwa	Humid Continental	Humid with severe, dry winter, hot summer
Dwb	Humid Continental	Humid with severe, dry winter, warm summer
Dwc	Subarctic	Severe, dry winter, cool summer
Dwd	Subarctic	Severe, very cold and dry winter, cool summer
ET	Tundra	Polar tundra, no true summer
EF	Ice Cap	Perennial ice
H	Complex Zone	Can encompass any of the above classifications due to the mountainous terrain.

Microclimates

> Nothing that is can pause or stay;
> The moon will wax, the moon will wane,
> The mist and cloud will turn to rain,
> The rain to mist and cloud again,
> Tomorrow be today.
> —Henry Wadsworth Longfellow

What is a microclimate? In answering this question, scale must first be talked about. For example, let's take a look at flow of air within a very small environment: the emission of smoke from a chimney. This flow represents one of the smallest spatial subdivisions of atmospheric motion, or microscale weather.

On a more realistic, but still relatively small scale, we must consider the geographical, biological, and man-made features that make local climate different from the general climate. This local climatic pattern is called a *microclimate*; it includes the climate of a small area such as a house, cave, city, or valley that may be different from that in the general region.

What are the elements or conditions that cause local or microclimates? Location, location, location—and local conditions—are the main ingredients making up a microclimate. Let's look at one example.

Large inland lakes moderate temperature extremes and climatic differences between their windward and leeward sides. For example, Seattle, on the windward side of Lake Washington, and the city of Bellevue, on the leeward side only about nine miles east, have microclimatic differences (although modest). Microclimatic differences exist in temperature fluctuations, precipitation levels, wind speed, and in relative humidity.

Even more dramatic differences can be seen in such parameters when a comparison is made between a city such as Milwaukee, on the windward side of Lake Michigan, and Grand Haven, on the leeward side, only 85 miles east.

Other examples of microclimates exist and these are listed below:

- Near the ground
- Over open land areas
- In woodlands or forested areas
- In valley regions
- In hillside regions
- In urban areas
- In sea-side locations

In the following sections we take a closer look at these microclimates: at their nature, causative factors, and their geographical/topographical locations.

Microclimates near the Ground

Nowhere in the atmosphere are climatic differences as distinct as they are near the ground. For instance, when you go to the beach on a warm summer day, you no doubt have noticed that the grass and water are much cooler to your feet than the sand. So, you may ask, what is it about this area near the ground that produces a microclimate with such major differences?

It's the interface (or activity zone) between the atmosphere and the ground surface (sandy shore) that causes the stark difference in temperature variability. Energy is reaching the sandy beach from the sun and from the atmosphere (though to a much lesser extent). The energy is either reflected and then returned to the atmosphere in a different form or is absorbed and stored in the sandy surface as heat.

Ground level energy absorption is very sensitive to the nature of the ground surface. Ground surface color, wetness, cover (vegetation), and topography are conditions that all affect the interaction between the ground and the atmosphere. Consider a snow-covered ground, for example. Clean snow reflects solar radiation, so the surface remains cool and the snow fails to melt. However, dirty snow absorbs more radiation, heats up, and is likely to melt. If the snowy area is shielded by vegetation, the vegetation, too, may protect the snow from the heat of the sun.

A surface cover such as clean snow has the ability to reflect solar radiation because of its high albedo. *Albedo* (the ratio between the light reflected from a surface and the total light falling on it) always has a value less than or equal to 1. An object with a high albedo, near 1, is very bright, while a body with a low albedo, near 0, is dark. For example, freshly fallen snow typically has an albedo that is between 75% and 90%; that is, 75% to 95% of the solar radiation that is incident on snow is reflected. At the other extreme, the albedo of a rough, dark surface, such as a green forest, may be as low as 5%. The albedos of some common surfaces are listed in Table 7.3. The portion of insolation not reflected is absorbed by the Earth's surface, warming it. This means Earth's albedo plays an important part in the Earth's radiation balance and influences the mean annual temperature and the climate on both local and global scales.

TABLE 7.3
The ALBEDO of Some Surface Types in % Reflected

Surface	ALBEDO
Water (low sun)	10–100
Water (high sun)	3–10
Grass	16–26
Glacier ice	20–40
Deciduous forest	15–20
Coniferous forest	5–15
Old snow	40–70
Fresh snow	75–95
Sea ice	30–40
Blacktopped tarmac	5–10
Desert	25–30
Crops	15–25

Microclimates over Open Land Areas

Many different properties of ground layer or soil type influence conditions in the thin layer of atmosphere just above it. Light-colored soils do not absorb energy as efficiently as do organically rich darker soils. Another important factor is soil moisture. Wet soils are normally dark, but moist soil (because water has a large heat capacity) requires a great deal of energy to raise its temperature. A moist soil warms up more slowly than a dry one.

Soil is a heterogeneous mixture of various particles. In between the soil particles is a large amount of air—air that is a poor conductor of heat. The larger the amount of air between the soil particles, the slower the heat transfers through the soil. As demonstrated in our example of the sandy beach, on a hot sunny day the heat is trapped in the upper layers, so the surface layers warm up more rapidly and become extremely hot. Water conducts heat more readily than air, so soils that contain some moisture are able to transmit warmth away from the surface more easily than dry soils. This is not always the case, however. If the soil contains too much water, the large heat capacity of the water will prevent the soil from warming despite heat being conducted from the surface.

Microclimates in Woodlands or Forested Areas

When making microclimate comparisons between open land areas and forested areas (commonly referred to as a forest climate), the differences are quite apparent. Forested areas, for example, are generally warmer in winter than the open areas, while open land is warmer in summer than forested areas. The forest climate has reduced wind speeds, while the open land area has higher wind speeds. The forest climate has higher relative humidity, while the open area has lower relative humidity. In the forest climate, water storage capacity is higher and evaporation rates are lower, while in the open land area water storage capacity is lower and evaporation is higher.

Microclimates in Valleys and Hillside Regions

Heavy, cold air flows downhill, forming cold pockets in valleys. Frost is much more common there, so orchards of apples and oranges and vines of grapes are planted on hillsides to ensure frost drainage when cold spells come.

Probably the best way in which to describe the microclimate in a typical valley region is to compare and contrast it with a hillside environment.

In a typical valley region, the daily minimum temperature is much lower than that in a hillside area. The daily and annual temperature range for a valley is much larger than that of a hillside area. In a valley region, more frost occurs than in a hillside region. Windspeed at night is lower in a valley than on a hillside, and morning fog is more prevalent and lasts longer in a valley region.

Microclimates in Urban Areas

The microclimate in an urban area as compared to that of the countryside is usually quite obvious. A city, for example, is usually characterized by having haze and smog, higher temperatures, lower wind speed, and reduced radiation. The countryside, on the other hand, is characterized by clear, clean air, lower temperatures, and high wind speeds and radiation.

These different microclimatic conditions should come as no surprise to anyone, especially when you consider what happens when a city is built. Instead of a mixture of soil or vegetation, the surface layer is covered with concrete, brick, glass, and stone surfaces ranging to heights of several hundred feet. These materials have vastly different physical properties from soil and trees. They shed and carry away water, absorb heat, block and channel the passage of winds, and present albedo levels significantly different from those of the natural world. All of these factors (and more) work to alter the climate conditions in the area.

Did You Know?

Urban areas have added roughness features and different thermal characteristics due to the presence of man-made elements. The thermal influence dominates the influence of the frictional components. Building materials such as brick and concrete absorb and hold heat more efficiently than soil and vegetation found in rural areas. After the sun sets, the urban area continues to radiate heat from buildings, paved surfaces, etc. Air warmed by the urban complex rises to create a dome over the city. It is called the *heat island effect*. The city emits heat all night. Just when the urban area begins to cool, the sun rises and begins to heat the urban complex again. Generally, city areas never revert to stable conditions because of the continual heating that occurs (EPA 2005).

Microclimates in Seaside Locations

The major climatic feature associated with seaside locations is the sea breeze. Sea breezes are formed by the different responses to heating of water and land. For example, if we have a bright, sunny morning with little wind, the ground surface warms rapidly as it absorbs short-wave radiation. Most of this heat is retained at the surface, although some will be transferred through the soil. As a result, the temperature of the ground surface increases and some of the heat warms the air above. When the sun sets, the surface starts to cool rapidly, because there is little store of heat in the soil. Thus, we find that land surfaces are characterized by high day (and summer) temperatures and low night (and winter) temperatures.

Now let's take a look at the response of the sea, which is very different. Solar energy (sunshine) is able to penetrate through the water to a certain level. Much solar energy has to be absorbed to raise its temperature. Through wave action and convection, the warm surface water is mixed with cooler deeper water. With enough solar energy and time, the top several feet of water forms an active layer where temperature change is slow. Slight warming occurs during the day and slight cooling at night. This means that the sea is normally cooler than the land by day and warmer by night.

Did You Know?

Have you ever noticed how much cooler it is in the shade than in direct sunlight? Of course! Or how much hotter it feels to stand on pavement as opposed to a grassy patch of land? As mentioned, temperature differences within a small area are indications of microclimates: very small-scale climate conditions. The following are a few examples of microclimatic variation:

- Dense, cold air sinking into the bottom of a valley can make the valley floor 20 degrees Celsius colder than a slope only 100 meters higher.
- Winter sunshine can heat the south-facing side of a tree (and the habitable cracks and crevices within it) to as high as 30 degrees Celsius, while the temperature only a few centimeters away from the tree is below freezing in the a high latitude case, i.e., not in the Pacific.
- The air temperature in a corn or wheat field can vary by 10 degrees Celsius from the soil to the top of the canopy.

Frogs, beetles, and other small animals experience temperature changes on even small scales (e.g., pockets of coolness formed by crevices in tree bark, the shade of a leaf or moist soil beneath a rock). Such small-scale temperature variations might seem unimportant, but they help set the distribution and abundance (DOE 2010).

The higher temperature over the land by day generates a weak low-pressure area. As this intensifies during daytime heating, a flow of cool, more humid air spreads inland from the sea, gradually changing in strength and direction during the day. At night the reverse occurs, with circulation of air from the cooler land to the warmer sea, though as the temperature difference is usually less, the land breeze is weak. Even large lakes can show a breeze system of this nature.

References and Recommended Reading

DOE. 2010. *Microclimate.* US Department of Energy. Accessed 10/18/10 @ http://education.arm.gov/teacher-lounge/lessons/microclimate.

EPA. 2005. *Basic Air Pollution Meteorology.* APTI Course SI:409. Accessed 1/9/08 @ www.epa.gov/apti

NOAA. 2010. *Koppen Climate Zones.* Accessed 10/16/10 @ http://www.shr.noaa.gov/jetstream/global/climate.htm.

Spellman, F.R. 2008. *The Science of Air.* Boca Raton, FL: CRC Press.

Spellman, F.R., and Whiting, N. 2006. *Environmental Science and Technology: Concepts and Applications,* 2nd ed. Lanham, MD: Government Institutes.

IV

WATER: EARTH'S BLOOD

Perhaps the profession of doing good may be full, but everybody should be kind at least to himself. Take a course in good water and air, and in the eternal youth of Nature you may renew your own. Go quietly, alone; no harm will befall you. Some have strange, morbid fears as soon as they find themselves with Nature even in the kindest and wildest of her solitudes, like very children afraid of the mother—as if God were dead the devil were King.

—John Muir, 1888; Mt. Shasta, *Picturesque California*, 165; 82.

8

Water on Earth

To the Reader

I N READING THIS SECTION, you are going to spend some time in following a drop of water on its travels. As Arabella B. Buckley wondered in 1878, "If I dip my finger in this basin of water and lift it up again, I bring with it a small glistening drop out of the water below and hold it before you. Tell me, have you any idea where this drop has been? What changes it has under gone, and what work it has been doing during all the long ages water has lain on the face of the earth? (quoted in Wick 1997)." Would it be long ages? Yes. Absolutely. Consider the water that weeps—literally weeps from Weeping Rock, Zion National Park, Utah. The weeps of wet stuff are relatively old ... well, sort of old, in the recent scheme of things, anyway ... about 4,000 years old. That 4,000 year old first drop of water, delivered as a weep today, entered from Echo Canyon above, a slot canyon, into the spring line between Kayenta and Navajo layered sandstone and slowly but inexorably seeped its way downward (remember: nothing on Earth is exempt from the heavy hand of gravity) through the **permeable** rock. Eventually, as with all movement on Earth, resistance is met in the form of Zion red rock shale. Shale is **impermeable**. Thus, downward flow of that drop of water shifted to lateral flow along its shale barricade. Water does not stay put in any one place forever. That first drop of water, joined by many of its brothers and sisters and cousins and others seeps horizontally and diagonally or whichever way it could until it reached daylight; a place where it transformed itself from seep to weep. Thus, to the delight to all those who witness, Weeping Rock weeps (see Figure 22).

FIGURE 22.   Weeping Rock, Zion National Park, Utah
Photo by Frank R. Spellman

Those lucky enough to taste the weeps probably hear that refrain from that great song: "cool cool water." Moreover, they now know what tasting sweet nectar of the gods must be all about . . . that feeling of being outside of one's self (ecstasy) . . . the 4,000 year wait well worth it.
                                                   —F.R. Spellman, 2007, *The Science of Water*

Introduction

When color photographs of the Earth as it appears from space were first published, it was a revelation: they showed our planet to be astonishingly beautiful. We were taken by surprise. What makes the earth so beautiful is its abundant water. The great expanses of vivid blue ocean with swirling, sunlit clouds above them should not have caused surprise, but the reality exceeded everybody's expectations. The pictures must have brought home to all who saw them the importance of water to our planet (E.C. Pielou, preface to *Fresh Water* 1998).

Water Is Special, Strange, and Different

Whether we characterize it as ice, rainbow, steam, frost, dew, soft summer rain, as fog, as flood or avalanche, or as stimulating as a stream or cascade, water is special—water is strange—water is different.

Water is the most abundant inorganic liquid in the world; moreover, it occurs naturally anywhere on earth. Literally awash with it, life depends on it, and yet water is so very different.

Water is scientifically different. With its rare and distinctive property of being denser as a liquid than as a solid, it is different. Water is different in that it is the only chemical compound found naturally in solid, liquid, gaseous states. Water is sometimes called the universal solvent. This is a fitting name, especially when you consider that water is a powerful reagent, which is capable in time of dissolving everything on earth.

Water is different. It is usually associated with all the good things on earth. For example, water is associated with quenching thirst, with putting out fires, and with irrigating the earth. The question is: Can we really say emphatically, definitively that water is associated with only those things that are good?

Not really!

Remember, water is different; nothing, absolutely nothing, is safe from it.

Water is different. This unique substance is odorless, colorless, and tasteless. Water covers 71% of the earth completely. Even the driest dust ball contains 10–15% water.

Water and life—life and water—inseparable.

The prosaic becomes wondrous as we perceive the marvels of water.

Three hundred twenty-six million cubic miles of water cover earth but only 3% of this total is fresh, with most locked up in polar ice caps, glaciers, in lakes; it flows through soil and in river and stream systems back to an ever increasingly saltier sea (only 0.027% is available for human consumption). Water is different.

Standing at a dripping tap, water is so palpably wet; one can literally hear the drip-drop-plop.

Water is special—water is strange—water is different—more importantly, water is critical to our survival, yet we abuse it, discard it, fowl it, curse it, dam it, and ignore it. At least this is the way we view the importance of water at this moment in time . . . however, because water is special, strange, and different, the dawn of tomorrow is pushing for quite a different view.

Along with being special, strange, and different, water is also a contradiction, a riddle.

How? Consider the Chinese proverb that states "water can both float and sink a boat."

Water's presence everywhere feeds these contradictions. Lewis (1996) points out that "water is the key ingredient of mother's milk and snake venom, honey and tears" (p. 90).

Leonardo da Vinci gave us insight into more of water's apparent contradictions:

Water is sometimes sharp and sometimes strong, sometimes acid and sometimes bitter;

Water is sometimes sweet and sometimes thick or thin;

Water sometimes brings hurt or pestilence, sometimes health-giving, sometimes poisonous.

Water suffers changes into as many natures as are the different places through which it passes.

Water, as with the mirror that changes with the color of its object, so it alters with the nature of the place, becoming: noisome, laxative, astringent, sulfurous, salt, incarnadined, mournful, raging, angry, red, yellow, green, black, blue, greasy, fat or slim.

Water sometimes starts a conflagration, sometimes it extinguishes one.

Water is warm and is cold.

Water carries away or sets down.

Water hollows out or builds up.

Water tears down or establishes.

Water empties or fills.

Water raises itself or burrows down.

Water spreads or is still.

Water is the cause at times of life or death, or increase of privation, nourishes at times and at others does the contrary.

Water, at times has a tang, at times it is without savor.

Water sometimes submerges the valleys with great flood.

In time and with water, everything changes.

Water's contradictions can be summed up by simply stating that though the globe is awash in it, water is no single thing, but an elemental force that shapes our existence. Da Vinci's last contradiction, "In time and with water, everything changes" concerns us most in this text.

Many of da Vinci's water contradictions are apparent to most observers. But with water there are other factors that do not necessarily stand out, that are not always so apparent. This is made clear by the following example—what you see on the surface of a water body is not necessarily what lies beneath.

Still Water

Consider a river pool, isolated by fluvial processes and time from the main stream flow. We are immediately struck by one overwhelming impression: It appears so still . . . so very still . . . still enough to soothe us. The river pool provides a kind of poetic solemnity, if only at the pool's surface. No words of peace, no description of silence or motionlessness can convey the perfection of this place, in this moment stolen out of time.

We ask ourselves, "The water is still, but does the term 'still' correctly describe what we are viewing . . . is there any other term we can use besides still—is there any other kind of still?"

Yes, of course, we know many ways to characterize still. For sound or noise, "still" can mean inaudible, noiseless, quiet, or silent. With movement (or lack of movement), still can mean immobile, inert, motionless, or stationary. At least this is how the pool appears to the casual visitor on the surface. The visitor sees no more than water and rocks.

The rest of the pool? We know very well that a river pool is more than just a surface. How does the rest of the pool (the subsurface, for example) fit the descriptors we tried to use to characterize its surface? Maybe they fit, maybe they don't. In time, we will go beneath the surface, through the liquid mass, to the very bottom of the pool to find out. For now, remember that images retained from first glances are almost always incorrectly perceived, incorrectly discerned, and never fully understood.

On second look we see that the fundamental characterization of this particular pool's surface is correct enough. Wedged in a lonely riparian corridor—formed by riverbank on one side and sandbar on the other—between a youthful, vigorous river system on its lower end and a glacier-and artesian-fed lake on its headwater end, almost entirely overhung by mossy old Sitka spruce, the surface of the large pool, at least at this particular location, is indeed still. In the proverbial sense, the pool's surface is as still and as flat as a flawless sheet of glass.

The glass image is a good one, because like perfect glass, the pool's surface is clear, crystalline, unclouded, definitely transparent, yet perceptively deceptive as well. The water's clarity, accentuated by its bone-chilling coldness, is apparent at close range. Further back, we see only the world reflected in the water—the depths are hidden and unknown. Quiet and reflective, the polished surface of the water perfectly reflects in mirror-image reversal the spring greens of the forest at the pond's edge, without the slightest ripple. Up close, looking straight into the bowels of the pool we are struck by the water's transparency. In the motionless depths, we do not see a deep, slow-moving reach with muddy bottom typical of a river or stream pool; instead, we clearly see the warm variegated tapestry of blues, greens, blacks stitched together with threads of fine, warm-colored sand that carpets the bottom, at least twelve feet below. Still waters can run deep.

No sounds emanate from the pool. The motionless, silent water doesn't, as we might expect, lap against its bank or bubble or gurgle over the gravel at its edge. Here, the river pool, held in temporary bondage, is patient, quiet, waiting, withholding all signs of life from its surface visitor.

Then the reality check: The present stillness, like all feelings of calm and serenity, could be fleeting, momentary, temporary, you think. And you would be correct, of course, because there is nothing still about a healthy river pool. At this exact moment, true clarity is present, it just needs to be perceived . . . and it will be.

* * *

We toss a small stone into the river pool, and watch the concentric circles ripple outward as the stone drops through the clear depths to the pool bottom. For a brief instant, we are struck by the obvious: the stone sinks to the bottom, following the laws of gravity, just as the river flows according to those same inexorable laws—downhill in its search for the sea. As we watch, the ripples die away; leaving as little mark as the usual human lifespan creates in the waters of the world, then disappears as if it had never been. Now the river water is as before, still. At the pool's edge, we look down through the massy depth to the very bottom—the substrate.

We determine that the pool bottom is not flat or smooth, but instead is pitted and mounded occasionally with discontinuities. Gravel mounds alongside small corresponding indentations—small, shallow pits—make it apparent to us that gravel was removed from the indentations and piled into slightly higher mounds. From our topside position, as we look down through the cool, quiescent liquid, the exact height of the mounds and the depth of the indentations are difficult for us to judge; our vision is distorted through several feet of water.

However, we can detect near the low gravel mounds (where female salmon buried their eggs, and where their young grow until they are old enough to fend for themselves), and actually through the gravel mounds, movement— water flow—an upwelling of groundwater. This water movement explains our ability to see the variegated color of pebbles. The mud and silt that would normally cover these pebbles has been washed away by the water's subtle, inescapable movement. Obviously, in the depths, our still water is not as still as it first appeared.

The slow, steady, inexorable flow of water in and out of the pool, along with the up-flowing of groundwater through the pool's substrate and through the salmon redds (nests) is only a small part of the activities occurring within the pool, including the air above it, the vegetation surrounding it, and the damp bank and sandbar forming its sides.

Let's get back to the pool itself. If we could look at a cross-sectional slice of the pool, at the water column, the surface of the pool may carry those animals that can literally walk on water. The body of the pool may carry rotifers and protozoa and bacteria—tiny microscopic animals—as well as many fish. Fish will also inhabit hidden areas beneath large rocks and ledges, to escape predators. Going down further in the water column, we come to the pool bed. This is called the **benthic** zone, and certainly the greatest number of creatures live here, including **larvae** and nymphs of all sorts, worms, leeches, flatworms, clams, crayfish, dace, brook lampreys, sculpins, suckers, and water mites.

We need to go down even farther, down into the pool bed, to see the whole story. How far this goes and what lives here, beneath the water, depends on whether it is a gravelly bed or a silty or muddy one. Gravel will allow water, with its oxygen and food, to reach organisms that live underneath the pool. Many of the organisms that are found in the benthic zone may also be found underneath, in the hyporheal zone.

But to see the rest of the story we need to look at the pool's outlet, and where its flow enters the main river. This is the **riffles**—shallow places where water runs fast and is disturbed by rocks. Only organisms that cling very well, such as net-winged midges, caddisflies, stoneflies, some mayflies, dace, and sculpins can spend much time here, and the plant life is restricted to **diatoms** and small algae. Riffles are a good place for mayflies, stoneflies, and caddisflies to live because they offer plenty of gravel in which to hide.

* * *

At first, we struggled to find the "proper" words to describe the river pool. Eventually, we settled on "Still Waters." We did this because of our initial impression, and because of our lack of understanding—lack of knowledge. Even knowing what we know now, we might still describe the river pool as still waters. However, in reality, we must call the pool what it really is: a dynamic habitat. This is true, of course, because each river pool has its own biological community, all members interwoven with each other in complex fashion, all depending on each other. Thus, our river pool habitat is part of a complex, dynamic ecosystem. On reflection, we realize, moreover, that anything dynamic certainly can't be accurately characterized as "still"—including our river pool.

* * *

Maybe you have not had the opportunity to observe a river pool like the one described above. Maybe such an opportunity does not interest you. However, the author's point can be made in a different manner.

Take a moment out of your hectic schedule and perform an action most people never think about doing. Hold a glass of water and think about the substance within the glass—about the substance you are getting ready to drink. You are aware that the water inside a drinking glass is not one of those items people usually spend much thought on, unless they are tasked with providing the drinking water—or dying of thirst.

As mentioned earlier, water is special, strange, and different. Some of us find water fascinating—a subject worthy of endless interest, because of its unique behavior, limitless utility, and ultimate and intimate connection with our existence. Perhaps you might agree with Tom Robbins, whose description of water follows.

> Stylishly composed in any situation—solid, gas or liquid—speaking in penetrating dialects understood by all things—animal, vegetable or mineral—water travels intrepidly through four dimensions, *sustaining* (Kick a lettuce in the field and it will yell "Water!") *destroying* (The Dutch boy's finger remembered the view from Ararat) and *creating* (It has even been said that human beings were invented by water as a device for transporting itself from one place to another, but that's another story). Always in motion, ever-flowing (whether at stream rate or glacier speed), rhythmic, dynamic, ubiquitous, changing and working its changes, a mathematics turned wrong side out, a philosophy in reverse, the ongoing odyssey of water is irresistible. *(Even Cowgirls Get The Blues*, pp. 1–2, 1976).

As Robbins said, water is always in motion. The one most essential characteristic of water is that it is dynamic: Water constantly evaporates from sea, lakes, and soil and transpires from foliage; is transported through the atmosphere; falls to earth; runs across the land; and filters down to flow along rock strata into **aquifers**. Eventually water finds its way to the sea again—indeed, water never stops moving.

A thought that might not have occurred to most people as they look at our glass of water is, "Who has tasted this same water before us?" Before us? Absolutely. Remember, water is almost a finite entity. What we have now is what we have had in the past. Every drop of water that exists on Earth today was here in one form or another that day our planet came into being! Think about it. The same water consumed by Cleopatra, Aristotle, da Vinci, Napoleon, Joan of Arc (and several billion other folks who preceded us), we are drinking now—because water is dynamic (never at rest), and because water constantly cycles and recycles, as discussed in another section.

Water never goes away, disappears, or vanishes (thankfully); it always returns in one form or another. As Dove (1989) points out "all water has a perfect memory and is forever trying to get back to where it was."

Setting the Stage

The availability of a water supply adequate in terms of both quantity and quality is essential to our very existence. One thing is certain: History has shown that the provision of an adequate quantity of quality potable water has been a matter of major concern since the beginning of civilization.

Water—especially clean, safe water—we know we need it to survive—we know a lot about it—however, the more we know the more we discover we don't know.

Modern technology has allowed us to tap potable water supplies and to design and construct elaborate water distribution systems. Moreover, we have developed technology to treat used water (wastewater); that is, water we foul, soil, pollute, discard, and flush away.

Have you ever wondered where the water goes when you drain the sink, tub, or flush the toilet? No. Probably not.

An entire technology has developed around treating water and wastewater. Along with technology, of course, technological experts have been developed. These experts range from environmental/structural/civil engineers to environmental scientists, geologists, hydrologists, chemists, biologists, and others.

Along with those who design and construct water and wastewater treatment works, there is a large cadre of specialized technicians, spread worldwide who operate water and wastewater treatment plants. These operators are tasked, obviously, with either providing a water product that is both safe and palatable for consumption and/or with treating (cleaning) a waste stream before it is returned to its receiving body (usually a river or stream). It is important to point out that not only are water practitioners who treat potable and used water streams responsible for ensuring quality, quantity, and reuse of their product, they are also tasked with, because of the events of 9/11, protecting this essential resource from terrorist acts.

The fact that most water practitioners know more about water than the rest of us comes as no surprise. For the average person, knowledge of water usually extends to knowing no more than that water is good or bad; it is terrible tasting, just great, wonderful, clean and cool and sparkling, or full of scum/dirt/rust/chemicals, great for the skin or hair, very medicinal, and so on. Thus, to say the water "experts" know more about water than the average person is probably an accurate statement.

At this point the reader is probably asking: What does all this have to do with anything? Good question.

What it has to do with water is quite simple. We need to accept the fact that we simply do not know what we do not know about water.

Then the question shifts to—why would you want to know anything about water in the first place? Another good question.

To start with, let's talk a little about the way in which we view water.

Earlier brief mention was made about the water contents of a simple drinking water glass. Let's face it, drinking a glass of water is something that normally takes little effort and even less thought. Trouble is our view of water and its importance is relative.

The situation could be different—even more relative, however. For example, consider the young woman who is an adventurer; an outdoors person. She likes to jump into her four-wheel-drive vehicle and head out for new adventure. On this particular day she decides to drive through Death Valley, California—one end to another and back on a seldom-used dirt road. She has done this a few times before. During her transit of this isolated region, she decides to take a side road that seems to lead to the mountains to her right.

She travels along this isolated, hardpan road for approximately 50 miles— then the engine in her four-wheel-drive vehicle quits. No matter what she does, the vehicle will not start. Eventually, the vehicle's battery dies; she had cranked on it too much.

Realizing that the vehicle was not going to start, she also realized she was alone and deep inside an inhospitable area. What she did not know was that the nearest human being was about 60 miles to the west.

She had another problem—a problem more pressing than any other. She did not have a canteen or container of water—an oversight on her part. Obviously, she told herself, this is not a good situation.

What an understatement this turned out to be.

Just before noon, on foot, she started back down the same road she had traveled. She reasoned she did not know what was in any other direction other than the one she had just traversed. She also knew the end of this side road intersected the major highway that bisected Death Valley. She could flag down a car or truck or bus; she would get help, she reasoned.

She walked—and walked—and walked some more. "Gee, if it wasn't so darn hot," she muttered to herself, to sagebrush, to scorpions, to rattlesnakes and to cacti. The point is it was hot; about 107°F.

She continued on for hours, but now she was not really walking; instead, she was forcing her body to move along. Each step hurt. She was burning up. She was thirsty. How thirsty was she? Well, right about now just about anything liquid would do, thank you very much!

Later that night, after hours of walking through that hostile land, she couldn't go on. Deep down in her heat-stressed mind, she knew she was in serious trouble. Trouble of the life-threatening variety.

Just before passing out, she used her last ounce of energy to issue a dry pathetic scream.

This scream of lost hope and imminent death was heard—but only by the sagebrush, the scorpions, Gila monsters, rattlesnakes, and cacti—and by the vultures that were now circling above her parched, dead remains. The vultures were of no help, of course. They had heard these desperate, hopeless screams before. They were indifferent; they had all the water they needed; their food supply wasn't all that bad either.

The preceding case sheds light on a completely different view of water. Actually, it is a very basic view that holds: We cannot live without it.

Historical Perspective

An early human, wandering alone from place to place, hunting and gathering to subsist, probably would have had little difficulty in obtaining drinking water, because such a person would—and could—only survive in an area where drinking water was readily available with little travail.

The search for clean, fresh, and palatable water has been a human priority from the very beginning. The authors take no risk in stating that when humans first walked the Earth, many of those steps were in the direction of water.

When early humans were alone or in small numbers, finding drinking water was a constant priority, to be sure, but for us to imagine today just how big a priority finding drinking water became as the number of humans proliferated is difficult.

Eventually communities formed, and with their formation came the increasing need to find clean, fresh, and palatable drinking water, and also to find a means of delivering it from the source to the point of use.

Archeological digs are replete with the remains of ancient water systems (Man's early attempts to satisfy that never-ending need). Those digs (spanning the history of the last 20 or more centuries) testify to this. For well over 2000 years, piped water supply systems have been in existence. Whether the pipes were fashioned from logs or clay or carved from stone or other materials is not the point—the point is they were fashioned to serve a vital purpose, one universal to the community needs of all humans: to deliver clean, fresh, and palatable water to *where* it was needed.

These early systems were not arcane. Today, we readily understand their intended purpose. As we might expect, they could be rather crude, but they were reasonably effective, though they lacked in two general properties we take for granted today.

First, of course, they were not pressurized, but instead relied on gravity flow, since the means to pressurize the mains was not known at the time—and even if such pressurized systems were known, they certainly would not have been used to pressurize water delivered via hollowed-out logs and clay pipe.

The second general property early civilizations lacked that we do not lack today (in the industrialized world that is) is sanitation. Remember, to know the need for something exists (in this case, the ability to sanitize, to disinfect water supplies), the nature of the problem must be defined. Not until the middle of the 1800s (after countless millions of deaths from waterborne disease over the centuries) did people realize that a direct connection between contaminated drinking water and disease existed. At that point, sanitation of water supply became an issue.

When the relationship between waterborne diseases and the consumption of drinking water was established, evolving scientific discoveries led the way toward the development of technology for unit processing of water and ultimate disinfection. Drinking water standards were developed by health authorities, scientists, and sanitary engineers.

With the current lofty state of effective technology that we in the U.S. and the rest of the developed world enjoy today, we could sit on our laurels, so to speak, and assume that because of the discoveries developed over time [and at the cost of countless people who died (and still die) from waterborne-diseases], that all is well with us—that problems related to providing us with clean, fresh, palatable drinking water are problems of the past.

Are they really problems of the past? Have we solved all the problems related to ensuring that our drinking water supply provides us with clean, fresh, and palatable drinking water? Is the water delivered to our tap as clean, fresh, and palatable as we think it is . . . as we hope it is? Does anyone really know?

What we do know is that we have made progress. We have come a long way from the days of gravity flow water delivered via mains of logs and clay or stone . . . Many of us on this planet have come a long way from the days of cholera epidemics.

However, to obtain a definitive answer to those questions, perhaps we should ask those who boiled their water for weeks on end in Sydney, Australia, in the fall of 1998. Or better yet, we should speak with those who drank the "city water" in Milwaukee in 1993, or in Las Vegas, Nevada—those who suffered and survived the onslaught of *Cryptosporidium*, from contaminated water out of their tap.

Or if we could, we should ask these same questions of a little boy named Robbie, who died of acute lymphatic leukemia, the probable cause of which is far less understandable to us: toxic industrial chemicals, unknowingly delivered to him via his local water supply.

So Necessary, but So Easy to Abuse

If water is so precious, so necessary for sustaining life, then two questions arise: (1) Why do we ignore water? (2) Why do we abuse it (pollute or waste it)? We ignore water because it is so common, so accessible, so available, so unexceptional (unless you are lost in the desert without a supply of it). Again, why do we pollute and waste water? There are several reasons.

You might be asking yourself: Is water pollution really that big of a deal? Simply stated, yes; it is. Man has left his footprint (in the form of pollution) on the environment, including on our water sources. Man has a bad habit of doing this. What it really comes down to is "out of sight out of mind" thinking. Or when we abuse our natural resources in any manner, maybe we think to ourselves: "Why worry about it. Let someone else sort it all out."

Finally, before moving on with the rest of the text, it should be pointed out the view held throughout this work is that water is special, strange, and different— and very vital. This view is held for several reasons, but the most salient factor driving this view is the one that points to the fact that on this planet, *water is life.*

References and Recommended Reading

DeZuane, J. 1997. *Handbook of Drinking Water Quality*, 2nd ed., New York: John Wiley and Sons.

Dove, R. 1989. *Grace Notes*. New York: Norton.

Gerba, C.P. 1996. "Risk Assessment." In *Pollution Science*, eds., Pepper, Gerba and Brusseau. San Diego: Academic Press.

Hammer, M.J., and Hammer, M.J., Jr. 1996. *Water and Wastewater Technology*, 3rd ed., Englewood Cliffs, NJ: Prentice Hall.

Harr, J. 1995. *A Civil Action*. New York: Vintage Books.

Lewis, S.A. 1996. *The Sierra Club Guide to Safe Drinking Water*. San Francisco: Sierra Club Books.

Metcalf and Eddy, Inc. 1991. *Wastewater Engineering: Treatment, Disposal, Reuse*. 3rd ed., New York: McGraw-Hill, Inc.

Meyer, W.B. 1996. *Human Impact on Earth*. New York: Cambridge University Press.

Nathanson, J.A. 1997. *Basic Environmental Technology: Water Supply, Waste Management, and Pollution Control*. Upper Saddle River, NJ: Prentice Hall.

Pielou, E.C. 1998. *Fresh Water*. Chicago: University of Chicago.

Robbins, T. 1976. *Even Cowgirls Get the Blues*. Boston: Houghton Mifflin Company.

Spellman, F.R. 2007. *The Science of Water*, 2nd ed. Boca Raton: CRC Press.

Wick, W. 1997. *A Drop of Water*. New York: Scholastic Press.

9

All about Water

Introduction

UNLESS YOU ARE THIRSTY, in real need of refreshment, when you look upon that glass of water we mentioned earlier, you might ask—well, what could be more boring? The curious might wonder what are the physical and chemical properties of water that make it unique and necessary for living things. Again, when you look at that glass of water, smell it, taste it—well, what could be more boring? Pure water is virtually colorless and has no taste or smell. But the hidden qualities of water make it a most interesting subject (USGS 2010).

When the uninitiated become initiated to the wonders of water, one of the first surprises is that the total quantity of water on earth is much the same now as it was more than three or four billion years ago, when the 320+ million cubic miles of it were first formed. Ever since then, the water reservoir has gone round and round, building up, breaking down, cooling, and then warming. Water is very durable, but it remains difficult to explain, because it has never been isolated in a completely undefiled state.

Remember, water is special, strange, and different.

Have you ever wondered what the nutritive value of water is? Well, the fact is water has no nutritive value. It has none; yet it is the major ingredient of all living things. Consider yourself, for example. Think of what you need to survive—just to survive. Food? Air? PS-3? MTV? iPod? Cell phone? Water? Naturally, the focus of this section is on water. Water is of major importance to all living things; in some organisms, up to 90% of their body weight comes from water. Up to 60% of the human body is water, the brain is composed of 70% water, and the lungs are nearly 90% water. About 83% of our blood is water, which helps digest our food, transport waste, and control body temperature. Each day humans must replace 2.4 liters of water, some through drinking and the rest taken by the body from the foods eaten.

Did You Know?

Unlike most liquids, water expands (gets less dense) when it freezes. Think of ice—it is one of the few items that float as a solid. If it didn't, then lakes would freeze from the bottom up (that would mean we'd have to wear wet suits when ice skating!), and some lakes way up north would be permanent blocks of ice.

There wouldn't be any you, me, or Luci the dog without the existence of an ample liquid water supply on Earth. The unique qualities and properties of water are what make it so important and basic to life. The cells in our bodies are full of water. The excellent ability of water to dissolve so many substances allows our cells to use valuable nutrients, minerals, and chemicals in biological processes.

Water's "stickiness" (from surface tension) plays a part in our body's ability to transport these materials all through ourselves. The **carbohydrates** and **proteins** that our bodies use as food are metabolized and transported by water in the bloodstream. No less important is the ability of water to transport waste material out of our bodies.

Water is used to fight forest fires. Yet we use water spray on coal in a furnace to make it burn better.

Condensation is water coming out of the air. Water that forms on the outside of a cold glass or on the inside of a window in winter is liquid water condensing from water vapor in the air. Air contains water vapor (humidity). In cold air, water vapor condenses faster than it evaporates. So, when the

Did You Know?

Water has the highest surface tension among common liquids (mercury is higher). Surface tension is the ability of a substance to stick to itself (cohere). That is why water forms drops, and also why when you look at a glass of water, the water "rises' where it touches the glass (the "meniscus"). Plants are happy that water has a high surface tension because they use capillary action to draw water from the ground up through their roots and stems.

warm air touches the outside of your cold glass, the air next to the glass gets chilled, and some of the water in that air turns from water vapor to tiny liquid water droplets. The cloud in the sky and the "cloud" you see when you exhale on a cold day are condensed water-vapor particles.

Did You Know?

It is a myth that clouds form because cold air cannot hold as much water vapor as warm air!

Chemically, water is hydrogen oxide. It turns out, however, on more advanced analysis to be a mixture of more than thirty possible compounds. In addition, all of its physical constants are abnormal (strange).

At a temperature of 2,900°C some substances that contain water cannot be forced to part with it. And yet others that do not contain water will liberate it when even slightly heated.

When liquid, water is virtually incompressible; as it freezes, it expands by an eleventh of its volume.

Water isn't known as the "Universal Solvent" for nothing! It can dissolve more substance than any other liquid. This is lucky for us . . . what if all the sugar in your soft drink ended up as a pile at the bottom of the glass? The water you see in rivers, lakes, and the ocean may look clear, but it actually contains many dissolved elements and minerals, and because these elements are dissolved, they can move with water over the surface of the Earth.

Contrary to popular belief rainwater is not the purest form of water; actually, distilled water is "purer." Rainwater contains small amounts of dissolved minerals that have been blown into the air by winds. Rainwater contains tiny particles of dust and dissolved gasses, such as carbon dioxide and sulfur dioxide (acid rain). That doesn't mean rainwater isn't very clean—normally only about 1/100,000th of the weight of rain comes from these substances.

Did You Know?

In regard to rainwater, the distillation process is responsible for its formation. Distilled water comes from water vapor condensing in a closed container (such as a glass jar). Rain is produced by water vapor evaporating from the earth and condensing in the sky. Both the closed jar and the Earth (via its atmosphere) are "closed systems," where water is neither added or lost.

For the above stated reasons, and for many others, we can truly say that water is special, strange, and different.

Characteristics of Water

To this point many things have been said about water; however, it has not been said that water is plain. This is the case because nowhere in nature is plain water to be found. Here on earth, with a geologic origin dating back over three to five billion years, water found in even its purest form is composed of many constituents. You probably know water's chemical description is H_2O—that is one atom of oxygen bound to two atoms of hydrogen. The hydrogen atoms are "attached" to one side of the oxygen atom, resulting in a water molecule having a positive charge on the side where the hydrogen atoms are and a negative charge on the other side, where the oxygen atom is. Since opposite electrical charges attract, water molecules tend to attract each other, making water kind of "sticky"—the hydrogen atoms (positive charge) attracts the oxygen side (negative charge) of a different water molecule.

✓ **Important Point:** all these water molecules attracting each other mean they tend to clump together. This is why water drops are, in fact, drops! If it wasn't for some of Earth's forces, such as gravity, a drop of water would be ball shaped—a perfect sphere. Even if it doesn't form a perfect sphere on Earth, we should be happy water is sticky.

Along with H_2O molecules, hydrogen (H^+), hydroxyl (OH^-), sodium, potassium, and magnesium, there are other ions and elements present. Additionally, water contains dissolved compounds including various carbonates, sulfates, silicates, and chlorides. As mentioned, rainwater, often assumed to be the equivalent of distilled water, is not immune to **contamination** that is collected as it descends through the atmosphere. The movement of water across the face of land contributes to its contamination, taking up dissolved gases, such as carbon dioxide and oxygen, and a multitude of organic substances and minerals leached from the soil. Don't let that crystal clear lake or pond fool you. These are not filled with water alone but are composed of a complex medium of chemical ingredients far exceeding the brief list presented here; it is a special medium in which highly specialized life can occur.

How important is water to life? To answer this question all we need do is to take a look at the common biological cell: it easily demonstrates the importance of water to life.

Living cells comprise a number of chemicals and organelles within a liquid substance, the cytoplasm, and the cell's survival may be threatened by changes in the proportion of water in the cytoplasm. This change in proportion of water in the cytoplasm can occur through desiccation (evaporation), oversupply, or the loss of either nutrients or water to the external environment. A cell that is unable to control and maintain homeostasis (i.e., the correct equilibrium/proportion of water) in its cytoplasm may be doomed—it may not survive.

✓ **Important Point:** As mentioned, water is called the "universal solvent" because it dissolves more substances than any other liquid. This means that wherever water goes, either through the ground or through our bodies, it takes along valuable chemicals, minerals, and nutrients.

Inflammable Air + Vital Air = Water

In 1783, in England, Henry Cavendish (a brilliant chemist and physicist) was "playing with" electric current. Specifically, Cavendish was passing electric current through a variety of substances to see what appended. Eventually, he got around to water. He filled a tube with water and sent his current through it. The water vanished.

To say that Cavendish was flabbergasted by the results of this experiment would be a mild understatement. "The tube has to have a leak in it," he reasoned.

He repeated the experiment again—same result.

Then again—same result.

The fact is he made the water disappear again and again. Actually, what Cavendish had done was convert the liquid water to its gaseous state—into an invisible gas.

When Cavendish analyzed the contents of the tube, he found it contained a mixture of two gases, one of which was *inflammable air* and the other was a heavier gas. This heavier gas had only been discovered a few years earlier by his colleague Joseph Priestley (English chemist and clergyman) who, finding that it kept a mouse alive and supported combustion, called it *vital air*.

Just Two H's and One O

Cavendish had been able to separate the two main constituents that make up water. All that remained was for him to put the ingredients back together again. He accomplished this by mixing a measured volume of inflammable air with different volumes of its vital counterpart, and setting fire to both. He found that most mixtures burned well enough, but when the proportions were precisely two to one, there was an explosion and the walls of his test tubes were covered with liquid droplets. He quickly identified these as water.

Cavendish made an announcement: Water was not water. Moreover, water is not just an odorless, colorless, and tasteless substance that lay beyond reach of chemical analysis. Water is not an element in its own right, but a compound of two independent elements, one a supporter of combustion and the other combustible. When united, these two elements become the preeminent quencher of thirst and flames.

It is interesting to note that a few year later, the great French genius, Antoine Lavoisier, tied the compound neatly together by renaming the ingredients *hydrogen*—"the water producer"—and *oxygen*. In a fitting tribute to his guillotined corpse (he was a victim of the revolution), his tombstone came to carry a simple and telling epitaph, a fitting tribute to the father of a new age in chemistry—*just two H's and one O*.

Somewhere Between 0° and 105°

We take water for granted now. Every high-school level student knows that water is a chemical compound of two simple and abundant elements. And yet scientists continue to argue the merits of rival theories on the structure of water. The fact is we still know little about water. For example, we don't know how water works.

Part of the problem lies with the fact that no one has ever seen a water molecule. It is true that we have theoretical diagrams and equations. We also have a disarmingly simple formula—H_2O. The reality however, is that water is very complex. X-rays, for example, have shown that the atoms in water are intricately laced.

It has been said over and over again that water is special, strange, and different. Water is also almost indestructible. Sure, we know that electrolysis can separate water atoms, but we also know that once they get together again they must be heated up to more than 2,900°C to separate them again.

Water is also idiosyncratic. This can be seen in the way in which the two atoms of hydrogen in a water molecule (see Figure 23) take up a very precise and strange (different) alignment to each other. Not all angles of 45°, 60°, or 90°—oh no, not water. Remember, water is different. The two hydrogen atoms always come to rest at an angle of approximately 105° from each other, making all diagrams of their attachment to the larger oxygen atom look like Mickey Mouse ears on a very round head (see Figure 23; remember that everyone's favorite mouse is mostly water, too).

This 105° relationship makes water lopsided, peculiar, and eccentric—it breaks all the rules. You're not surprised are you?

One thing is certain, however; this 105° angle is crucial to all life as we know it. Thus, the answer to defining why water is special, strange, different—and vital, lies somewhere between 0°–105°.

Water's Physical Properties

Water has several unique physical properties. These properties are:

- Water is unique in that it is the only natural substance that is found in all three states—liquid, solid (ice), and gas (steam)—at the temperatures

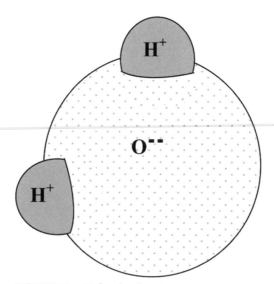

FIGURE 23. Molecule of water

normally found on Earth. Earth's water is constantly interacting, changing, and in movement.

- Water freezes at 32°F and boils at 212°F (at sea level, but 186.4°F at 14,000 feet). In fact, water's freezing and boiling points are the baseline with which temperature is measured: 0° on the Celsius scale is water's freezing point, and 100° is water's boiling point. Water is unusual in that the solid form, ice, is less dense than the liquid form, which is why ice floats.
- Water has a high specific heat index. This means that water can absorb a lot of heat before it begins to get hot. This is why water is valuable to industries and in your car's radiator as a coolant. The high specific heat index of water also helps regulate the rate at which air changes temperature, which is why the temperature change between seasons is gradual rather than sudden, especially near the oceans.
- Water has a very high surface tension. In other words and as previously mentioned, water is sticky and elastic, and tends to clump together in drops rather than spread out in a thin film. Surface tension is responsible for capillary action (discussed in detail later), which allows water (and its dissolved substances) to move through the roots of plants and through the thin blood vessels in our bodies.
- Here's a quick rundown of some of water's properties:
  - Weight: 62.416 pounds per cubic foot at 32°F
  - Weight: 61.998 pounds per cubic foot at 100°F
  - Weight: 8.33 pounds per gallon, 0.036 pounds per cubic inch
  - Density: 1 gram per cubic centimeter (cc) at 39.2°F, 0.95865 gram per cc at 212°F
  - 1 gallon = 4 quarts = 8 pints = 128 ounces = 231 cubic inches
  - 1 liter = 0.2642 gallons = 1.0568 quarts = 61.02 cubic inches
  - 1 million gallons = 3.069 acre-feet = 133,685.64 cubic feet

---

Did You Know?

The boiling point of water gets lower as you go up in altitude. At beach level, water boils at 212°F. But at 5,000 feet, about where Denver is located, water boils at 202.9°F, and up at 10,000 feet it boils at 193.7°F. This is because as the altitude gets higher, the air pressure (the weight of all that air above you) becomes less. Because there is less pressure pushing on a pot of water at a higher altitude, it is easier for the water molecules to break their bonds and attraction to each other and, thus, it boils more easily.

Capillary Action

If we were to mention the term capillary action, or capillarity, to the man or woman on the street, they might instantly nod their heads and respond that their bodies are full of them—that capillaries are the tiny blood vessels that connect the smallest arteries and the smallest of the veins. This is true, of course. But in the context of water science, capillary action is something different than capillary action in the human body.

Even if you've never heard of capillary action, it is still important in your life. Capillary action is important for moving water (and all of the things that are dissolved in it) around. It is defined as the movement of water within the spaces of a porous material due to the forces of adhesion, cohesion, and surface tension.

Surface tension is a measure of the strength of the water's surface film. The attraction between the water molecules creates a strong film, which among other common liquids is only surpassed by that of mercury. This surface tension permits water to hold up substances heavier and denser than itself. A steel needle carefully placed on the surface of that glass of water will float. Some aquatic insects such as the water strider rely on surface tension to walk on water.

Capillary action occurs because water is sticky; thanks to the forces of cohesion (water molecules like to stay closely together) and adhesion (water molecules are attracted to and stick to other substances). So, water tends to stick together, as in a drop, and it sticks to glass, cloth organic tissues, and soil. Dip a paper towel into a glass of water and the water will "climb" onto the paper towel. In fact, it will keep going up the towel until the pull of gravity is too much for it to overcome.

Water Cycle

Among water's most indispensable qualities is that it is Earth's only self-renewing vital resource. Earth's self-renewing water cycle, or natural *water cycle* or *hydrological cycle*, is the means by which water in all three forms—solid, liquid, and vapor—circulates through the biosphere. The water cycle is all about describing how water moves above, on, and through the Earth. Much more water, however, is "in storage" for long periods of time than is actually moving through the cycle. The storehouses for the vast majority of all water on Earth are the oceans. It is estimated that of the 332,500,000 cubic miles of the world's water supply, about 321,000,000 cubic miles is stored in oceans. That is about 96.5%. It is also estimated that the oceans supply about 90% of the evaporated water that goes into the water cycle.

Water—lost from the Earth's surface to the atmosphere either by evaporation from the surface of lakes, rivers, and oceans or through the transpiration of plants—forms clouds that condense to deposit moisture on the land and sea. Evaporation from the oceans is the primary mechanism supporting the surface-to-atmosphere portion of the water cycle. Note, however, that a drop of water may travel thousands of miles between the time it evaporates and the time it falls to Earth again as rain, sleet, or snow. The water that collects on land flows to the ocean in streams and rivers or seeps into the earth, joining groundwater. Even groundwater eventually flows toward the ocean for recycling (see Figure 24). The cycle constantly repeats itself; a cycle without end.

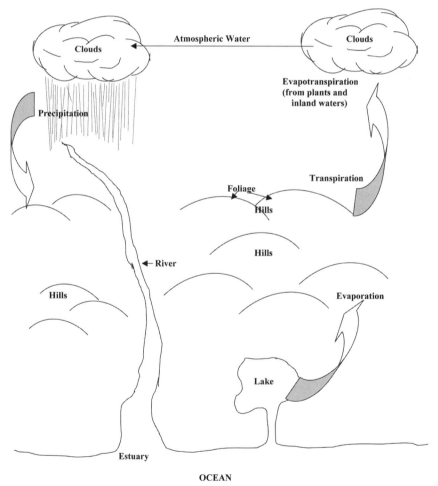

FIGURE 24. Water cycle

✓ **Note:** How long water that falls from the clouds takes to return to the atmosphere varies tremendously. After a short summer shower, most of the rainfall on land can evaporate into the atmosphere in only a matter of minutes. A drop of rain falling on the ocean may take as long as 37,000 years before it returns to atmosphere and some water has been in the ground or caught in glaciers for millions of years.

✓ **Important Point:** Only about 2% of the water absorbed into plant roots is used in photosynthesis. Nearly all of it travels through the plant to the leaves, where transpiration to the atmosphere begins the cycle again.

Specific Water Movements

The water cycle shown in Figure 24 illustrates continuous movement of Earth's water in very simple terms. However, it is important to point out that the actual movement of water on Earth is much more complex. As mentioned earlier, there are three different methods of transport involved in this water movement: *evaporation, precipitation,* and *run-off.*

Evaporation of water is a major factor in hydrologic systems. Evaporation is a function of temperature, wind velocity, and relative humidity. Evaporation (or vaporization) is, as the name implies, the formation of vapor. Dissolved constituents such as salts remain behind when water evaporates. Evaporation of the surface water of oceans provides most of the water vapor. It should be pointed out, however, that water can also vaporize through plants, especially from leaf surfaces. As pointed out earlier, this process is called *evapotranspiration.* Plant transpiration is pretty much an invisible process—since the water is evaporating from the leaf surfaces, you don't just go out and see the leaves "breathe." During a growing season, a leaf will transpire many times more water than its own weight. A large oak tree can transpire 40,000 gallons (151,000 liters) per year (USGS 2006).

USGS (2006) points out that the amount of water that plants transpire varies greatly geographically and over time. There are a number of factors that determine transpiration rates:

- **Temperature**—Transpiration rates go up as the temperature goes up, especially during the growing season, when the air is warmer due to stronger sunlight and warmer air masses.
- **Relative Humidity**—As the relative humidity of the air surrounding the plant rises the transpiration rate falls. It is easier for water to evaporate into dryer air than into more saturated air.

- **Wind and Air Movement**—Increased movement of the air around a plant will result in a higher transpiration rate.
- **Soil-moisture Availability**—When moisture is lacking, plants can begin to senesce (i.e., premature aging, which can result in leaf loss) and transpire less water.
- **Type of Plant**—Plants transpire water at different rates. Some plants which grow in arid regions, such as cacti and succulents, conserve precious water by transpiring less water than other plants.

✓ **Interesting Point:** It may surprise you that ice can also vaporize without melting first. However, this *sublimation* process is slower than vaporization of liquid water.

Evaporation rates are measured with evaporation pans. These evaporation pans provide data that indicate the atmospheric evaporative demand of an area and can be used to estimate (1) the rates of evaporation from ponds, lakes, and reservoirs, and (2) evapotranspiration rates. It is important to note that several factors affect the rate of pan evaporation. These factors include the type of pan, type of pan environment, method of operating the pan, exchange of heat between pan and ground, solar radiation, air temperature, wind, and temperature of the water surface (Jones 1992).

Recall that *precipitation* includes all forms in which atmospheric moisture descends to earth—rain, snow, sleet, and hail. Before precipitation can occur, the water that enters the atmosphere by vaporization must first condense into liquid (clouds and rain) or solid (snow, sleet, and hail) before it can fall. This vaporization process absorbs energy. This energy is released in the form of heat when the water vapor condenses. You can best understand this phenomena when you compare it to what occurs when water that evaporates from your skin absorbs heat, making you feel cold.

Note: The annual evaporation from ocean and land areas is the same as the annual precipitation.

Run-off is the flow back to the oceans of the precipitation that falls on land. This journey to the oceans is not always unobstructed—flow back may be intercepted by vegetation (from which it later evaporates), a portion is held in depressions, and a portion infiltrates into the ground. A part of the infiltrated water is taken up by plant life and returned to the atmosphere through evapotranspiration, while the remainder either moves through the ground or is held by capillary action. Eventually, water drips, seeps, and flows its way back into the oceans.

Assuming that the water in the oceans and ice caps and glaciers is fairly constant when averaged over a period of years, the water balance of the Earth's surface can be expressed by the relationship: Water lost = Water gained (Turk and Turk 1988).

Sources of Water

Approximately 40 million cubic miles of water cover or reside within the Earth. The oceans contain about 97% of all water on Earth. The other 3% is freshwater: (1) snow and ice on the surface of Earth contains about 2.25% of the water; (2) usable groundwater is approximately 0.3%; and (3) surface freshwater is less than 0.5%.

In the U.S., for example, average rainfall is approximately 2.6 feet (a volume of 5,900 cubic kilometers). Of this amount, approximately 71% evaporates (about 4,200 cubic cm), and 29% goes to stream flow (about 1,700 cubic km).

Beneficial freshwater uses include manufacturing, food production, domestic and public needs, recreation, hydroelectric power production, and flood control. Stream flow withdrawn annually is about 7.5% (440 cubic km). Irrigation and industry use almost half of this amount (3.4% or 200 cubic km per year). Municipalities use only about 0.6% (35 cubic km per year) of this amount.

Historically, in the U.S., water usage is increasing (as might be expected). For example, in 1975, 40 billion gallons of freshwater were used. In 1990, the total increased to 455 billion gallons.

The primary sources of freshwater include the following:

1. Captured and stored rainfall in cisterns and water jars
2. Groundwater from springs, artesian wells, and drilled or dug wells
3. Surface water from lakes, rivers, and streams
4. Desalinized seawater or brackish groundwater
5. Reclaimed wastewater

Current federal drinking water regulations actually define three distinct and separate sources of freshwater. They are surface water, groundwater, and groundwater under the direct influence of surface water (GUDISW). This last classification is the result of the Surface Water Treatment Rule (SWTR). The definition of the conditions that constitute GUDISW, while specific, is not obvious.

Potable Water Source

Because of huge volume and flow conditions, the quality of natural water cannot be modified significantly within the body of water. Accordingly, humans must augment Nature's natural purification processes with physical, chemical, and biological treatment procedures. Essentially, this quality control approach is directed to the water withdrawn, which is treated, from a source for a specific use.

Potable Water

Potable water is water fit for human consumption and domestic use, which is sanitary and normally free of minerals, organic substances, and toxic agents in excess of reasonable amounts for domestic usage in the area served, and normally adequate in quantity for the minimum health requirements of the persons served.

In regard to a potential potable water supply, the key words, as previously mentioned, are "quality and quantity." Obviously, if we have a water supply that is unfit for human consumption, we have a quality problem. If we do not have an adequate supply of quality water, we have a quantity problem.

Surface Water

Where do we get our potable water from? From what water source is our drinking water provided? To answer these questions, we would most likely turn to one of two possibilities: our public water is provided by a groundwater or surface water source because these two sources are, indeed, the primary sources of most water supplies.

From the earlier discussion of the hydrologic or water cycle, we know that from whichever of the two sources we obtain our drinking water, the source is constantly being replenished (we hope) with a supply of freshwater. This water cycle phenomenon was best summed up by Heraclitus of Ephesus, who said, "You could not step twice into the same rivers; for other waters are ever flowing on to you."

In this section, one of the drinking water practitioner's (and humankind in general) primary duties—to find and secure a source of potable water for human use—is discussed.

Location! Location! Location!

In the real estate business, location is everything. The same can be said when it comes to sources of water. In fact, the presence of water defines "location" for communities. Although communities differ widely in character and size, all have the common concerns of finding water for industrial, commercial, and residential use. Freshwater sources that can provide stable and plentiful supplies for a community don't always occur where we wish. Simply put, on land, the availability of a regular supply of potable water is the most important factor affecting the presence—or absence—of many life-forms. A map of the world immediately shows us that surface waters are not uniformly distributed over the Earth's surface. U.S. land holds rivers, lakes, and streams on only about 4% of its surface. The heaviest populations of any life-forms, including humans, are found in regions of the U.S. (and the rest of the world) where potable water is readily available, because lands barren of water simply won't support large populations. One thing is certain: if a local supply of potable water is not readily available, the locality affected will seek a source. This is readily apparent (absolutely crystal clear), for example, when one studies the history of water "procurement" for the communities located within the Los Angeles basin.

✓ **Important Point:** The volume of freshwater sources depends on geographic, landscape and temporal variations, and on the impact of human activities.

How Readily Available Is Potable Water?

Approximately 326 million cubic miles of water comprise Earth's entire water supply. Of this massive amount of water—though providing us indirectly with freshwater through evaporation from the oceans—only about 3% is fresh. Even most of the minute percentage of freshwater Earth holds is locked up in polar ice caps and in glaciers. The rest is held in lakes, in flows through soil, and in river and stream systems. Only 0.027% of Earth's freshwater is available for human consumption (see Table 9.1 for the distribution percentages of Earth's water supply).

We see from Table 9.1 that the major sources of drinking water are from surface water, groundwater, and from groundwater under the direct influence of surface water (i.e., springs or shallow wells).

Surface waters are not uniformly distributed over the Earth's surface. In the U.S., for example, only about 4% of the landmass is covered by rivers, lakes, and streams. The volumes of these freshwater sources depend on geographic, landscape, and temporal variations, and on the impact of human activities.

TABLE 9.1
World Water Distribution

Location	Percent of Total
Land areas	
Freshwater lakes	0.009
Saline lakes and inland seas	0.008
Rivers (average instantaneous volume)	0.0001
Soil moisture	0.005
Groundwater (above depth of 4000 m)	0.61
Ice caps and glaciers	2.14
Total: Land areas	**2.8**
Atmosphere (water vapor)	0.001
Oceans	97.3
Total all locations (rounded)	**100**

Source: USGS, 2006.

Again, surface water is that water that is open to the atmosphere and results from overland flow (i.e., runoff that has not yet reached a definite stream channel). Put a different way, surface water is the result of **surface runoff.**

For the most part, however, surface (as used in the context of this text) refers to water flowing in streams and rivers, as well as water stored in natural or artificial lakes, man-made impoundments such as lakes made by damming a stream or river; springs that are affected by a change in level or quantity; shallow wells that are affected by precipitation; wells drilled next to or in a stream or river; rain catchments; and/or muskeg and tundra ponds.

Specific sources of surface water include:

1. Rivers
2. Streams
3. Lakes
4. Impoundments (man-made lakes made by damming a river or stream)
5. Very shallow wells that receive input via precipitation
6. Springs affected by precipitation (flow or quantity directly dependent upon precipitation)
7. Rain catchments (**drainage basins**)
8. Tundra ponds or muskegs (peat bogs)

Surface water has advantages as a source of potable water. Surface water sources are usually easy to locate—unlike groundwater, finding surface water does not take a geologist or hydrologist. Normally, surface water is not tainted with minerals precipitated from the earth's strata.

Ease of discovery aside, surface water also presents some disadvantages: surface water sources are easily contaminated (polluted) with microorganisms that can cause waterborne diseases (anyone who has suffered from "hiker's disease" or "hiker's diarrhea" can attest to this), and from chemicals that enter from surrounding runoff and upstream discharges. **Water rights** can also present problems.

As mentioned, most surface water is the result of surface runoff. The amount and flow rate of this surface water is highly variable, which comes into play for two main reasons: (1) human interferences (influences) and (2) natural conditions. In some cases, surface water runs quickly off land surfaces. From a water resources standpoint, this is generally undesirable, because quick runoff does not provide enough time for the water to infiltrate the ground and recharge groundwater aquifers. Surface water that quickly runs off land also causes erosion and flooding problems. Probably the only good thing that can be said about surface water that runs off quickly is that it usually does not have enough contact time to increase in mineral content. Slow surface water off land has all the opposite effects.

Drainage basins collect surface water and direct it on its gravitationally influenced path to the ocean. The drainage basin is normally characterized as an area measured in square miles, acres, or sections. Obviously, if a community is drawing water from a surface water source, the size of its drainage basin is an important consideration.

Surface water runoff, like the flow of electricity, flows or follows the path of least resistance. Surface water within the drainage basin normally flows toward one primary watercourse (river, stream, brook, creek, etc.), unless some manmade distribution system (canal or pipeline) diverts the flow.

✓ **Important Point:** Many people probably have an overly simplified idea that precipitation falls on the land, flows overland (runoff), and runs into rivers, which then empty into the oceans. That is "overly simplified" because rivers also gain and lose water to the ground. Still, it is true that much of the water in rivers comes directly from runoff from the land surface, which is defined as surface runoff.

Surface water runoff land surfaces depend on several factors, which include:

- **Rainfall duration**—Even a light, gentle rain, if it lasts long enough, can, with time, saturate soil and allow runoff to take place.

- **Rainfall intensity**—With increases in intensity, the surface of the soil quickly becomes saturated. This saturated soil can hold no more water; as more rain falls and water builds up on the surface, it creates surface runoff.
- **Soil moisture**—The amount of existing moisture in the soil has a definite impact on surface runoff. Soil already wet or saturated from a previous rain causes surface runoff to occur sooner than if the soil were dry. Surface runoff from frozen soil can be up to 100% of snowmelt or rain runoff because frozen ground is basically impervious.
- **Soil composition**—The composition of the surface soil directly affects the amount of runoff. For example, hard rock surfaces, obviously, result in 100% runoff. Clay soils have very small void spaces that swell when wet; the void spaces close and do not allow infiltration. Coarse sand possesses large void spaces that allow easy flow through of water, which produces the opposite effect, even in a torrential downpour.
- **Vegetation cover**—Groundcover limits runoff. Roots of vegetation and pine needles, pine cones, leaves, and branches create a porous layer (a sheet of decaying natural organic substances) above the soil. This porous "organic" sheet readily allows water into the soil. Vegetation and organic waste also act as cover to protect the soil from hard, driving rains, which can compact bare soils, close off void spaces, and increase runoff. Vegetation and ground cover work to maintain the soil's infiltration and water-holding capacity, and also work to reduce soil moisture evaporation.
- **Ground slope**—When rain falls on steeply sloping ground, up to 80+% may become surface runoff. Gravity moves the water down the surface more quickly than it can infiltrate the surface. Water flow off flat land is usually slow enough to provide opportunity for a higher percentage of the rainwater to infiltrate the ground.
- **Human influences**—Various human activities have a definite impact on surface water runoff. Most human activities tend to increase the rate of water flow. For example, canals and ditches are usually constructed to provide steady flow, and agricultural activities generally remove ground cover that would work to retard the runoff rate. On the opposite extreme, man-made dams are generally built to retard the flow of runoff.

Paved streets, tarmacs, paved parking lots, and buildings are impervious to water infiltration, greatly increasing the amount of stormwater runoff from precipitation events. These man-made surfaces (which work to hasten the flow of surface water), often cause flooding to occur, sometimes with devastating consequences. In badly planned areas, even relatively light precipitation can cause local flooding. Impervious surfaces not only present flooding problems, they also do not allow water to percolate into the soil to recharge groundwater supplies—often another devastating blow to a location's water supply.

Groundwater Supply

Unbeknownst to most of us, our Earth possesses an unseen ocean. This ocean, unlike the surface oceans that cover most of the globe, is freshwater: the groundwater that lies contained in aquifers beneath Earth's crust. This gigantic water source forms a reservoir that feeds all the natural fountains and springs of earth. But how does water travel into the aquifers that lie under Earth's surface?

Groundwater sources are replenished from a percentage of the average approximately three feet of water that falls to earth each year on every square foot of land. Water falling to earth as precipitation follows three courses. Some runs off directly to rivers and streams (roughly six inches of that three feet), eventually working back to the sea. Evaporation and transpiration through vegetation takes up about two feet. The remaining six inches seeps into the ground, entering and filling every interstice, each hollow and cavity. Gravity pulls water toward the center of the Earth. That means that water on the surface will try to seep into the ground below it. Although groundwater comprises only 1/6 of the total, (1,680,000 miles of water), if we could spread out this water over the land, it would blanket it to a depth of 1,000 feet.

Groundwater

As mentioned, part of the precipitation that falls on land infiltrates the land surface, percolates downward through the soil under the force of gravity, and becomes groundwater. Groundwater, like surface water, is extremely important to the hydrologic cycle and to our water supplies. Almost half of the people in the U.S. drink public water from groundwater supplies. Overall, more water exists as groundwater than surface water in the U.S., including the water in the Great Lakes. But sometimes, pumping it to the surface is not economical, and in recent years, pollution of groundwater supplies from improper disposal has become a significant problem. We find groundwater in saturated layers called aquifers under the earth's surface. Three types of aquifers exist: unconfined, confined, and springs.

Aquifers are made up of a combination of solid material such as rock and gravel and open spaces called pores. Regardless of the type of aquifer, the groundwater in the aquifer is in a constant state of motion. This motion is caused by gravity or by pumping.

The actual amount of water in an aquifer depends upon the amount of space available between the various grains of material that make up the aquifer. The amount of space available is called **porosity**. The ease of movement through an aquifer is dependent upon how well the pores are connected. For example, clay can hold a lot of water and has high porosity, but the pores are not connected, so water moves through the clay with difficulty. The ability of an aquifer to allow water to infiltrate is called permeability.

The aquifer that lies just under the earth's surface is called the zone of saturation, an unconfined aquifer (see Figure 25). The top of the zone of saturation is the water table. An unconfined aquifer is only contained on the bottom and is dependent on local precipitation for recharge. This type of aquifer is often called a water table aquifer.

Unconfined aquifers are a primary source of shallow well water (see Figure 25). These wells are shallow (and not desirable as a public drinking water source). They are subject to local contamination from hazardous and toxic materials—fuel and oil, and septic tanks and agricultural runoff providing increased levels of nitrates and microorganisms. These wells may be classified as groundwater under the direct influence of surface water (GUDISW), and therefore require treatment for control of microorganisms.

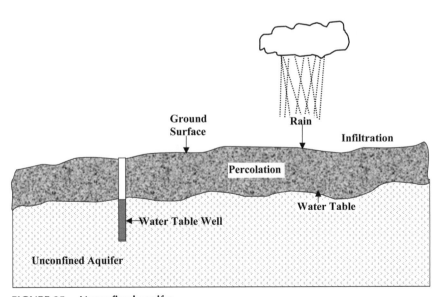

FIGURE 25. Unconfined aquifer

A confined aquifer is sandwiched between two impermeable layers that block the flow of water. The water in a confined aquifer is under hydrostatic pressure. It does not have a free water table (see Figure 26).

Confined aquifers are called artesian aquifers. Wells drilled into artesian aquifers are called artesian wells and commonly yield large quantities of high quality water. An artesian well is any well where the water in the well casing would rise above the saturated strata. Wells in confined aquifers are normally referred to as deep wells and are not generally affected by local hydrological events.

A confined aquifer is recharged by rain or snow in the mountains where the aquifer lies close to the surface of the Earth. Because the **recharge area** is some distance from areas of possible contamination, the possibility of contamination is usually very low. However, once contaminated, confined aquifers may take centuries to recover.

Groundwater naturally exits the Earth's crust in areas called springs. The water in a spring can originate from a water table aquifer or from a confined

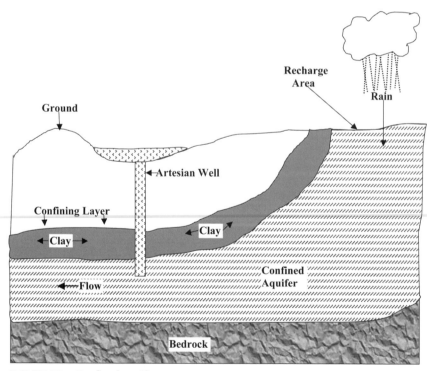

FIGURE 26.   Confined aquifer

aquifer. Only water from a confined spring is considered desirable for a public water system.

Groundwater Quality

Generally, groundwater possesses high chemical, bacteriological, and physical quality. When pumped from an aquifer composed of a mixture of sand and gravel, if not directly influenced by surface water, groundwater is often used without filtration. It can also be used without disinfection if it has a low Coliform count. However, as mentioned, groundwater can become contaminated. When septic systems fail, saltwater intrudes, improper disposal of wastes occurs, improperly stockpiled chemicals leach, underground storage tanks lead, hazardous materials spill, fertilizers and pesticides are misplaced, and when mines are improperly abandoned, groundwater can become contaminated.

To understand how an underground aquifer becomes contaminated, you must understand what occurs when pumping is taking place within the well. When groundwater is removed from its underground source (i.e., from the water-bearing stratum) via a well, water flows toward the center of the well. In a water table aquifer, this movement causes the water table to sag toward the well. This sag is called the **cone of depression**. The shape and size of the cone depends on the relationship between the pumping rate and the rate at which water can move toward the well. If the rate is high, the cone is shallow, and its growth stabilizes. The area that is included in the cone of depression is called the cone of influence, and any contamination in this zone will be drawn into the well.

GUDISW

Groundwater under the direct influence of surface water (GUDISW) is not classified as a groundwater supply. A supply designated as GUDISW must be treated under the state's surface water rules rather than the groundwater rules.

The Surface Water Treatment Rules (SWTR) of the Safe Drinking Water Act requires each site to determine which groundwater supplies are influenced by surface water (i.e., when surface water can infiltrate a groundwater supply and could contaminate it with *Giardia*, viruses, turbidity, and organic material from the surface water source). To determine whether a groundwater supply is under the direct influence of surface water, USEPA has developed procedures that focus on significant and relatively rapid shifts in water quality characteristics, including turbidity, temperature, and pH. When these shifts can be closely correlated with rainfall or other surface water conditions, or when certain indicator organisms associated with surface water are found, the source is said to be under the direct influence of surface water.

Almost all groundwater is in constant motion through the pores and crevices of the aquifer in which it occurs. The water table is rarely level; it generally follows the shape of the ground surface. Groundwater flows in the downhill direction of the sloping water table. The water table sometimes intersects low points of the ground, where it seeps out into springs, lakes, or streams.

Usual groundwater sources include wells and springs that are not influenced by surface water or local hydrologic events.

As a potable water source, groundwater has several advantages over surface water. Unlike surface water, groundwater is not easily contaminated. Groundwater sources are usually lower in bacteriological contamination than surface waters. Groundwater quality and quantity usually remains stable throughout the year. In the United States, groundwater is available in most locations.

As a potable water source, groundwater does present some disadvantages compared to surface water sources. Operating costs are usually higher, because groundwater supplies must be pumped to the surface. Any contamination is often hidden from view. Removing any contaminants is very difficult. Groundwater often possesses high mineral levels, and thus an increased level of hardness, because it is in contact longer with minerals. Near coastal areas, groundwater sources may be subject to salt water intrusion.

✓ **Important Point:** Groundwater quality is influenced by the quality of its source. Changes in source waters or degraded quality of source supplies may seriously impair the quality of the groundwater supply.

Prior to moving onto water use it is important to point out that our freshwater supplies are constantly renewed through the hydrologic cycle, but the balance between the normal ratio of freshwater to salt water is not subject to our ability to change. As our population grows and we move into lands without ready freshwater supplies, we place ecological strain upon those areas, and on their ability to support life.

Communities that build in areas without adequate local water supply are at risk in the event of emergency. Proper attention to our surface and groundwater sources, including remediation, pollution control, and water reclamation and reuse can help to ease the strain, but technology cannot fully replace adequate local freshwater supplies, whether from surface or groundwater sources.

References and Recommended Reading

Angele, F.J., Sr. 1974. *Cross Connections and Backflow Protection*, 2nd ed. Denver: American Water Association.

Jones, F.E. 1992. *Evaporation of Water*. Chelsea, MI: Lewis Publishers.

Lewis, S.A. 1996. *The Sierra Club Guide To Safe Drinking Water*. San Francisco: Sierra Club Books.

McGhee, T.J. 1991. *Water Supply and Sewerage*, 6th ed. New York: McGraw-Hill, Inc.

Meyer, W.B. 1996. *Human Impact on Earth*. New York: Cambridge University Press.

Peavy, H. S., et al. 1985. *Environmental Engineering*. New York: McGraw-Hill, Inc.

Pielou, E. C. 1998. *Fresh Water*. Chicago: University of Chicago Press.

Powell, J.W. 1904. *Twenty-second Annual Report of the Bureau of American Ethnology to the Secretary of the Smithsonian Institution, 1900–1901*. Washington, DC: Government Printing Office.

Spellman, F.R. 2003. *Handbook of Water and Wastewater Treatment Plant Operations*. Boca Raton, FL: Lewis Publishers.

Turk, J., and Turk, A. 1988. *Environmental Science*, 4th ed. Philadelphia: Saunders College Publishing.

USEPA. 2006. *Watersheds*. Accessed 12/06 @ http://www.epa.gov/owow/watershed/whatis.html.

USGS. 2004. *Estimated Use of Water in the United States in 2000*. Washington, DC: U.S. Geological Survey.

USGS. 2006. *Water Science in Schools*. Washington, DC: U.S. Geological Survey.

USGS. 2010. *Water: Challenge Questions*. Accessed 10/24/10 @ http://ga.water.usgs.gov/edse/sc3.html.

10

Earth's Oceans and Their Margins

The ocean, whose tides respond, like women's menses, to the pull of the moon, the ocean which corresponds to the amniotic fluid in which human life begins, the ocean on whose surface vessels (personified as female) can ride but in whose depth sailors meet their death and monsters conceal themselves . . . it is unstable and threatening as the earth is not; it spawns new life daily, yet swallows up lives; it is changeable like the moon, unregulated, yet indestructible and eternal.

—Adrienne Rich, *Of Woman Born*, 1976

Oceans

EARTH'S OCEANS—Atlantic, Pacific, Arctic, Indian, and Southern—are the storehouses of Earth's saline water. Oceans cover about 71% of Earth's surface. Average depth of Earth's oceans is about 3,800 m, with the greatest ocean depth recorded at 11,036 m in the Mariana Trench. At the present time, the oceans contain a volume of about 1.35 billion cubic kilometers (96.5% of Earth's total water supply), but the volume fluctuates with the growth and melting of glacial ice.

Composition of ocean water has remained constant in composition throughout geologic time. The major constituents dissolving in ocean water (from rivers and precipitation and the result of weathering and degassing of the mantle by volcanic activity) are composed of about 3.5%, by weight, of dissolved salts including chloride (55.07%), sodium (30.62%), sulfate (7.72%), magnesium

(3.68%), calcium (1.17%), potassium (1.10%), bicarbonate (0.40%), bromine, (0.19%) and strontium (0.02%).

The most significant factor related to ocean water that everyone is familiar with is the salinity of the water—how salty it is. *Salinity*, a measure of the amount of dissolved ions in the oceans, ranges between 33 and 37 parts per thousand. Often the concentration is the amount (by weight) of salt in water, as expressed in "parts per million" (ppm). Water is saline if it has a concentration of more than 1,000 ppm of dissolved salts; ocean water contains about 35,000 ppm of salt (USGS 2007). Chemical precipitation, absorption onto clay minerals, and plants and animals prevents seawater from containing even higher salinity concentrations. However, salinity does vary in the oceans because surface water evaporates, rain and stream water is added, and ice forms or thaws.

Did You Know?

Salinity is higher in mid-latitude oceans because evaporation exceeds precipitation. Salinity is also higher in restricted areas of the oceans like the Red Sea (up to 40 parts per thousand). Salinity is lower near the equator because precipitation is higher and is lower near the mouths of major rivers because of the input of fresh water.

Along with salinity another important property of seawater includes temperature. The temperature of surface seawater varies with latitude, from near 0°C near the poles to 29°C near the equator. Some isolated areas can have temperatures up to 37°C. Temperature decreases with ocean depth.

Did You Know?

Most scientists agree that the atmosphere and the oceans accumulated gradually over millions and millions of years with the continual 'degassing' of the Earth's interior. According to this theory, the oceans formed from the escape of water vapor and other gases from the molten rocks of the Earth to the atmosphere surrounding the cooling planet. After the Earth's surface had cooled to a temperature below the boiling point of water, rain began to fall—and continued to fall for centuries. As the water drained into the great hollows in the Earth's surface, the primeval ocean came into existence. The forces of gravity prevented the water from leaving the planet (NOAA 2010).

Ocean Floors

The bottoms of the oceans' basins (ocean floors) are marked by mountain ranges, plateaus, and other relief features similar to (although not as rugged as) those on the land.

As shown in Figure 27, the floor of the ocean has been divided into four divisions: the continental shelf, continental slope, continental rise, and deep-sea floor or abyssal plain.

- **Continental Shelf**—is the flooded, nearly flat true margins of the continents. Varying in width to about 40 miles and in depth to approximately 650 feet, continental shelves slope gently outward from the shores of the continents (see Figure 27). Continental shelves occupy approximately 7.5% of the ocean floor.
- **Continental Slope**—is a relatively steep slope descending from the continental shelf (see Figure 27) rather abruptly to the deeper parts of the ocean. These slopes occupy about 8.5% of the ocean floor.
- **Continental Rise**—is a broad gentle slope below the continental slope containing sediment that has accumulated along parts of the continental slope.
- **Abyssal Plain**—is a sediment-covered deep-sea plain about 12,000–18,000 feet below sea level. This plane makes up about 42% of the ocean floor.

The deep ocean floor does not consist exclusively of the abyssal plain. In places there are areas of considerable relief. Among the more important such features are:

- **Seamounts**—are isolated mountain-shaped elevations more than 3,000 feet high.

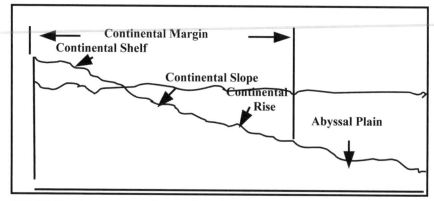

FIGURE 27.   Cross section of ocean floor showing major elements of topography

- **Mid-oceanic Ridge**—are submarine mountains, extending more than 37,000 miles through the oceans, generally 10,000 feet above the abyssal plain.
- **Trench**—is a deep, steep-sided trough in an abyssal plain.
- **Guyot**—is a seamount that is flat-topped and was once a volcano. They rise from the ocean bottom and usually are covered by 3000 to 6000 feet of water.

Did You Know?

In the beginning, the primeval seas were probably only slightly salty. But over time, as rain fell to the Earth and ran over the land, breaking up rocks and transporting their minerals to the ocean, the ocean has become saltier. Raining replenishes freshwater in rivers and streams, so they don't taste salty. However, the water in the ocean collects all of the salt and minerals from all of the rivers that flow into it. It is estimated that the rivers and streams flowing from the United States alone discharge 225 million tons of dissolved solids and 513 million tons of suspended sediment annually to the ocean. Throughout the world, rivers carry an estimated four billion tons of dissolved salts to the ocean annually. About the same tonnage of salt from ocean water probably is deposited as sediment on the ocean bottom and thus, yearly gains may offset yearly losses. In other words, the ocean today probably has a balanced salt input and output (and so the ocean is no longer getting saltier) (NOAA 2010).

Ocean Tides, Currents, and Waves

Water is the master sculptor of Earth's surfaces. The ceaseless, restless motion of the sea is an extremely effective geologic agent. Besides shaping inland surfaces, water sculpts the coast. Coasts include sea cliffs, shores, and beaches. Seawater set in motion erodes cliffs, transports eroded debris along shores, and dumps it on beaches. Therefore most coasts retreat or advance. In addition to the unceasing causes of motion—wind, density of sea water, and rotation of the Earth—the chief agents in this process are tides, currents, and waves.

Tides

The periodic rise and fall of the sea (once every twelve hours and twenty-six minutes) produces the tides. Tides are due to the gravitational attraction of the moon, and to a lesser extent, the sun on the Earth. The moon has a larger effect on tides and causes the Earth to bulge toward the moon. It is interesting to note that at the same time the moon causes a bulge on Earth, a bulge occurs on the opposite side of the Earth due to inertial forces (further explanation is beyond the scope of this text). The effect of the tides is not too noticeable in the open sea, the difference between high and low tide amounting to about two feet. The tidal range may be considerably greater near shore, however. It may range from less than two feet to as much as fifty feet. The tidal range will vary according to the phase of the moon and the distance of the moon from the Earth. The type of shoreline and the physical configuration of the ocean floor will also affect the tidal range.

The times, heights, and extents of both the inflow and outflow of the tidal waters that support a number of different aspects of our daily lives are measured on a continuous basis. Navigating ships safely through shallow water ports, intracoastal waterways, and estuaries (where rivers meet the sea) required knowledge of the time and height of the tides as well as the speed and direction of the tidal currents. Mariners need accurate data because the depths and widths of the channels and the increase of marine traffic leave very little room for error.

Tidal data is critical to engineers for harbor engineering projects, to fishing, recreational boating, and surfing. Commercial and recreation fishermen use their knowledge of the tides and tidal currents to help them improve their catches. Depending on the species and water depth in a particular area, fish may concentrate during ebb or flood tidal currents (NOAA 2010).

Currents

The oceans have localized movements of masses of seawater called ocean currents. These are the result of drift of the upper 50 to 100 m of the ocean due to drag by wind. Thus, surface ocean currents generally follow the same patterns as atmospheric circulation with the exception that atmospheric currents continue over the land surface while ocean currents are deflected by the land. Along with wind action, current may also be caused by tides, variation in salinity of the water, rotation of the earth, and concentrations of turbid or muddy water.

Winds drive currents that are at or near the ocean's surface. These currents are generally measured in meters per second or in knots (1 knot = 1.85 kilometers per hour or 1.15 miles per hour). Winds drive currents near coastal areas on a localized scale and in the open ocean on a global scale.

The rise and fall of the tide create a current in the oceans, near the shore, and in bays and estuaries along the coast. These are called *tidal currents*. Tidal currents are the only type of currents that change in a very regular pattern and can be predicted for future dates.

Temperature changes in water affect water density which, in turn, causes currents—these currents cause seawater to circulate vertically. The *thermohaline circulation* process is driven by the density differences in water due to temperature (thermo) and salinity (haline) variations in different parts of the ocean.

Did You Know?

Currents affect the Earth's climate by driving warm water from the Equator and cold water from the poles around the Earth. The warm Gulf Stream, for instance, brings milder winter weather to Bergen, Norway, than to New York, much farther south. It keeps the Norwegian coast an incredible 6.1 degrees Celsius (43 degrees Fahrenheit) warmer than other places equally far north. In regard to the Gulf Stream, it is an intense, warm ocean current in the western North Atlantic Ocean. It moves north along the coast of Florida and then turns eastward off of North Carolina, flowing northeast across the Atlantic. Off the Atlantic seaboard of the United States, the Gulf Stream flows at a rate nearly 300 times faster than the typical flow of the Amazon River. The velocity of the current is faster near the surface, with the maximum speed typically about 5.6 miles per hour (nine kilometers per hour). The average speed of the Gulf Stream, however, is four miles per hour (6.4 kilometers per hour). The current slows to a speed of about one mile per hour (1.6 kilometers per hour) as it widens to the north.

Waves

Waves, varying greatly in size, are produced by the friction of wind on open water. Wave height and power depend upon wind strength and fetch—the amount of unobstructed ocean over which the wind has blown. In a wave, water travels in loops. Essentially an up-and done movement of the water, the diameter of the loops decreases with depth. The diameter of loops at the surface is equal to wave height (h). Breakers are formed when the wave comes into shallow water near the shore. The lower part of the wave is retarded by the ocean bottom, and the top, having greater momentum, is hurled forward causing the wave to break. These breaking waves may do great damage to coastal property as they race across coastal lowlands driven by winds of gale or hurricane velocity.

Coastal Erosion, Transportation, and Deposition

The geologic work of the sea, like previously discussed geologic agents, consists of erosion, transportation, and deposition. The sea accomplishes its work of coastal landform sculpting largely by means of waves and wave-produced currents; their effect on the seacoast may be quite pronounced. The coast and accompanying coastal deposits and landform development represent a balance between wave energy and sediment supply.

Wave Erosion

Waves attack shorelines and erode by a combination of processes. The resistance of the rocks composing the shoreline and the intensity of wave action to which it is subjected are the factors that determine how rapidly the shore will be eroded. Wave erosion works chiefly by hydraulic action, corrosion, and attrition. As waves strike a sea cliff, *hydraulic action* crams air into rock crevices putting tremendous pressure on the surrounding rock; as waves retreat, the explosively expanding air enlarges cracks and breaks off chunks of rock (*scree*). Chunks hurled by waves against the cliff break off more scree (via a sandpapering action)—a process called *corrasion*. When the sea rubs and grinds rocks together, forming scree that is thrown into the cliffs, reducing broken rocks to pebbles and sand grains—this process is called *attrition* (Lambert 2007).

Several features are formed by marine erosion—different combinations of wave action, rock type, and rock beds produce these features. Some of the more typical erosion-formed features of shorelines are discussed below.

- **Sea Cliffs or Wave Cut Cliffs**—are formed by wave erosion of underlying rock followed by the caving-in of the overhanging rocks. As waves eat farther back inland, they leave a wave-cut beach or platform. Such cliffs are essentially vertical and are common at certain localities along the New England and Pacific coasts of North America.
- **Wave-cut Benches**—are the result of wave action not having enough time to lower the coastline to sea level. Because of the resistance to erosion, a relatively flat wave-cut bench develops. If subsequent uplift of the wave-cut bench occurs, it may be preserved above sea level as a wave-cut bench.
- **Headlands**—are finger-like projections of resistant rock extending out into the water. Indentations between headlands are termed *coves*.

- **Sea Caves, Sea Arches, and Stacks**—are formed by continued wave action on a sea cliff. Wave action hollows out cavities or caves in the sea cliffs. Eventually, waves may cut completely through a headland to form a sea arch; if the roof of the arch collapses the rock left separated from the headland is called a stack.

Marine Transportation

Waves and currents are important transporting agents. Rip currents and undertow carry rock particles back to the sea, and long-shore currents will pick up sediments (some of it in solution), moving them out from shore into deeper water. Materials carried in solution or suspension may drift seaward for great distances and eventually be deposited far from shore. During the transportation process sediments undergo additional erosion, becoming reduced in size.

Marine Deposition

Marine deposition takes place whenever currents and waves suffer reduced velocity. Some rocks are thrown up on the shore by wave action. Most of the sediments thus deposited consist of rock fragments derived from the mechanical weathering of the continents, and they differ considerably from terrestrial or continental deposits. Due to input of sediments from rivers, deltas may form; due to beach drift such features as spits and hooks, bay barriers, and tombolos may form. Depositional features along coasts are discussed below.

- **Beaches**—are transitory coastal deposits of debris which lie above the low-tide limit in the shore zone.
- **Barrier Islands**—are long narrow accumulations of sand lying parallel to the shore and separated from the shore by a shallow lagoon.
- **Spits and Hooks**—are elongated, narrow embankments of sand and pebble extending out into the water but attached by one end to the land.
- **Tombolos**—are bars of sand or gravel connecting an island with the mainland or another island.
- **Wave-built Terraces**—are structures built up from sediments deposited in deep water beyond a wave-cut terrace.
- **Deltas**—form where sediment supply is greater than ability of waves to remove sediment.

Did You Know?

The ocean is blue because water absorbs colors in the red part of the light spectrum. Like a filter, this leaves behind colors in the blue part of the light spectrum for us to see. The ocean may also take on green, red, or other hues as light bounces off of floating sediments and particles in the water. Most of the ocean, however, is completely dark. Hardly any light penetrates deeper than 20 meters (656 feet), and no light penetrates deeper than 2,000 meters (3,280 feet).

References and Recommended Reading

Gross, G.M. 1995. *Oceanography: A View of the Earth.* Englewood Cliffs, NJ: Prentice-Hall, Inc.

Lambert, D. 2007. *The Field Guide to Geology.* New York: Checkmark Books.

NOAA. 2010. *Why Do We Have Oceans?* Accessed 10/27/10 @ http://oceanservice.noaa .gov/facts.

Pinet, P.R. 1996. *Invitation to Oceanography.* St. Paul, MN: West Publishing Company.

USGS. 2007. *The Water Cycle: Water Storage in Oceans.* Accessed 7/11/08 @ http://ga .water.usgs.gov/edu/watercycleoceans.html.

11

Running Waters

Little brook, sing a song of a leaf that sails along,
Down the golden braided center of your current swift and strong.

—J.W. Riley (1849–1916)

Stream Genesis and Structure

EARLY IN THE SPRING on a snow and ice-covered high alpine meadow the time and place of the water cycle continues. The cycle's main component, water, has been held in reserve—literally frozen, for the long dark winter months, but with longer, warmer spring days, the sun is higher, more direct, and of longer duration, and the frozen masses of water respond to the increased warmth. The melt begins with a single drop, then two, then more and more. As the snow and ice melts, the drops join a chorus that continues unending; they fall from their ice-bound lip to the bare rock and soil terrain below.

The terrain the snow-melt strikes is not like glacial till, the unconsolidated, heterogeneous mixture of clay, sand, gravel, and boulders, dug-out, ground-out, and exposed by the force of a huge, slow, and inexorably moving glacier. Instead, this soil and rock ground is exposed to the falling drops of snowmelt because of a combination of wind and the tiny, enduring force exerted by drops of water as over season after season they collide with the thin soil cover, exposing the intimate bones of the earth.

Gradually, the single drops increase to a small rush—they join to form a splashing, rebounding, helter-skelter cascade, and many separate rivulets that trickle, and then run their way down the face of the granite mountain. At an indented ledge halfway down the mountain slope, a pool forms whose beauty, clarity, and sweet iciness provides the visitor with an incomprehensible, incomparable gift—a blessing from earth.

The mountain pool fills slowly, tranquil under the blue sky, reflecting the pines, snow, and sky around and above it, an open invitation to lie down and drink, and to peer into the glass-clear, deep waters, so clear that it seems possible to reach down over fifty feet and touch the very bowels of the mountain. The pool has no transition from shallow margin to depth; it is simply deep and pure. As the pool fills with more melt water, we wish to freeze time, to hold this place and this pool in its perfect state forever, it is such a rarity to us in our modern world. But this cannot be—Mother Nature calls, prodding, urging—and for a brief instant, the water laps in the breeze against the outermost edge of the ridge, then a trickle flows over the rim. The giant hand of gravity reaches out and tips the overflowing melt onward and it continues the downward journey, following the path of least resistance to its next destination, several thousand feet below.

When the overflow, still high in altitude, but with its rock-strewn bed bent downward, toward the sea, meets the angled, broken rocks below, it bounces, bursts, and mists its way against steep, V-shaped walls that form a small valley, carved out over time by water and the forces of the earth.

Within the valley confines, the melt water has grown from drops to rivulets to a small mass of flowing water. It flows through what is at first a narrow opening, gaining strength, speed, and power as the V-shaped valley widens to form a U-shape. The journey continues as the water mass picks up speed and tumbles over massive boulders, and then slows again.

At a larger but shallower pool, waters from higher elevations have joined the main body—from the hillsides, crevices, springs, rills, mountain creeks. At the influent pool-sides all appears peaceful, quiet, and restful, but not far away, at the effluent end of the pool, gravity takes control again. The overflow is flung over the jagged lip, and cascades downward several hundred feet, where the waterfall again brings its load to a violent, mist-filled meeting.

The water separates and joins again and again, forming a deep, furious, wild stream that calms gradually as it continues to flow over lands that are less steep. The waters widen into pools overhung by vegetation, surrounded by tall trees. The pure, crystalline waters have become progressively discolored on their downward journey, stained brown-black with humic acid, and literally filled with suspended sediments; the once-pure stream is now muddy.

The mass divides and flows in different directions, over different landscapes. Small streams divert and flow into open country. Different soils work to retain or speed the waters, and in some places the waters spread out into shallow swamps, bogs, marshes, fens, or mires. Other streams pause long enough to fill deep depressions in the land and form lakes. For a time, the water remains and pauses in its journey to the sea. But this is only a short-term pause, because lakes are only a short-term resting place in the water cycle. The water will eventually move on, by evaporation, or seepage into groundwater. Other portions of the water mass stay with the main flow, and the speed of flow changes to form a river, which braids its way through the landscape, heading for the sea. As it changes speed and slows the river bottom changes from rock and stone to silt and clay. Plants begin to grow, stems thicken, and leaves broaden. The river is now full of life and the nutrients needed to sustain life. But the river courses onward, its destiny met when the flowing rich mass slows its last and finally spills into the sea (Spellman & Whiting 1998).

As related above, each stream has its own history, which can be revealed through the eyes of those who follow it from its beginning to where it empties its water into a large stream, lake, pond, or the sea.

Streams

As mentioned, about 71% of the Earth's surface is covered by water. Most of the Earth's water is in the oceans (97.2%). Less than 3% of the water on Earth is located on or beneath the continents, and most of that (2.15%) is in the form of ice. About 0.001% of the Earth's water is in the atmosphere, and only 0.0001% is contained in the world's river systems. Despite the low proportion of water in streams, running water:

1. serves as the primary source of drinking, industrial, and irrigation water;
2. is the most important erosional agent on Earth;
3. is used for transportation;
4. supplies about 8% of the electricity used in North America;
5. is an important part of the water cycle; running waters carry most of the water that goes from the land to the sea;
6. is the site of most population centers because they provide a major source of water and transportation.

Did You Know?

Streams are bodies of running water that carry rock particles (sediment loads) and dissolved ions and flow downslope along a clearly defined path, called a *channel*. Thus, streams may vary in width from a few inches to several miles.

Characteristics of Stream Channels

A standard rule of thumb states: flowing waters (rivers and streams) determine their own channels, and these channels exhibit relationships attesting to the operation of physical laws—laws that are not, as of yet, fully understood. The development of stream channels and entire drainage networks, and the existence of various regular patterns in the shape of channels, indicate that streams are in a state of dynamic equilibrium between erosion (sediment loading) and deposition (sediment deposit), and governed by common hydraulic processes. However, because channel geometry is four dimensional with a long profile, cross-section, depth, and slope profile, and because these mutually adjust over a time scale as short as years and as long as centuries or more, cause and effect relationships are difficult to establish. Other variables that are presumed to interact as the stream achieves its graded state include width and depth, velocity, size of sediment load, bed roughness, and the degree of braiding (**sinuosity**).

Stream Profiles

Mainly because of gravity, most streams exhibit a downstream decrease in gradient along their length. Beginning at the headwaters, the steep gradient becomes less so as one proceeds downstream, resulting in a concave longitudinal profile. Though diverse geography provides for almost unlimited variation, a lengthy stream that originates in a mountainous area (such as the one described in the chapter opening) typically comes into existence as a series of springs and rivulets; these coalesce into a fast-flowing, turbulent mountain stream, and the addition of tributaries results in a large and smoothly flowing river that winds through the lowlands to the sea.

When studying a stream system of any length, it becomes readily apparent (almost from the start of such studies) that what we are studying is a body of flowing water that varies considerably from place to place along its length. For example, a common variable—the results of which can be readily seen—

is whenever discharge increases, causing corresponding changes in the stream's width, depth, and velocity. In addition to physical changes that occur from location to location along a stream's course, there is a legion of biological variables that correlate with stream size and distance downstream. The most apparent and striking changes are in steepness of slope and in the transition from a shallow stream with large boulders and a stony substrate to a deep stream with a sandy substrate.

The particle size of bed material at various locations is also variable along the stream's course. The particle size usually shifts from an abundance of coarser material upstream to mainly finer material in downstream areas.

Did You Know?

A fundamental rule for stream flow (similar to the rule for the flow of electrons in electricity) is: Flowing water will always seek the path of least resistance.

Sinuosity

Unless forced by man in the form of heavily regulated and channelized streams, straight channels are uncommon. Stream flow creates distinctive landforms composed of straight (usually in appearance only), **meandering**, and braided channels, channel networks, and flood plains. Simply put: flowing water will follow a sinuous course. The most commonly used measure is the sinuosity index (SI).

$$SI = \frac{\text{Channel distance}}{\text{Down valley distance}}$$

or

$$SI = \frac{\text{Actual path-length}}{\text{Shortest path-length}}$$

Sinuosity equals 1 in straight channels and more than 1 in sinuous channels. Meandering is the natural tendency for alluvial channels and is usually defined as an arbitrarily extreme level of sinuosity, typically an SI greater than 1.5. Many variables affect the degree of sinuosity, however, and so SI values range from near unity in simple, well-defined channels to four in highly meandering channels (Gordon et al. 1992).

It is interesting to note that even in many natural channel sections of a stream course that appear straight, meandering occurs in the line of maximum water or channel depth (known as the **thalweg**). Keep in mind that streams have to meander, that is how they renew themselves. By meandering, they wash plants and soil from the land into their waters, and these serve as nutrients for the plants in the rivers. If rivers aren't allowed to meander, if they are channelized, the amount of life they can support will gradually decrease. That means less fish, ultimately—and fewer bald eagles, herons, and other fishing birds (Spellman 1996).

Meander flow follows a predictable pattern and causes regular regions of erosion and deposition. The streamlines of maximum velocity and the deepest part of the channel lie close to the outer side of each bend and cross over near the point of inflection between the banks. A huge elevation of water at the outside of a bend causes a helical flow of water towards the opposite bank. In addition, a separation of surface flow causes a back eddy. The result is zones of erosion and deposition, and explains why point bars develop in a downstream direction in depositional zones (Morisawa 1968).

Did You Know?

Meandering channels can be highly convoluted or merely sinuous but maintain a single thread in curves having definite geometric shape. Straight channels are sinuous but apparently random in occurrence of bends. Braided channels are those with multiple streams separated by bars and islands (Leopold et al. 1964; Leopold 1994).

Bars, Riffles, and Pools

Implicit in the morphology and formation of meanders are bars, riffles, and pools. Bars develop by deposition in slower, less competent flow on either side of the sinuous mainstream. Onward moving water, depleted of bed load, regains competence and shears a pool in the meander—reloading the stream for the next bar. Alternating bars migrate to form riffles.

As stream flow continues along its course a pool-riffle sequence is formed. Basically the riffle is a mound or hillock and the pool is a depression.

Floodplain

Stream channels influence the shape of the valley floor through which they course. This self-formed, self-adjusted flat area near to the stream is the floodplain, which loosely describes the valley floor prone to periodic inundation during over-bank discharges. What is not commonly known is that valley flooding is a regular and natural behavior of the stream. Many people learn about this natural phenomenon the hard way—that is, whenever their farms, towns, streets, and homes become inundated by a river or stream that is doing nothing more than following its "natural" periodic cycle—conforming to the Master Plan designed by the Master Planner: Mother Nature.

Did You Know?

Floodplain rivers are found where regular floods form lateral plains outside the normal channel which seasonally become inundated, either as a consequence of greatly increased rainfall or snowmelt.

Water Flow in a Stream

Most elementary students learn early in their education process that water on Earth flows downhill (gravity)—from land to the sea. However, they may or may not be told that water flows downhill toward the sea by various routes.

For the moment the "route" (channel, conduit, or pathway) we are concerned with is the surface water route taken by surface runoff. Surface runoff is dependent on various factors. For example, climate, vegetation, topography, geology, soil characteristics, and land-use determine how much surface runoff occurs compared with other pathways.

The primary source (input) of water to total surface runoff, of course, is precipitation. This is the case even though a substantial portion of all precipitation input returns directly to the atmosphere by evapotranspiration. Evapotranspiration is a combination process, as the name suggests, whereby water in plant tissue and in the soil evaporates and transpires to water vapor in the atmosphere.

Probably the easiest way to understand precipitation's input to surface water runoff is to take a closer look at this precipitation input.

Again, a substantial portion of precipitation input returns directly to the atmosphere by evapotranspiration. It is also important to point out that when precipitation occurs, some rainwater is intercepted or blocked or caught by vegetation where it evaporates, never reaching the ground or being absorbed by plants. A large portion of the rainwater that reaches the surface on ground, in lakes, and streams also evaporates directly back to the atmosphere. Although plants display a special adaptation to minimize transpiration, plants still lose water to the atmosphere during the exchange of gases necessary for photosynthesis. Notwithstanding the large percentage of precipitation that evaporates, rain- or melt-water that reach the ground surface follow several pathways in reaching a stream channel or groundwater.

Soil can absorb rainfall to its infiltration capacity (i.e., to its maximum intake rate). During a rain event, this capacity decreases. Any rainfall in excess of infiltration capacity accumulates on the surface. When this surface water exceeds the depression storage capacity of the surface, it moves as an irregular sheet of **overland flow**. In arid areas, overland flow is likely because of the low permeability of the soil. Overland flow is also likely when the surface is frozen and/or when human activities have rendered the land surface less permeable. In humid areas, where infiltration capacities are high, overland flow is rare.

In rain events where the infiltration capacity of the soil is not exceeded, rain penetrates the soil and eventually reaches the groundwater—from which it discharges to the stream slowly and over a long period of time. This phenomenon helps to explain why stream flow through a dry weather region remains constant; the flow is continuously augmented by groundwater. This type of stream is known as a **perennial stream**, as opposed to an intermittent one, because the flow continues during periods of no rainfall.

Streams that course their way in channels through humid regions are fed water via the water table, which slopes toward the stream channel. Discharge from the water table into the stream accounts for flow during periods without precipitation, and also explains why this flow increases, even without tributary input, as one proceeds downstream. Such streams are called gaining or effluent, as opposed to losing or influent streams that lose water into the ground. It is interesting to note that the same stream can shift between gaining and losing conditions along its course because of changes in underlying strata and local climate.

Stream Water Discharge

The current velocity (speed) of water (driven by gravitational energy) in a channel varies considerably within a stream's cross section owing to friction with the bottom and sides, with sediment, with obstructions (rocks and logs, etc.) and the atmosphere, and to sinuosity (bending or curving). Highest velocities, obviously, are found where friction is least, generally at or near the surface and near the center of the channel. In deeper streams current velocity is greatest just below the surface due to the friction with the atmosphere; in shallower streams current velocity is greatest at the surface due to friction with the bed. Velocity decreases as a function of depth, approaching zero at the substrate surface. A general and convenient rule of thumb is that the deepest part of the channel occurs where the stream velocity is the highest. Additionally, both width and depth of a stream increase downstream because discharge (the amount of water passing any point in a given time) increases downstream. As discharge increases the cross-sectional shape will change, with the stream becoming deeper and wider. Velocity is important to discharge because discharge (m³/sec) = cross-sectional Area [width x average depth] (m²) x Average Velocity (m/sec).

$$Q = A \times V$$

A stream is constantly seeking balance. This can be seen whenever the amount of water in a stream increases, the stream must adjust its velocity and cross-sectional area to reach balance. Discharge increases as more water is added through precipitation, tributary streams, or from groundwater seeping into the stream. As discharge increases, generally width, depth, and velocity of the stream also increase.

Transport of Material (Load)

Streams possess both potential and kinetic energy. As water moves downstream, potential energy is converted into kinetic energy. Most of the kinetic energy is released as frictional heat, but a small amount remains to transport eroded material.

Water flowing in a channel may exhibit laminar flow (parallel layers of water shear over one another vertically), or turbulent flow (complex mixing) (see Figure 28). In streams, laminar flow is uncommon, except at boundaries where flow is very low and in groundwater. Thus the flow in streams generally is turbulent. Turbulence exerts a shearing force that causes particles to move along the stream bed by pushing, rolling, and skipping, referred to as bed load. This same shear causes turbulent eddies that entrain particles in suspension

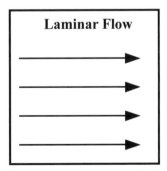
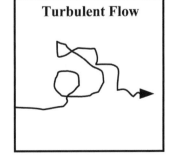

FIGURE 28. Laminar and turbulent flow

(called the suspended load—particles size under 0.06 mm). Entrainment is the incorporation of particles when stream velocity exceeds the entraining velocity for a particular particle size.

Did You Know?

Entrainment is a natural extension of erosion and is vital to the movement of stationary particles in changing flow conditions. Remember, all sediments ultimately derive from erosion of basin slopes, but the immediate supply usually derives from the stream channel and banks, while the bedload comes from the streambed itself and is replaced by erosion of bank regions.

The entrained particles in suspension (suspended load) also include fine sediment, primarily clays, silts, and fine sands that require only low velocities and minor turbulence to remain in suspension. These are referred to as wash load (under 0.002 mm), because this load is "washed" into the stream from banks and upland areas (Gordon et al. 1992; Spellman 1996).

Thus the suspended load includes the wash load and coarser materials (at lower flows). Together, the suspended load and bed load constitute the solid load. It is important to note that in bedrock streams the bed load will be a lower fraction than in alluvial streams where channels are composed of easily transported material.

A substantial amount of material is also transported as the dissolved load. Solutes (ions) are generally derived from chemical weathering of bedrock and soils, and their contribution is greatest in sub-surface flows, and in regions of limestone geology.

The relative amount of material transported as solute rather than solid load depends on basin characteristics, lithology (i.e., the physical character of rock) and hydrologic pathways. In areas of very high runoff, the contribution of solutes approaches or exceeds sediment load, whereas in dry regions, sediments make up as much as 90% of the total load.

Deposition occurs when stream competence (i.e., the largest particle that can be moved as bedload and the critical erosion—competent—velocity is the lowest velocity at which a particle resting on the streambed will move) falls below a given velocity. Simply stated: the size of the particle that can be eroded and transported is a function of current velocity.

Sand particles are the most easily eroded. The greater the mass of larger particles (e.g., coarse gravel) the higher the initial current velocities must be for movement. However, smaller particles (silts and clays) require even greater initial velocities because of their cohesiveness and because they present smaller, streamlined surfaces to the flow. Once in transport, particles will continue in motion at somewhat slower velocities than initially required to initiate movement, and will settle at still lower velocities.

Particle movement is determined by size, flow conditions, and mode of entrainment. Particles over 0.02 mm (medium-coarse sand size) tend to move by rolling or sliding along the channel bed as traction load. When sand particles fall out of the flow, they move by saltation or repeated bouncing. Particles under 0.06 mm (silt) move as suspended load and particles under 0.002 (clay), indefinitely, as wash load. A considerable amount of particle sorting takes place because of the different styles of particle flow in different sections of the stream (Richards 1982; Likens 1984).

Unless the supply of sediments becomes depleted the concentration and amount of transported solids increases. However, discharge is usually too low, throughout most of the year, to scrape or scour, shape channels, or move significant quantities of sediment in all but sand-bed streams, which can experience change more rapidly. During extreme events, the greatest scour occurs and the amount of material removed increases dramatically.

Sediment inflow into streams can both be increased and decreased as a result of human activities. For example, poor agricultural practices and deforestation greatly increase erosion.

Man-made structures such as dams and channel diversions can, on the other hand, greatly reduce sediment inflow.

Groundwater

Unbeknownst to most of us, Earth possesses an unseen ocean of water. This ocean, unlike the surface oceans that cover most of the globe, is freshwater: the groundwater that lies contained in aquifers beneath Earth's crust. This gigantic freshwater water source (makes up about 1% of the water on Earth; about 35 times the amount of water in lakes and streams) forms a reservoir that feeds all the natural fountains and springs of Earth. But how does water travel into the aquifers that lie under Earth's surface?

Did You Know?

Groundwater occurs almost everywhere beneath the land surface. The widespread occurrence of potable groundwater is the reason that it is used as a source of water supply by about one-half the population of the United States, including almost all of the population that is served by domestic water-supply systems.

Groundwater sources are replenished from a percentage of the average approximately three feet of water that falls to earth each year on every square foot of land. Water falling to earth as precipitation follows three courses. Some runs off directly to rivers and streams (roughly six inches of that three feet), eventually working back to the sea. Evaporation and transpiration through vegetation takes up about two feet. The remaining six inches seeps into the ground, entering and filling every interstice, each hollow and cavity. Gravity pulls water toward the center of the Earth. That means that water on the surface will try to seep into the ground below it. Although groundwater comprises only 1/6 of the total (1,680,000 miles of water), if we could spread out this water over the land, it would blanket it to a depth of 1,000 feet.

As mentioned, part of the precipitation that falls on land infiltrates the land surface, percolates downward through the soil under the force of gravity, and becomes groundwater. Groundwater, like surface water, is extremely important to the hydrologic cycle and to our water supplies. Almost half of the people in the U.S. drink public water from groundwater supplies. Overall, more water exists as groundwater than surface water in the U.S., including the water in the Great Lakes. But sometimes, pumping it to the surface is not economical, and in recent years, pollution of groundwater supplies from improper disposal has become a significant problem.

We find groundwater in saturated layers called aquifers under the Earth's surface. Three types of aquifers exist: unconfined, confined, and springs.

Aquifers are made up of a combination of solid material such as rock and gravel and open spaces called pores. Regardless of the type of aquifer, the groundwater in the aquifer is in a constant state of motion. This motion is caused by gravity or by pumping.

The actual amount of water in an aquifer depends upon the amount of space available between the various grains of material that make up the aquifer. The amount of space available is called porosity. The ease of movement through an aquifer is dependent upon how well the pores are connected. For example, clay can hold a lot of water and has high porosity, but the pores are not connected, so water moves through the clay with difficulty. The ability of an aquifer to allow water to infiltrate is called permeability.

As described earlier, the aquifer that lies just under the earth's surface is called the zone of saturation, an unconfined aquifer. The top of the zone of saturation is the water table. An unconfined aquifer is only contained on the bottom and is dependent on local precipitation for recharge. This type of aquifer is often called a water table aquifer.

Unconfined aquifers are a primary source of shallow well water. These wells are shallow (and not desirable as a public drinking water source). They are subject to local contamination from hazardous and toxic materials—fuel and oil, and septic tanks and agricultural runoff providing increased levels of nitrates and microorganisms. These wells may be classified as groundwater under the direct influence of surface water (GUDISW), and therefore require treatment for control of microorganisms.

As mentioned, a confined aquifer is sandwiched between two impermeable layers that block the flow of water. The water in a confined aquifer is under hydrostatic pressure. It does not have a free water table.

Confined aquifers are called artesian aquifers. Wells drilled into artesian aquifers are called artesian wells and commonly yield large quantities of high quality water. An artesian well is any well where the water in the well casing would rise above the saturated strata. Wells in confined aquifers are normally referred to as deep wells and are not generally affected by local hydrological events.

A confined aquifer is recharged by rain or snow in the mountains where the aquifer lies close to the surface of the earth. Because the recharge area is some distance from areas of possible contamination, the possibility of contamination is usually very low. However, once contaminated, confined aquifers may take centuries to recover.

Groundwater naturally exits the earth's crust in areas called springs. The water in a spring can originate from a water table aquifer or from a confined aquifer. Only water from a confined spring is considered desirable for a public water system.

Almost all groundwater is in constant motion through the pores and crevices of the aquifer in which it occurs. The water table is rarely level; it generally follows the shape of the ground surface. Groundwater flows in the downhill direction of the sloping water table. The water table sometimes intersects low points of the ground, where it seeps out into springs, lakes, or streams.

Usual groundwater sources include wells and springs that are not influenced by surface water or local hydrologic events.

As a potable water source, groundwater has several advantages over surface water. Unlike surface water, groundwater is not easily contaminated. Groundwater sources are usually lower in bacteriological contamination than surface waters. Groundwater quality and quantity usually remain stable throughout the year. In the United States, groundwater is available in most locations.

As a potable water source, groundwater does present some disadvantages compared to surface water sources. Operating costs are usually higher, because groundwater supplies must be pumped to the surface. Any contamination is often hidden from view. Removing any contaminants is very difficult. Groundwater often possesses high mineral levels, and thus an increased level of hardness, because it is in contact longer with minerals. Near coastal areas, groundwater sources may be subject to salt water intrusion.

Geologic Activity of Groundwater

Groundwater contributes to geologic activity in a number of ways, including:

- **Dissolution**—Because water is the main agent of chemical weathering (nothing is safe from water), groundwater too is an active weathering agent that in part or in total (limestone) leaches ions from rock.
- **Chemical Cementation and Replacement**—Water not only carries ions away from rock but also brings chemical agents into masses of rock or rock structures. When some of the water-aided transported chemical enters rocks or rocky masses, many act as cement and bind sedimentary rocks together. In a similar water-aided transport manner water transports replacement molecules that work to fossilize organic substances or petrify wood.
- **Caves and Caverns**—When large areas of limestone underground are dissolved by the action of groundwater these cavities can become caves or caverns (caves with many interconnected chambers) once the water table is lowered. Once formed a cave is open to the atmosphere and water percolating it can precipitate new material such as the common cave decorations like stalactites (grow and hang from the ceiling), stalagmites (grow from the floor upward).
- **Sinkholes**—these are common in areas underlain by limestone, carbonate rock, salt beds, or rocks that can naturally be dissolved by groundwater circulating through them (USGS 2006).
- **Karst Topography**—In limestone terrains where dissolution is the main type of weathering, groundwater may work to form caves and sinkholes, and their collapse and coalescence may result in this highly irregular topography.

Did You Know?

Velocities of groundwater flow generally are low and are orders of magnitude less than velocities of streamflow. The movement of groundwater normally occurs as slow seepage through the pore spaces between particles of unconsolidated earth material or through networks of fractures and solution openings in consolidated rocks. A velocity of 1 ft/day or greater is a higher rate of movement for groundwater, and groundwater velocity can be as low as 1 ft/yr or 1 ft/decade. In contrast, velocities of streamflow generally are measured in feet per second. A velocity of 1 ft/sec equals about 16 miles per day.

References and Recommended Reading

Giller, P.S., and Jalmqvist, B. 1998. *The Biology of Streams and Rivers.* Oxford: Oxford University Press.

Gordon, N.D., McMahon, T.A., and Finlayson, B.L. 1992. *Stream Hydrology: An Introduction for Ecologists.* Chichester: Wiley.

Leopold, L.B. 1994. *A View of the River.* Cambridge, MA: Harvard University Press.

Leopold, L.B., Wolman, M.G., and Miller, J.P. 1964. *Fluvial Processes in Geomorphology.* San Francisco: W.H. Freeman and Co.

Likens, W.M. 1984. Beyond the Shoreline: A Watershed Ecosystem Approach. *Vert. Int. Ver. Theor. Aug Liminol.* 22: 1–22.

Morisawa, M. 1968. *Streams: Their Dynamics and Morphology.* New York: McGraw-Hill.

Mueller, J. 1968. An Introduction to the Hydraulic and Topographic Sinuosity Indexes. *Annals of the Association of American Geographers* 58: 371–385.

Richards, K. 1982. *Rivers: Form and Processes in Alluvial Channels.* London: Mehuen.

Spellman, F.R. 1996. *Stream Ecology and Self-Purification.* Lancaster, PA: Technomic Publishing Company.

Spellman, F.R., and Whiting, N. 1998. *Environmental Science & Technology.* Rockville, MD: Government Institutes.

USGS. 2006. *Sinkholes.* Accessed 7/06/08 @ http://ga.wwater.usgs.gov/edu/earthqwsink holes.htm.

12

Still Waters

Let yourself be open and life will be easier. A spoon of salt in a glass of water makes the water undrinkable. A spoon of salt in a lake is almost unnoticed.

—Buddha

I'm an old-fashioned guy. . . . I want to be an old man with a beer belly sitting on a porch, looking at a lake or something.

—Johnny Depp

STILL WATERS are ephemeral waterbodies that are one of the signs of the youth of our landscape. In areas to the south and west, where the glaciers departed less recently, the various depressions left by the ice have drained away or been eutrophied (filled in by) erosion from the surrounding uplands.

Lentic Habitat

When we look at a globe of the Earth we quickly perceive that ours is a water planet. Water covers most of Earth's surface (about three quarters of it). With all of Earth's water, the irony is that if all of Earth's 325 trillion gallons were squeezed into a gallon container and we poured off what was not drinkable (polluted, salty, or frozen) we would be left with one drop. Within this one drop a very small percentage of its whole represents all the freshwater contained in all of the Earth's lakes.

Lakes and ponds are found in many parts of the world and many of them are of great importance to human welfare. The total volume of freshwater contained in lakes in thousands of cubic kilometers is 91. More specifically, freshwater contained in lakes represents 0.007% of Earth's total water and 0.26% of Earth's total freshwater supply (Spellman 2008).

Lakes and ponds range in size from just a few square feet (ponds are generally 2 to 8 hectares) to thousands of square miles. Scattered throughout the earth, many of the first lakes evolved during the Pleistocene Ice Age. Lakes are found at all altitudes. Lake Titicaca (Peru and Chile) is 12,500 feet above sea level. At the other extreme, the Dead Sea in Israel and Jordan is almost 1300 feet below sea level. Many ponds are seasonal, just lasting a couple of months, such as sessile pools, while lakes last many years (none—even the Great Lakes—will last forever).

Lakes and ponds are divided into four different "zones" which are usually determined by depth and distance from the shoreline. The four distinct zones—littoral, limnetic, profundal, and benthic—are shown in Figure 29. Miller (1998) points out that each zone provides a variety of ecological **niches** for different species of plant and animal life.

The littoral zone is the top-most zone near the shores of the lake or pond with light penetration to the bottom. It provides an interface zone between the land and the open water of lakes. This zone contains rooted vegetation such as grasses, sedges, rushes, water lilies, and waterweeds, and a large variety of organisms. The littoral zone is further divided into concentric zones, with

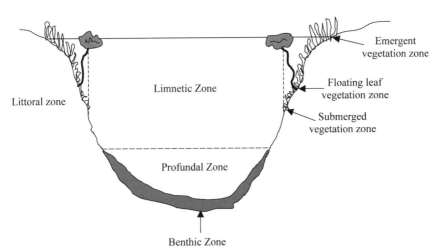

FIGURE 29.   Vertical section of a pond showing major zones of life

Source: Modified from Enger, Kormelink, Smith, and Smith, 1989, *Environmental Science: An Introduction.* Dubuque, Iowa: Wm. C. Brown Publications, p.77

one group replacing the other as the depth of water changes. Figure 29 also shows these concentric zones: emergent vegetation, floating leaf vegetation, and submerged vegetation zones, proceeding from shallow to deeper water.

The littoral zone is the warmest zone since it is the area that light hits, contains flora such as rooted and floating aquatic plants, and contains a very diverse community, which can include several species of algae (like diatoms), grazing snails, clams, insects, crustaceans, fishes, and amphibians. The aquatic plants aid in providing support by establishing excellent habitats for photosynthetic and **heterotrophic** (requires organic food from the environment) microflora as well as many zooplankton and larger **invertebrates** (Wetzel 1983).

From Figure 30 it can be seen that the limnetic zone is the open-water zone up to the depth of effective light penetration; that is, the open water away from the shore. The community in this zone is dominated by minute suspended organisms, the **plankton**, such as **phytoplankton** (plants) and zooplankton (animals), and some consumers such as insects and fish. Plankton are small organisms that can feed and reproduce on their own and serve as food for small chains.

Did You Know?

Without plankton in the water, there would not be any living organisms in the world, including humans.

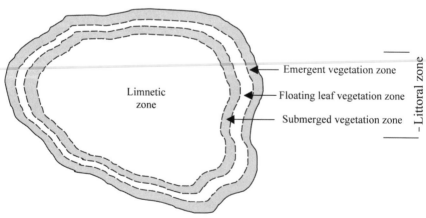

FIGURE 30.   View looking down on concentric zones that make up the littoral zone

In the limnetic zone, the population density of each species is quite low. The rate of photosynthesis is equal to the rate of respiration; thus, the limnetic zone is at compensation level. Small shallow ponds do not have this zone; they have a littoral zone only. When all lighted regions of the littoral and limnetic zones are discussed as one, the term euphotic is used for both, designating these zones as having sufficient light for photosynthesis and the growth of green plants to occur.

The small plankton do not live for a long time. When they die, they fall into the deep-water part of the lake/pond, the profundal zone. The profundal zone, because it is the bottom or deep water region, is not penetrated by light. This zone is primarily inhabited by heterotroph adapted to its cooler, darker water and lower oxygen levels.

The final zone, the benthic zone, is the bottom region of the lake. It supports scavengers and decomposers that live on sludge. The decomposers are mostly large numbers of bacteria, fungi, and worms, which live on dead animal and plant debris and wastes that find their way to the bottom.

Classification of Lakes

It is our natural tendency to classify things. In regards to the classification of lakes, Kevern et al. (1999) classify lakes in three ways. One classification is based on productivity of the lake (or its relative richness). This is the trophic basis of classification. A second classification is based on the times during the year that the water of a lake becomes mixed and the extent to which the water is mixed. And a third classification is based on the fish community of lakes.

In the following section we present a somewhat different classification scheme than the one just described. That is, lakes are classified based on eutrophication, special types of lakes, and impoundments.

Eutrophication is a natural aging process that results in organic material being produced in abundance due to a ready supply of nutrients accumulated over time. Through natural succession (i.e., the process by which biological communities replace each other in a relatively predictable sequence), eutrophication causes a lake ecosystem to turn into a bog and eventually to a terrestrial ecosystem. Eutrophication has received a great amount of publicity lately. In recent years, humans have accelerated the eutrophication of many surface waters by the addition of wastes containing nutrients. This accelerated process is called cultural eutrophication. Sources of human wastes and pollution are sewage, agricultural runoff, mining, industrial wastes, urban runoff, leaching from cleared land, and landfills.

Classification Based on Eutrophication

Lakes can be classified into three types based on their eutrophication stage.

1. **Oligotrophic Lakes** (few foods)—are young, deep, crystal-clear water, nutrient-poor lakes with little biomass productivity. Only a small quantity of organic matter grows in an oligotrophic lake; the phytoplankton, the zooplankton, the attached algae, the **macrophytes** (aquatic weeds), the bacteria, and the fish are all present as small populations. It's like planting corn in sandy soil, not much growth (Kevern et al. 1999). Lake Superior is an example from the Great Lakes.

2. **Mesotrophic Lakes**—It is hard to draw distinct lines between oligotrophic and eutrophic lakes, and often the term mesotrophic is used to describe a lake which falls somewhere between the two extremes. Mesotrophic lakes develop with the passage of time. Nutrients and sediments are added through runoffs, and the lake becomes more productive biologically. There is a great diversity of species with very low populations at first, but a shift towards higher and higher populations with fewer and fewer species. Sediments and solids contributed by runoffs and organisms make the lake shallower. At an advanced mesotrophic stage, a lake has undesirable odors and colors in certain parts. Turbidity increases and the bottom has organic deposits. Lake Ontario has reached this stage.

3. **Eutrophic Lakes** (good foods)—are lakes with a large or excessive supply of nutrients. As the nutrients continue to be added, large algal blooms occur, fish types change from sensitive to more pollution-tolerant ones, and biomass productivity becomes very high. Populations of a small number of species become very high. The lake takes on undesirable characteristics such as offensive odors, very high turbidity, and a blackish color. This high level of turbidity can be seen in studies of Lake Washington in Seattle, Washington. Laws (1993) reports that Secchi depth (measure of turbidity of water) measurements made in Lake Washington from 1950 to 1979 show an almost fourfold reduction in water clarity. Along with the reduction in turbidity, the lake becomes very shallow. Lake Erie is at this stage. Over a period of time, a lake eventually becomes filled with sediments as it evolves into a swamp and finally into a land area.

Special Types of Lakes

Odum (1971) refers to several special lake types:

1. **Dystrophic** (like bog lakes)—develop from the accumulation of organic matter from outside of the lake. In this case the **watershed** is often for-

ested and there is an input of organic acids (e.g., humic acids) from the breakdown of leaves and evergreen needles. There follows a rather complex series of events and processes resulting finally in a lake that is usually low in pH (acid) and often is moderately clear, but color ranges from yellow to brown. Dissolved solids, nitrogen, phosphorus, and calcium are low and humic matter is high. These lakes are sometimes void of fish fauna; other organisms are limited. When fish are present, production is usually poor. They are typified by the bog lakes of northern Michigan.

2. **Deep Ancient Lakes**—contain animals found nowhere else (endemic fauna), for example, Lake Baikal in Russia.

3. **Desert Salt Lakes**—are specialized environments like the Great Salt Lake, Utah, where evaporation rates exceed precipitation rates, resulting in salt accumulation.

4. **Volcanic Lakes**—are lakes on volcanic mountain peaks as in Japan and the Philippines.

5. **Chemically Stratified Lakes**—are stratified due to different densities of water caused by dissolved chemicals; examples include Big Soda Lake in Nevada. They are meromictic, which means partly mixed.

6. **Polar Lakes**—are lakes in the polar regions, their surface water temperature mostly below 4°C.

7. **Marl Lakes**—are different in that they generally are very unproductive (Kevern et al. 1999), yet they may have summer-time depletion of **dissolved oxygen** in the bottom waters and very shallow Secchi disk depths, particularly in the late spring and early summer. These lakes gain significant amounts of water from springs which enter at the bottom of the lake. When rainwater percolates through the surface soils of the drainage basin, the leaves, grass, and other organic materials incorporated in these soils are attacked by bacteria. These bacteria extract the oxygen dissolved in the percolating rainwater and add carbon dioxide. The resulting concentrations of carbon dioxide can get quite high and when they interact with the water, carbonic acid is formed.

Impoundments (Shut-Ins)

These are artificial lakes made by trapping water from rivers and watersheds. They vary in their characteristics according to the region and nature of drainage. They have high turbidity and a fluctuating water level. The biomass productivity, particularly of benthos, is generally lower than that of natural lakes (Odum 1971).

References and Recommended Reading

Kevern, N.R., King, D.L., and Ring, R. 1999. Lake Classification Systems, Part I, *The Michigan Riparian*, p. 1.

Laws, E.A. 1993. *Aquatic Pollution: An Introductory Text.* New York: Wiley & Sons, 59.

Miller, G.T. 1998. *Environmental Science: An Introduction.* Belmont, CA: Wadsworth Publishing Company, p. 77.

Odum, E.P. 1971. *Fundamentals of Ecology.* Philadelphia: Saunders College Publishing, p. 312–313.

Spellman, F.R. 2008. *The Science of Water*, 2nd ed. Boca Raton, FL: CRC Press.

USGS. 2008. *Earth's Water: Lakes and reservoirs.* Accessed 09/14/08 @ http://ga.water .usgs.gov/edu/earthlakes.html.

Wetzel, R.G. 1983. *Limnology.* New York: Harcourt Brace Jovanovich College Publishers.

V

EARTH'S LIFE FORMS

The true beauty of nature is her amplitude; she exists neither for nor because of us, and possesses a staying power that all our nuclear arsenals cannot threaten (much as we can easily destroy our puny selves).

—Stephen Jay Gould, *Bully for Brontosaurus*, 1991

13

Biosphere

We do not inherit the earth from our ancestors, we borrow it from our children.

—Navajo Proverb

Introduction

E ARLIER WE MENTIONED that Earth is composed of shells beginning with the inner core and outer core and progressing outward to the mantle and finally the crust. We can also say that there is an additional shell, the absolute outer shell, called the **biosphere**. This absolute outer shell includes land, surface rocks, water, air, and the atmosphere. This is the shell where life occurs, and which biotic processes in turn alter or transform. Moreover, it is the shell that consists of the global ecological system. Within the ecological system, all living beings and their relationship, including their interaction with the elements of the lithosphere, hydrosphere, and atmosphere are integrated.

Thanks to the life-giving qualities of air and water, earth is populated by countless species of plants (plants make up over nine-tenths of the total biomass) and animals. Estimates indicate that the entire Earth contains over 75 billion tons of biomass (life—of which 0.6% is accounted for by humans), which lives within various environments comprising the biosphere (U.S. Census Bureau 2006). Most of the planet's life is found from three meters below the ground to thirty meters above it and in the top 200 meters of the oceans and seas.

In regard to the life-forms that make up the biosphere, life as we know it is only known to exist on Earth. The origin of life is still a poorly understood process, but it is thought to have occurred about 3.9 to 3.5 billion years ago. Once life had appeared, the process of **evolution** by natural selection resulted in the development of ever-more diverse life forms. Species that were unable to adjust to the changing environment and **competition** from other life forms became extinct.

What Is Life?

Have you ever asked what life is? What does it mean to be alive? Have you ever tried to define life? If so, how did you define it? If these questions strike you as odd, consider them for a moment (they are almost as difficult as defining the origin of life). Of course we all have an intuitive sense of what life is, but, if you had difficulty, as is probably the case, with answering these questions, you are not alone. These questions are open to debate and have been from the beginning of time. One thing is certain; life is not a simple concept and it is difficult if not impossible to define.

Along with the impossibility of defining life definitively, it is not always an easy thing to tell the difference between living, dead, and non-living things. Prior to the seventeenth century, many people believed that nonliving things could spontaneously turn into living things. For example, it was believed that piles of straw could turn into mice. Obviously, that is not the case. There are some very general rules to follow when trying to decide if something is living, dead, or non-living. Scientists have identified seven basic characteristics of life. Keep in mind that for something to be described as living, that something must display *all* seven of these characteristics (i.e., "characteristic" is plural). Although many of us have many different opinions about what "living" means, the following characteristics were designated "characteristics of living things" with the consensus of the scientific community.

- **Living things are composed of cells.** Living things exhibit a high level of organization, with multicellular organisms being subdivided into cells, and cells into organelles, and organelles into molecules, etc.
- **Living things reproduce.** All living organisms reproduce, either by sexual or asexual means.
- **Living things respond to stimuli.** All living things respond to stimuli in their environment.
- **Living things maintain homeostasis.** All living things maintain a state of internal balance in terms of temperature, pH, water concentrations, etc.

- **Living things require energy.** Some view life as a struggle to acquire energy (from sunlight, inorganic chemicals, or another organism), and release it in the process of forming adenosine triphosphate (ATP). The conventional view is that living organisms require energy, usually in the form of ATP. They use this energy to carry out energy-requiring activities such as metabolism and locomotion.
- **Living things display heredity.** Living organisms inherit traits from the parent organisms that created them.
- **Living things evolve and adapt.** All organisms have the ability to adapt or adjust to their surroundings. An example of this is they might be adapting to environmental change resulting in an increased ability to reproduce.

Again, if something follows one or just a few of the characteristics listed above, it does not necessarily mean that it is living. To be considered alive, an object must exhibit *all* of the characteristics of living things. A good example of a nonliving object that displays at least one characteristic for living is sugar crystals growing on the bottom of a syrup dispenser. On the other hand, there is a stark exception to the characteristics above. For example, mules cannot reproduce because they are sterile. Another nonliving object that exhibits many of the characteristics of life is a flame. Think about it, a flame:

- respires
- requires nutrition
- reproduces
- excretes
- grows
- moves
- is irritable
- is organized

We all know that a flame is not alive, but how do we prove that to the skeptic? The best argument we can make is:

1. Non-living materials never replicate using **DNA** and RNA (hereditable materials).
2. Non-living material cannot carry out **anabolic** metabolism.

Life: An After-Thought

The reader, based on the previous section, should have little difficulty in understanding our dilemma in attempting to describe that which is indescribable.

In our nebulous attempts and amphorous uses of symphonic prose to describe what life is, we often forget that life is nothing more than living. Thus, the analysis shifts from trying to define life to trying to define living. Maybe the title of the previous section should not have been "What is Life?" but, instead, "What is Living?"

Living, what is it? The simple answer: living is a state of mind. The complex answer: What? Maybe a few examples will help us clear the air:

Life is the chirp of a bird.
Life is an ill child who is cured by accumulated knowledge and skill.
Life is the indescribable beauty of a flower.
Life is a high alpine meadow filled with wildflowers in full bloom.
Life is that look-around at nature that generates a smile
broader than an ocean.
Life is living—the taking in of it all.

Did You Know?

All four divisions of Earth can be and often are present in a single location. For example, a piece of soil will of course have mineral material from the lithosphere. Additionally, there will be elements of the hydrosphere present as moisture within the soil, the biosphere as insects and plants, and even the atmosphere as pockets of air between soil pieces.

Levels of Organization

All living things are organized into several basic levels of organization. The levels of organization from the smallest to the largest are:

- **Atoms**—are basic units of matter having a specific chemical property.
- **Molecules**—are the smallest multi-atom-containing unit having a specific chemical property.
- **Macromolecules**—occur when two or more molecules bond to form large molecules.
- **Organelles**—include many subcellular structures, usually found only in **eukaryotes**.
- **Cell**—is the basic unit of life.
- **Tissue**—is a group of similar cells acting as a structural or functional unit.

- **Organ**—is a group of dissimilar tissues that act together and have a specific function.
- **Organ System**—is a group of dissimilar organs that together have a specific function.
- **Organism**—contains cells and is capable of growing and replicating itself.

In addition to the levels of organization for living things, there are also levels of organization in the biosphere. From the smallest to largest group these are:

- **Species**—a group of organisms that resemble one another closely
- **Population**—all organisms of the same species in a specific space and time
- **Community**—all populations in a specific space and time
- **Ecosystem**—community plus abiotic environment in a specific space and time
- **Biome**—a major regional or global biotic community, such as a grassland or desert, characterized chiefly by the dominant forms of plant life and the prevailing climate.
- **Biosphere**—is part of a planet's outer shell—including air, land, surface rocks, and water—within which life occurs, and which biotic processes in turn alter or transform.

The Scientific Method

In order to conduct science, one must know the rules of the game. USEPA (2006) points out that the word "science" is derived from a Latin verb meaning "to know." Scientists use the *scientific method* to construct an accurate representation of the world through the testing of scientific theories. These theories enhance our understanding and knowledge (to know) of the world. A theory, such as Einstein's theory of relativity or Darwin's theory of natural selection, accounts for many facts and attempts to explain a great variety of phenomena. Such a unifying theory does not become widely accepted in science unless its prediction can withstand thorough and continuous testing by experiments and observations (Campbell 2004).

It is important to point out that the scientific method is to be used as a guide that can be modified. In some sciences, such as geology, and taxonomy, lab experiments are not necessarily performed. Instead, after formulating a hypothesis, additional observations and/or collections are made from different localities.

Sir Karl Raimund Popper (1902–1994) developed the theory of the scientific method in 1934 in his book *The Logic of Scientific Discovery*. Many scientists,

including biologists, chemists, ecologists, physicists, and environmental practitioners, use the scientific method to test new theories.

While there are several forms of the scientific method, the basic steps involved are:

- Initial observations and objectives
- Hypothesis formulation
- Data collection—(e.g., record observations)
- Analysis of the data (e.g., perform calculations) to test the hypothesis
- Summarization of results (create tables, graphics, etc.)
- Discussion of limitations and conclusions
- Identification of future research needs

The University of Maryland School of Medicine (Department of Pathol-
ogy) suggests an ideal example of the scientific method and its importance. A
researcher may observe that a large number of insects all have three pairs of
legs, including flies, beetles, grasshoppers, and wasps. A conclusion may be
drawn that all insects have three pairs of legs. Then after evaluating additional
insects including cockroaches, crickets, moths, and bees, a hypothesis might
be formed that all insects have three pairs of legs. However, a good scientist
would not stop there. New hypotheses should be formed to further test the
initial observation. Perhaps immature moths should be considered, as they
are insects too, so they should have three pairs of legs. However, findings
would determine that caterpillars or immature moths do not have any legs.
Then the generalization becomes reformulated into all adult insects have
three pairs of legs.

How Is the Scientific Method Used?

1. **Initial Observations and Objectives**—Scientists are usually curious
   about their surroundings and may notice something and want to understand more about it. This type of curiosity may lead them to observe
   their surroundings more carefully. As a result, when they document
   their observations, questions may arise which are then formulated into
   hypotheses.
2. **Hypothesis Formulation**—As observations are made, questions are
   formulated and scientists try to answer these questions, which leads to
   guesses or hypotheses. If observations do not support these statements,
   the hypothesis is rejected. A hypothesis must be stated in a way that can
   be tested by the scientific method. Reviewing similar studies performed
   by others can also be helpful in this step.

3. **Data Collection**—In order to test the hypotheses, data must be collected. Scientists must design a data collection plan which considers: what to collect? when to collect? and how many samples are necessary? Developing sample surveys involves determining appropriate sample size, monitoring frequency, and the need for repetition. Through the use of variables and controls, results can be determined and documented. Variables are those factors being tested in an experiment, which are usually compared to a control. A control is a known measure to which scientists can compare their results.

The scientist must also determine which equipment, supplies, or materials are necessary to complete the study. Prior to conducting an experiment it is important to document data collection methods. This step will ensure the quality of the experiment if someone else should reproduce it. Most scientific papers published in journals include a methods section documenting the way in which the experiment was performed. Attention must be paid to make sure that data collection methods are kept unbiased. Data can be represented by many formats. For example,

The Environmental Monitoring and Assessment Program Near Coastal (EMAP-NC) Program Plan for 1990: Estuaries (EPA/600/4-90/033) provides an example of applying the scientific method. Initially, observations were made regarding estuarine and coastal ecological conditions leading to the identification of the problem—humans. From this, a solution to the problem was proposed—the need for better environmental surveillance to assess the status and trends of the nation's ecological problems. To achieve this, a sampling design was developed. Since the problem identified estuarine and coastal wetlands, estuaries, coastal waters, and the Great Lakes as areas of potential concern, the scope of the project needed to be determined. Due to budget and time limitations, the EMAP program chose to focus its initial efforts on the state of estuaries along the nation's coastline (e.g., where to collect).

The sampling design for the EMAP-NC had three major elements including a regionalization scheme, a classification scheme, and a statistical design.

Specifically, the statistical design identified what parameters were to be sampled, when sampling should occur, and how many samples should be taken. Parameters to be sampled were identified including, but not limited to, benthic and fish species composition and biomass, habitat indicators including salinity, temperature, pH, dissolved oxygen, sediment characteristics, and water depth. Station locations were chosen using a randomly placed systematic grid and sampling times were identified to fall within an index period, which was chosen to be the summer.

4. **Analysis of the Data**—This step is necessary to prove or disprove a hypothesis by experimentation. The methods involved in testing/analyzing

the data are also important since an experiment should be repeated by others to ensure the quality of results. For instance, if two people on different sides of the country decide to perform the same experiment, they should end up with the same results. Statistics are then used to analyze the data. Descriptive statistics are a means of summarizing observational data through the calculation of a mean, mode, average, standard deviation, variance, etc. More advanced comparisons can also be completed. This step is very important as it transforms raw data into information, which can be used to report results in a user-friendly format. To further analyze the data, parametric or non-parametric statistical tests can be performed. Parametric tests assume that data are normally distributed. Generally, to ensure that a parametric test is appropriate, tests for normality and variance need to be completed to identify any necessary transformations (e.g., taking the log of the values) or if a non-parametric test is warranted. Non-parametric tests make fewer assumptions about the distributions of the data.

5. **Summarization of Results**—The presentation of the results is very important. Often scientists will rely heavily on graphics, tables, flow charts, maps, and diagrams to facilitate the interpretation of the results. Graphics can be used to model future predictions. Graphics (e.g., scatter plots) can also assist with the identification of relationships (e.g., correlations) between environmental parameters. If two variables are correlated, when one changes the other will do so in a related manner. This relationship can be either positive or negative. Often when a correlation is found, it is assumed that there is a "cause and effect" relationship between the variables. Although there should be some logical basis for relating variables, cause is not demonstrated with a statistical technique. When we say two variables are correlated, we can say that they are associated in some way. A written discussion documenting identifiable trends or correlations generally accompanies these graphics. This step is just the presentation of results; it does not include any interpretation.

6. **Discussion of Limitation and Conclusions**—This is the section where the hypothesis is accepted or rejected. Many scientists no longer try to define cause and effect parameters, but instead identify relationships between the data. In this manner, ideas can be formed about why certain results were found while identifying previous studies that may have had similar or contradicting results. It is important to reference all studies so that other scientists can refer to them if necessary.

7. **Identification of Future Research Needs**—This may include areas of related interest that should be studied to better understand the subject. This section may give information about limitations of the study, such as what items should be modified to try to reach the intended goal.

✓ **Important Point:** In addition to the 7 steps involved, there are three important factors to consider. First, throughout the entire process of the scientific method, quality assurance/quality control (QA/QC) is extremely important. The goal of QA/QC is to ensure that environmental data are of sufficient quantity and quality to support the intended use of the data. Second, methods and observations, statistical procedures, and results should be documented and reviewed to ensure a quality study. This step is vital to the third factor—the repeatability of the process. If two exact experiments are conducted, they should have the same findings. Additional testing ensures the quality of the results. Even if the experiment does not prove the hypothesis, it is still important. Having the knowledge that an experiment did not work is a step toward finding the answer.

CASE STUDY
Scientific Method: *Pfiesteria* and Fish Health

In a 2006 presentation, David Goshorn described *Pfiesteria* as "a very small, single-celled organism without flagella. It's got an extremely complex life cycle. Most of the time populations are benign, feeding on algae and bacteria, but some populations, not all, are capable of producing a toxin which can cause fish health problems." They can apparently also cause human health problems. Goshorn noted that, in 1997, "Maryland experienced four separate toxic outbreaks on three different Eastern Shore rivers. North Carolina has had its problems for quite some time."

Sea Grant Maryland (SGM 2006), in their *Fish Disease Information: The Scientific Method, Fish Health and Pfiesteria* reported (in regards to fish health in the Chesapeake Bay) that their "understanding of the situation is that fish have had lesions and that fish have died. *Pfiesteria piscicida* and *Pfiesteria*-like organisms have been cultured from water samples taken in the vicinity where fish with lesions have been observed or where fish kills have occurred. It was reported that *Pfiesteria*-like organisms have caused lesions and mortalities in laboratory exposures."

Well, sounds like the jury is back; a simple matter of cause and effect. *Pfiesteria* were present, fish had lesions, and fish died. Moreover, *Pfiesteria*-like organisms have caused lesions and mortalities in lab exposures. That seals the deal; a slam duck verdict. Right? Not so fast my environmental practitioner friends. We know better (hopefully) than to jump the gun, don't we? As SGM points out "as good scientists know, it's not that easy."

Notwithstanding that many of us are scientists or want-to-be scientists or those who just want to know more about science, there are some who might ask: "Why, in this case, is it not easy to conclude that fish lesions are caused by *Pfiesteria* or *Pfiesteria*-like organisms?" SGM points out that the "very same water which was cultured for *Pfiesteria* could have also grown bacteria." Therefore, in order to properly study the cause of the fish lesions and the possibility that *Pfiesteria* are the culprits, we need to follow Koch's postulates.

Robert Koch (1882, 1884, 1893), a German physician and bacteriologist, in the course of his studies of anthrax and tuberculosis, formulated four postulates (i.e., four criterions for judging whether a given bacteria is the cause of a given disease) in 1884 and refined and published them in 1890. To say that Koch's criteria brought some muchneeded scientific clarity to what was then a very confused field is an understatement. Koch's postulates are still used today.

Koch's postulates are as follows:

1. The organism must be found in all animals suffering from the disease, but not in healthy animals.
2. The organism must be isolated from a diseased animal and grown in pure culture.
3. The cultured organism should cause disease when introduced into a healthy animal.
4. The organism must be reisolated from the experimentally infected animal.

It should now be clear why it is difficult to establish a cause and effect relationship between fish lesions and *Pfiesteria*. SGM (2006) points out "*Pfiesteria* (or its toxin) has not yet been isolated from fish or fish lesions." The record shows that fish with lesions have been taken from waters where *Pfiesteria*-like organisms have not been identified (of course that does not mean that they were not present). Moreover, based on tests it has been shown that lesions on fish from lab exposures are not identical to those seen in some natural habitat-collected specimens. This and especially the non-specificity of the lesions means that the lesions could have been caused by a number of different causative agents (e.g., bacteria, virus, etc.). SGM reports that they were unable to isolate the infectious organism from the host, culture it, and reinfect another host to be able to observe if the same lesions occur.

So, does the above account, using Koch's postulates, prove or disprove the association of *Pfiesteria* and fish lesions? At the present time
we cannot definitively say for sure one way or other on the *Pfiesteria*-
fish lesion connection. Instead, we are at the hypothesis stage and
testing is on-going, a work in progress. The point is that eventually,
after the hypothesis is tested, challenged, revised, retested, rechal-
langed, and re-revised, a theory is developed. This theory still remains
subject to debate by all who question it; they may counter it with new
evidence. Ultimately, the theory may be accepted as fact.

✓ **Important Point:** It is important to point out that Koch's postulates have
their limitations and so may not always be the last word (Med Net 2006).
They may not hold if:

- The particular bacteria (such as the one that causes leprosy) cannot
  be "grown in pre culture" in the lab.
- There is no animal model of infection with that particular bacterium.

Source: Adapted from a Sea Grant Maryland (SGM)—funding provided by the
National Oceanic and Atmospheric Administration (NOAA)), work titled: *The Sci-
entific Method, Fish Health and Pfiesteria* (related to fish health in the Chesapeake
Bay), with the writings of William T. Keeton (1996) and a presentation by David
Goshorn, Ph.D. (2006).

Why Is the Scientific Method Important?

The scientific method has played an instrumental part in scientific research
for almost 500 years. From Galileo's experiment back in the 1590s to current
scientific research, the scientific method has contributed to the creation of vac-
cines and advancements in medicine and technology. Through the use of the
scientific method, scientific theories can be tested. A scientific theory is a logical
explanation of observed events. Once a scientific method has been tested and
widely accepted as true it becomes scientific law (e.g., Newton's law of gravity).
However, scientific methods must be continually examined for possible errors.
Openness to new innovative ideas and an organized approach of skepticism are
necessary to protect against collective bias in scientific results. The manner in
which the hypothesis is formed, as well as the way data is collected, analyzed,
and interpreted all need to be monitored for the potential introduction of bias.

The scientific method also ensures the quality of data for public use. This can be completed through submitting work for a peer review. A peer review is a critical review by technical experts without a vested interest in the particular investigation. Peer reviews confirm that the research has been conducted in a scientifically sound manner. For example, peer reviews are performed on all scientific articles being submitted to journals to check the quality of each study before it is released to the public. The USEPA always peer reviews their studies and data before they are released into the public domain.

References and Recommended Reading

American Heritage Dictionary of the English Language, 4th ed. 2000. Boston: Houghton Mifflin Company.

Campbell, N.A. 2004. *Biology: Concepts & Connections*, 4th CD-Rom ed. Benjamin-Cummings Publishing Company.

Goshorn, D. 2006. Proceedings—DELMARVA Coastal Bays Conference III: Tri-State Approaches to Preserving Aquatic Resources. USEPA.

Huxley, T.H. 1876. *Science & Education, Volume III, Collected Essays*. New York: D. Appleton & Company.

Jones, A. 1997. *Environmental Biology*. New York: Routledge.

Keeton, W.T. 1996. *Biological Science*. New York: R.S. Means Company.

King, R.M. 2003. *Biology Made Simple*. New York: Broadway Books.

Koch, R. 1882. *Uber die Atiologie der Tuberkulose*. In: *"Verhandlungen des Knogresses fur Innere Medizin*. Erster Kongress, Wiesbaden.

Koch R. 1893. *J. Hyg. Inf.* 14: 319–333.

Koch, R. 1884. *Mitt Kaiser Gesundh* 2: 1–88.

Larsson, K.A. 1993. Prediction of the pollen season with a cumulated activity method. *Grana* 32: 111–114.

Med Net. 2006. *Definition of Koch's Postulates*. Medicine Net.com.

SGM. 2006. *The Scientific Method, Fish Health and Pfiesteria*. University of Maryland: NOAA.

Spellman, F.R. 2009. *Biology for Non-Biologists*. Lanham, MD: Government Institutes.

Spellman, F.R., and Price-Bayer, J. 2011. *In Defense of Science: Why Scientific Literacy Matters*. Lanham, MD: Government Institutes.

Spellman, F.R., and Whiting, N.E. 2006. *Environmental Science and Technology*, 2nd ed. Rockville, MD: Government Institutes.

Spieksma, F.T. 1991. Aerobiology in the Nineties: Aerobiology and pollinosis. *International Aerobiology Newsletter* 34: 1–5.

US Census Bureau. 2006. World Population Information. Accessed 11/01/10 @ http://www.census.gov/ipc/www/world.html.

USEPA. 2006. What is the scientific method? Accessed 11/01/10 @ http://www.epa.gov/maia/html/scientific.html.

14

Ecosystems

Man has generally been preoccupied with obtaining as much "production" from the landscape as possible, by developing and maintaining early successional types of ecosystems, usually monocultures. But, of course, man does not live by food and fiber alone; he also needs a balanced CO_2-O_2 atmosphere, the climatic buffer provided by oceans and masses of vegetation, and clean (that is, unproductive) water for cultural and industrial uses. Many essential life-cycle resources, not to mention recreational and esthetic needs, are best provided man by the less 'productive' landscapes. In other words, the landscape is not just a supply depot but is also the *Oikos*—the home—in which we must live.

—Eugene P. Odum, 1969

Introduction

ON ONE of their late August holiday outings, a family of eighteen picnickers from a couple of small rural towns visited a local stream that coursed its way alongside and/or through one of the towns. This annual outing was looked upon with great anticipation for it was that one time each year when aunts, uncles, and cousins came together as one big family. The streamside setting was perfect for such an outing, but historically, until quite recently, the stream had been posted "DANGER—NO SWIMMING— CAMPING or FISHING!"

Because the picnic area was such a popular location for picnickers, swimmers, and fishermen, over the years several complaints about the polluted stream were filed with the County Health Department. The Health Department finally took action to restore the stream to a relatively clean condition: sanitation workers removed debris and old tires, and plugged or diverted end-of-pipe industrial outfalls upstream of the picnic area. After two years of continuous worker-aided stream clean-up and the stream's natural self-purification process, the stream was given a clean bill of health by the Health Department. The danger postings were removed.

Once the stream had been declared clean, fit for use by swimmer and fisher, with postings removed, it did not take long for the word to get out. Local folks and others made certain, at first opportunity, that they flocked to the restored picnic and swimming and fishing site alongside the stream.

During most visits to the restored picnic-stream area, visitors, campers, fishermen, and others were pleased with their cleaned-up surroundings. However, during late summer, when the family of eighteen and several others visited the restored picnic-stream area they found themselves swarmed over by thousands of speedy dragonflies and damselflies, especially near the bank of the stream. Soon they found the insects too much to deal with so they stayed clear of the stream. To themselves and to anyone that would listen the same complaint was heard over and over again: "What happened to our nice clean stream? With all those nasty bugs the stream is polluted again." So, when August arrived with its hordes of dragonfly-type insects, the picnickers, campers, swimmers, and fishermen avoided the place until the insects departed, until the human visitors thought the stream was clean again.

However, there is one local family that does not avoid the stream-picnic area in August; on the contrary, August is one of their favorite times to visit, camp, swim, take in nature, and fish—they usually have most of the site to themselves. The family is led by a local university professor of ecology, and she knows the truth about the picnic-stream area and the dragonflies and other insects. She knows that dragonflies and damselflies are **macroinvertebrate** indicator organisms; they only inhabit, grow, and thrive in and around streams that are clean and healthy—when dragonflies and damselflies are around they indicate non-polluted water. Further, the ecology professor knows that dragonflies are valued as predators, friends and allies in waging war against flies and controlling populations of harmful insects, such as mosquitoes. In regard to mosquitoes, dragonflies take the wrigglers in the water, and the adults, on swiftest wings (25–35 mph) that are hovering over streams and ponds laying their eggs.

The ecology professor's husband, an amateur poet, also understood the significance of the presence of the indicator insects and had no problem sharing the same area with them. He also viewed the winged insects differently, with the eye of a poet. He knew that the poets have been lavish in their attention to the dragonflies and have paid them delightful tributes. James Witcomb Riley (1849–1916) says:

> Till the dragon fly, in light gauzy armor
>   burnished bright,
> Came tilting down the waters in a wild,
>   bewildered flight.

The early great poets were not finished with the dragonfly. For example, Alfred, Lord Tennyson drew inspiration for one of his most beautiful poems from the two stages of dragonfly life. But perhaps James Russell Lowell's (1819–1891) poem, *The Fountain of Youth*, gives us the perfect description of these insects:

> In summer-noon flushes
> When all the wood hushes,
> Blue dragon-flies knitting
> To and fro in the sun,
> With sidelong jerk flitting,
> Sink down on the rushes.
> And, motionless sitting,
> Hear it bubble and run,
> Hear its low inward singing
> With level wings swinging
> On green tasseled rushes,
> To dream in the sun.

Probably the best way to understand ecology—to get a really good "feel" for it—or to get to the heart of what ecology is all about is to read the following by Rachel Carson (1962):

"We poison the caddis flies in a stream and the salmon runs dwindle and die. We poison the gnats in a lake and the poison travels from link to link of the food chain and soon the birds of the lake margins become victims. We spray our elms and the following springs are silent of robin song, not because we sprayed the robins directly but because the poison traveled, step by step, through the now familiar elm leaf-earthworm-robin cycle. These are matters of record, observable, part of the visible world around us. They reflect the web of life—or death— that scientists know as ecology."

As Carson points out, what we do to any part of our environment has an impact upon other parts. In other words, there is an interrelationship between the parts that make up our environment. Probably the best way to state this interrelationship is to define ecology definitively—that is, to define it as it is used in this text: "Ecology is the science that deals with the specific interactions that exist between organisms and their living and nonliving environment" (Tomera 1989).

Did You Know?

No ecosystem can be studied in isolation. If we were to describe ourselves, our histories, and what made us the way we are, we could not leave the world around us out of our description! So it is with streams: they are directly tied in with the world around them. They take their chemistry from the rocks and dirt beneath them as well as for a great distance around them (Spellman 1996).

Ecosystems

As mentioned, **ecosystem**, a contraction of "ecological" and "system," is a term introduced by Tansley to denote an area that includes all organisms therein and their physical environment. Specifically, an ecosystem is defined as the geographic area including all the living organisms, their physical surroundings, and the natural cycles that sustain them. All of these elements are interconnected (USFWS 2007). Simply, the ecosystem is the major ecological unit in nature. Elements of an ecosystem may include flora, fauna, lower life forms, water, and soil (Ecosystem 2007). "There is a constant interchange of the most various kinds within each system, not only between the organisms but between the organic and the inorganic" (Tansley 1935). Living organisms and their nonliving environment are inseparably interrelated and interact upon each other to create a self-regulating and self-maintaining system. To create a self-regulating and self-maintaining system, ecosystems are homeostatic, i.e., they resist any change through natural controls. These natural controls are important in ecology. This is especially the case since it is people through their complex activities who tend to disrupt natural controls.

Tansley regarded the ecosystem as not only the organism complex, but also the whole complex of physical factors forming what we call the environment. It was first applied to levels of biological organization represented by units such as community and the biome. Odum (1952) and Evans (1956) expanded the extent of the concept to include other levels of organization (USDA 1982).

As stated earlier, an ecosystem encompasses both the living and nonliving factors in a particular environment. The living or biotic part of the ecosystem is formed by two components: **autotrophic** and heterotrophic. The autotrophic (self-nourishing) component does not require food from its environment but can manufacture food from inorganic substances. For example, some autotrophic components (plants) manufacture needed energy through photosynthesis. Heterotrophic components, on the other hand, depend upon autotrophic components for food (Porteous 1992).

Did You Know?

Modern usage of the term ecosystem derives from the work done by Raymond Lindeman. Lindeman's central concepts were that of functional organization and ecological energy efficiency ratios. This approach is connected to ecological energetics and might also be thought of as environmental rationalism. It was subsequently applied by H.T. Odum in founding the transdiscipline known as systems ecology (Lindeman 1942).

The non-living or abiotic part of the ecosystem is formed by three components: inorganic substances, organic **compounds** (link biotic and abiotic parts), and climate regime. Figure 31 shows a simplified diagram depicting a few of the living and nonliving components of an ecosystem found in a freshwater pond.

An ecosystem is a cyclic mechanism in which biotic and abiotic materials are constantly exchanged through biogeochemical cycles. **Biogeochemical cycles** are defined as follows: *bio* refers to living organisms and *geo* to water, air, rocks, or solids. *Chemical* is concerned with the chemical composition of the Earth. Biogeochemical cycles are driven by energy, directly or indirectly from the sun. They will be discussed later.

The simplified freshwater pond shown in Figure 31 depicts an ecosystem where biotic and abiotic materials are constantly exchanged. Producers construct organic substances through photosynthesis and chemosynthesis. Consumers and decomposers use organic matter as their food and convert it into abiotic components. That is, they dissipate energy fixed by producers through food chains. The abiotic part of the pond in Figure 31 is formed of inorganic and organic compounds dissolved and in sediments such as carbon, oxygen, nitrogen, sulfur, calcium, hydrogen, and humic acids. The biotic part is represented by producers such as rooted plants and phytoplanktons. Fish, crustaceans, and insect larvae make up the consumers. Detrivores, which feed on organic detritus, are represented by mayfly nymphs. Decomposers make up

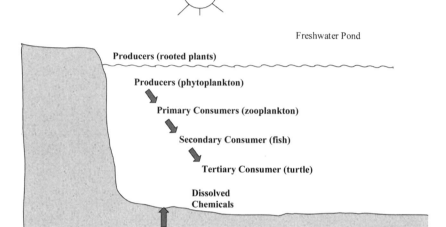

FIGURE 31.   Major components of a freshwater pond ecosystem

the final abiotic part. They include aquatic bacteria and fungi, which are distributed throughout the pond.

As stated earlier, an ecosystem is a cyclic mechanism. From a functional viewpoint, an ecosystem can be analyzed in terms of several factors. The factors important in this study include: biogeochemical cycles, energy, and food chains; these factors are discussed in detail later.

Types of Ecosystems

Individual ecosystems consist of physical, chemical, and biological components. As mentioned, the physical and chemical components are known as abiotic (not living environment) factors that influence living organisms in both terrestrial and aquatic ecosystems. The abiotic factors are:

For Terrestrial Ecosystems
Sunlight
Temperature
Precipitation
Wind
Latitude
Altitude
Fire frequency
Soil

For Aquatic Ecosystems
Light penetration
Water currents
Dissolved nutrient concentrations
Suspended solids
Salinity

The **biotic** (living environment) **factors** making up an ecosystem include the producers, consumers, and decomposers.

Abiotic and biotic factors combine to make up the following types of terrestrial and aquatic ecosystems:

- Estuaries
- Swamps and marshes
- Tropical rain forest
- Temperate forest
- Northern coniferous forest (taiga)
- Savanna
- Agricultural land
- Woodland and shrubland
- Temperate grassland
- Lake and streams
- Continental shelf
- Open ocean
- Tundra (arctic and alpine)
- Desert scrub
- Extreme desert

Biogeochemical Cycles

Some time ago I heard of an old man down on a hill farm in the South, who sat on his front porch as a newcomer to the neighborhood passed by. The newcomer to make talk said, "Mister, how does the land lie around here?" The old man replied, "Well, I don't know about the land alying; it's these real estate people that do the lying."

In a very real sense the land does not lie; it bears a record of what men write on it. In a larger sense a nation writes its record on the land, and civilization writes its record on the land—a record that is easy to read by those who understand the simple language of the land.

—W.C. Lowdermilk, 1938

Nutrient Cycles

All matter cycles; it is neither created nor destroyed. This is true whether the matter is an element, such as carbon, nitrogen, oxygen, or a molecule, such as water. Because the Earth is essentially a closed system with respect to matter, it can be said that all matter on Earth cycles. Ecologists study the flow of nutrients in ecosystems.

Ecosystem elements, like streams—and their larger cousins, rivers—are complex ecosystems that take part in the physical and chemical cycles (biogeochemical cycles) that shape our planet and allow life to exist. A **biogeochemical cycle** is composed of bioelements (chemical elements that cycle through living organisms), and occurs when there is an interaction between the biological and physical exchanges of bioelements. In a biogeochemical cycle, nutrient cycling and recycling through ecosystems results from the actions of geology, meteorology, and living things. Various nutrient biogeochemical cycles include (Spellman 1996):

- water cycle
- carbon cycle
- oxygen cycle
- nitrogen cycle
- **phosphorus cycle**
- sulfur cycle

Did You Know?

Contrary to an incorrect assumption, energy does not cycle through an ecosystem, chemicals do. The inorganic nutrients cycle through more than the organisms, however, they also enter into the oceans, atmosphere, and even rocks. Because these chemicals cycle through both the biological and the geological world, we call the overall cycles biogeochemical cycles.

Each chemical has its own unique cycle, but all of the cycles do have some things in common. Reservoirs are those parts of the cycle where the chemical is held in large quantities for long periods of time (e.g., the oceans for water and rocks for phosphorous). In exchange pools, on the other hand, the chemical is held for only a short time (e.g., the atmosphere; a cloud). The length of time a chemical is held in an exchange pool or a reservoir is termed its residence time. The biotic community includes all living organisms. This community may serve as an exchange pool (although for some chemicals like carbon, bound in certain tree species for a thousand years, it may seem more like a reservoir), and also serve to move chemicals (bioelements) from one stage of the cycle to another. For instance, the trees of the tropical rain forest bring water up from the forest floor to be transpired into the atmosphere. Likewise, coral organisms take carbon from the water and turn it into limestone rock. The energy for most of the transportation of chemicals from one place to another is provided either by the sun or by the heat released from the mantle and core of the Earth (Spellman 1996).

Did You Know?

Water is exchanged between the hydrosphere, lithosphere, atmosphere, and biosphere. The oceans are large reservoirs that store water; they ensure thermal and climatic stability.

In addition to these exchanges of nutrients (losses and gains), Abedon (1997) points out examples of other losses and gains:

- Minerals can be lost from ecosystems by the action of rain.
- Nutrients can also be carried into ecosystems by the action of wind or migrating animals.
- Movement of salmon up rivers is an example of how nutrients might be delivered into an upstream ecosystem (e.g., from the oceans back to terrestrial forests).
- A consequence of ecosystem disruption is an impaired ability to recycle nutrients which leads to nutrient loss and long term ecosystem impoverishment.
- In general, a disturbed habitat probably loses (rather than recycles) nutrients to a much greater degree than an undisturbed habitat where the action of human activities is not necessarily rapidly or readily reversible. A common consequence of human disturbance of ecosystems and the associated irreversible loss of nutrients is desertification.

In the case of chemical elements that cycle through living things; i.e., bioelements (ISWS 2005):

- All bioelements reside in compartments or defined spaces in nature.
- A compartment contains a certain quantity, or pool, of bioelements.
- Compartments exchange bioelements. The rate of movement of bioelements between two compartments is called the flux rate.
- The average length of time a bioelement remains in a compartment is called the mean residence time (MRT).
- The flux rate and pools of bioelements together define the nutrient cycle in an ecosystem.
- Ecosystems are not isolated from one another, and bioelements come into an ecosystem through meteorological, geological, or biological transport

mechanisms—meteorological (e.g., deposition in rain and snow, atmospheric gases)

* ecological (e.g., surface and subsurface drainage)
* biological (e.g., movement of organisms between ecosystems)

As a result, biogeochemical cycles can be:

* local

Smith (1974) categorizes biogeochemical cycles into two types, the gaseous and the sedimentary. Gaseous cycles include the carbon and nitrogen cycles. The main pool (or sink) of nutrients in the gaseous cycle is the atmosphere and the ocean. The sedimentary cycles include the sulfur and phosphorous cycles. The main sink for sedimentary cycles is soil and rocks of the Earth's crust.

Between twenty to forty elements of the Earth's ninety-two naturally occurring elements are ingredients that make up living organisms. The chemical elements carbon, hydrogen, oxygen, nitrogen, and phosphorus are critical in maintaining life as we know it on Earth. Odum (1971) points out that of the elements needed by living organisms to survive, oxygen, hydrogen, carbon, and nitrogen are needed in larger quantities than are some of the other elements. The point is—no matter what particular elements are needed to sustain life, these elements exhibit definite biogeochemical cycles. These biogeochemical cycles will be discussed in detail later. For now it is important to cover the life-sustaining elements in greater detail.

The elements needed to sustain life are products of the global environment. The global environment consists of three main subdivisions:

1. **Hydrosphere**—includes all the components formed of water bodies on the earth's surface.
2. **Lithosphere**—comprises the solid components, such as rocks.
3. **Atmosphere**—is the gaseous mantle that envelopes the hydrosphere and lithosphere.

To survive, organisms require inorganic metabolites from all three parts of the biosphere. For example, the hydrosphere supplies water as the exclusive source of needed hydrogen. Essential elements such as calcium, sulfur, and phosphorus are provided by the lithosphere. Finally, oxygen, nitrogen, and carbon dioxide are provided by the atmosphere.

Within the biogeochemical cycles, all the essential elements circulate from the environment to organisms and back to the environment. Because of the critical importance of elements in sustaining life, it may be easily understood why the biogeochemical cycles are readily and realistically labeled nutrient cycles. Through these biogeochemical or nutrient cycles, nature processes and reprocesses the critical life-sustaining elements in definite inorganic-organic cycles. Some cycles, such as carbon, are more perfect than others; that is, there is no loss of material for long periods of time. One major point to keep in mind, energy (explained later in Chapter 15) flows "through" an ecosystem, but nutrients are cycled and recycled.

Humans need most of these recycled elements to survive. Because we need almost all the elements in our complex culture, we have speeded up the movement of many materials so that the cycles tend to become imperfect or what Odum calls acyclic. Odum goes on to explain that our environmental impact on phosphorus demonstrates one example of a somewhat imperfect cycle.

We mine and process phosphate rock with such careless abandon that severe local pollution results near mines and phosphate mills. Then, with equally acute myopia we increase the input of phosphate fertilizers in agricultural systems without controlling in any way the inevitable increase in run-off output that severely stresses our waterways and reduces water quality through eutrophication. (Odum 1971)

As related above, in agricultural ecosystems, we often supply necessary nutrients in the form of fertilizer to increase plant growth and yield. In natural ecosystems, however, these nutrients are recycled naturally through each trophic (feeding) level. For example, the elemental forms are taken up by plants. The consumers ingest these elements in the form of organic plant material. Eventually, the nutrients are degraded to the inorganic form again. The following pages present and discuss the nutrient cycles for water, carbon, nitrogen, phosphorus, and sulfur.

Water Cycle

Recall that the water cycle describes how water moves through the environment and identifies the links between groundwater, surface water, and the atmosphere. As illustrated in Figure 32, water is taken from the Earth's surface to the atmosphere by evaporation from the surface of lakes, rivers, streams, and oceans. This evaporation process occurs when the sun heats water. The sun's heat energizes surface molecules, allowing them to break free of the attractive force binding them together, and then evaporate and rise as invisible vapor in the atmosphere. Water vapor is also emitted from plant leaves by a process called *transpiration*. Every day, an actively growing plant transpires five to ten times as much water as it can hold at once [e.g., one acre of corn gives off 3,000–4,000 gallons (11,400–15,100 liters) of water per day, and a large oak tree can transpire 40,000 gallons (151,000 liters) per year (USGS 2005)]. As water vapor rises, it cools and eventually condenses, usually on tiny particles of dust in the air. When it condenses, it becomes a liquid again or turns directly into a solid (ice, hail, or snow). These water particles then collect and form clouds. The atmospheric water formed in clouds eventually falls to earth as precipitation. The precipitation can contain contaminants from air pollution. The precipitation may fall directly onto surface waters, be intercepted by plants or structures, or fall onto the ground. Most precipitation falls in coastal areas or in high elevations. Some of the water that falls in high elevations becomes runoff water, the water that runs over the ground (sometimes collecting nutrients from the soil) to lower elevations to form streams, lakes, and fertile valleys.

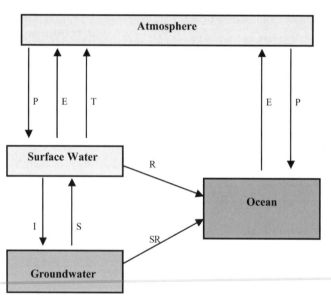

P = Precipitation
E = Evaporation
T = Transpiration
R = Surface Runoff
SR = Subsurface Runoff
I = Infiltration
S = Springs

FIGURE 32.   Water cycle box

Evapotranspiration is the sum of evaporation and plant transpiration. In regards to evapotranspiration and the water cycle, evapotranspiration is a significant water loss from a watershed. Types of land use and vegetation significantly affect evapotranspiration, and there the amount of water leaving a watershed. Evapotranspiration cannot be measured directly but may be estimated by creating an equation of the water balance of a watershed. The equation balances the change in water stored with the basin (S) with inputs and exports:

$$S = P - ET - Q - D \qquad (14.1)$$

where

S = water stored in watershed
P = input in precipitation
ET = missing flux
Q = streamflow
D = groundwater recharge

The water we see is known as *surface water*. Surface water can be broken down into five categories: oceans, lakes, rivers (streams), estuaries, and wetlands.

Water Reservoirs, Fluxes, and Residence Times

Earth's water reservoirs include the atmosphere, ocean, land surface (lakes and rivers), land subsurface (groundwater), and ice (glaciers). Water cycle fluxes include precipitation, evaporation, transpiration included in evaporation, surface runoff, subsurface runoff, infiltration, springs, and human use. Water cycle residence times are accumulated while in the atmosphere, ocean, streams and rivers, and groundwater. Table 14.1 lists quantities for water reservoirs, fluxes, and residence times.

Because the amount of rain and snow remains almost constant, but population and usage per person are both increasing rapidly, water is in short supply. In the United States alone, water usage is four times greater today than it was in 1900. In the home, this increased use is directly related to an increase in the number of bathrooms, garbage disposals, home laundries, and lawn sprinklers. In industry, usage has increased 13 times since 1900.

170,000+ small-scale suppliers provide drinking water to approximately 200+ million Americans by 60,000+ community water supply systems, and to nonresidential locations, such as schools, factories, and campgrounds. The rest of Americans are served by private wells. The majority of the drinking water used in the U.S. is supplied from groundwater. Untreated water drawn

TABLE 14.1
Water Reservoirs, Fluxes, and Residence Times

Reservoirs	km3	%
Atmosphere	12,700	.001
Ocean	1,230,000,000	97.2
Land surface		
Lakes	123,000	.009
Rivers & Streams	1,200	.0001
Land subsurface (groundwater)	4,000,000	.31
Ice (glaciers)	28,600,000	2.15

Fluxes		km3/yr
Precipitation	total	496,000
	Land	111,000
	Ocean	385,000
Evaporation (includes Transpiration)	total	496,000
	Land	71,000
	Ocean	425,000
Surface runoff		26,000
Subsurface runoff	liquid	12,000
	Ice	2,000
Infiltration		14,000
Springs		2,000
Human Use	total	3,000

Residence	Residence Time
Atmosphere	0.03 yr or 9 days
Ocean	2,900 years
Streams and rivers	0.05 yr or 17 days
Groundwater	330 years

from groundwater and surface waters, and used as a drinking water supply, can contain contaminants that pose a threat to human health.

Carbon Cycle

Carbon (the name comes from Latin language *carbo*, coal; carbon is one of the few elements known since antiquity) is an essential ingredient of all living things; it is the 15th most abundant element in the Earth's crust and the fourth most abundant element in the universe by mass after hydrogen, helium, and oxygen. There are several allotropes (i.e., a variant of a substance) of which the best known are graphite diamond, and amphorous carbon. It is the basic building block of the large organic molecules necessary for life. Inorganic forms of carbon (carbon dioxide, bicarbonate, and carbonate) strongly affect the acidity of soils and natural waters, the heat insulating capability of the atmosphere, and the rates of such key natural processes as photosynthesis, weathering, and biomineralization. In reduced form, carbon provides the elemental backbone for myriad organic molecules that comprise living organisms, soil humus, and fossil fuels (NSF 2000). Carbon is cycled into food chains from the atmosphere, as shown in Figure 33.

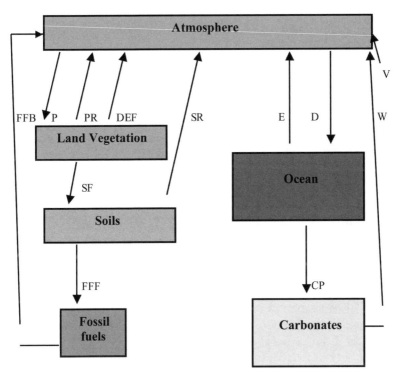

P = Photosynthesis
PR = Plant respiration
SR = Soil respiration
SF = Plants to soils
FFF = Fossil fuel formation
FFB = Fossil fuel burning
DEF = Deforestation
D = Dissolving
E = Exsolving
CP = Carbonate formation
W = Weathering
V = Volcanoes

FIGURE 33.   Carbon cycle box

The Carbon Cycle is based on carbon dioxide, which makes up only a small percentage of the atmosphere. From Figure 33 it can be seen that green plants obtain carbon dioxide (CO_2) from the air and, through photosynthesis, described by Asimov as the "most important chemical process on Earth," it produces the food and oxygen that all organisms live on (Asimov 1989). Part of the carbon produced remains in living matter and the other part is released as CO_2 in cellular respiration. Miller points out that the carbon dioxide released by cellular respiration in all living organisms is returned to the atmosphere (Miller 1988).

Did You Know?

About a tenth of the estimated 700 billion tons of carbon dioxide in the atmosphere is fixed annually by photosynthetic plants. A further trillion tons are dissolved in the ocean, more than half in the photosynthetic layer.

Some carbon is contained in buried dead animal and plant materials. Much of these buried plant and animal materials were transformed into fossil fuels. Fossil fuels, coal, oil, and natural gas, contain large amounts of carbon. When fossil fuels are burned, stored carbon combines with oxygen in the air to form carbon dioxide, which enters the atmosphere (Moran et al. 1986).

In the atmosphere, carbon dioxide acts as a beneficial heat screen as it does not allow the radiation of earth's heat into space. This balance is important. The problem is that as more carbon dioxide from burning is released into the atmosphere, balance can and is being altered. Odum (1983) warns that with the recent increases in consumption of fossil fuels "coupled with the decrease in the 'removal capacity' of the green belt is beginning to exceed the delicate balance." Massive increases of carbon dioxide into the atmosphere tend to increase the possibility of global warming. The consequences of global warming "would be catastrophic . . . and the resulting climatic change would be irreversible" (Abrahamson 1988).

Nitrogen Cycle

Nitrogen was discovered as a separable component of air, by Scottish physician Daniel Rutherford, in 1772. It is important to all life because it is a necessary nutrient. Nitrogen in the atmosphere or in the soil can go through many complex chemical and biological changes, be combined into living and nonliving material, and return to the soil or air in a continuing cycle. This is called the Nitrogen Cycle (Killpack & Buchholz 1993). The Nitrogen Cycle consists of the following processes and various states:

Processes	*States*
Fertilizer	N_2—elemental nitrogen is a gaseous form of nitrogen
Volatilization	NH_3—ammonia is a gaseous form of nitrogen
Animal Wastes	NO—nitric oxide is a gaseous form of nitrogen
Organic Matter	NH_4+—Ammonium is attracted to soil particles
Immobilization	N_2O—nitrous oxide is a gaseous form of nitrogen
Nitrification	NO_3—nitrate is not attracted to soil particles
Biological Fixation	
Mineralization	
Denitrification	
Crop Uptake	

The following are important nitrogen cycle terms.

- **Limiting Nutrient**—amount of an element necessary for plant life is in short supply.
- **Nitrogen Fixation**—chemical conversion to N_2 to NH_3 (ammonia) or NO_3 (nitrate).
- **Denitrification**—chemical conversion from nitrate (NO_3) back to N_2.

Today we readily apply nitrogen in the form of fertilizer to plants to ensure proper growth. Before fertilizer containing nitrogen was commercially available, how did agriculture survive? Early farmers had to rely on natural regeneration of fixed nitrogen:

- **Annual Floods**—bring fresh sediments (e.g., Nile Valley).
- **Slash/burn Agriculture**—Once the soil nutrients are depleted, move on to a new place.
- **Crop Rotation**—Certain crops (e.g., soybeans) are good at fixing nitrogen, others (e.g., corn) use it up; plant on alternate years.

The atmosphere contains 78% by volume of nitrogen, making it the largest pool of nitrogen. Moreover, as stated above, nitrogen is an essential element for all living matter and constitutes 1–3% dry weight of cells, yet nitrogen is not a common element on Earth. Although it is an essential ingredient for plant growth, it is chemically very inactive and before it can be incorporated by the vast majority of the biomass, it must be "fixed" (Porteous 1992).

Price (1984) describes the nitrogen cycle as an example "of a largely complete chemical cycle in ecosystems with little leaching out of the system." From the water/wastewater specialist's point of view, nitrogen and phosphorous are both commonly considered as limiting factors for productivity. Of the two, nitrogen is harder to control, but is found in smaller quantities in wastewater.

As stated earlier, nitrogen gas makes up about 78 percent of the volume of the earth's atmosphere. As such, it is useless to most plants and animals. Fortunately, nitrogen gas is converted into compounds containing nitrate ions, which are taken up by plant roots as part of the nitrogen cycle, shown in simplified form in Figure 34 and in box model form below.

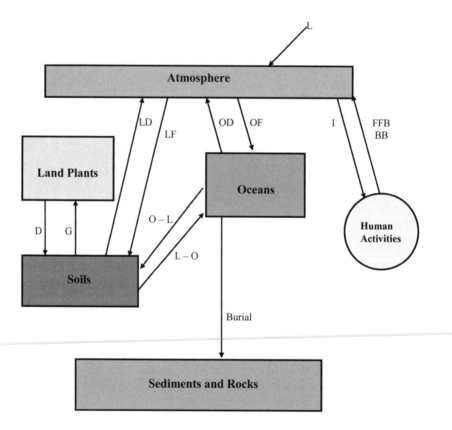

LF = Land Fixation   BB = Biomass Burning
LD = Land Denitrification   L = Lightning
OF = Oceanic Fixation   D = Decay
OD = Oceanic Denitrification G = Growth
I = Industrial Fixation   O – L = Ocean-to-Land
FFB = Fossil Fuel Burning   L – O = Land-to-Ocean

FIGURE 34. Nitrogen cycle box

Aerial nitrogen is converted into nitrates mainly by microorganisms, bacteria, and blue-green algae. Lightning also converts some aerial nitrogen gas into forms that return to the earth as nitrate ions in rainfall and other types of precipitation. From Figure 34 it can be seen that ammonia plays a major role in the nitrogen cycle. **Excretion** by animals and **anaerobic** decomposition of dead organic matter by bacteria produce ammonia. Ammonia, in turn, is converted by nitrification bacteria into nitrites and then into nitrates. This process is known as nitrification. Nitrification bacteria are **aerobic**. Bacteria that convert ammonia into nitrites are known as nitrite bacteria (Nitrosococcus and Nitrosomonas); they convert nitrites into nitrates and nitrate bacteria (Nitrobacter). In wastewater treatment, ammonia is produced in the sludge digester, nitrates in the aerobic sewage treatment process.

Nitrogen reservoirs and quantities in millions of metric tons are:

- Atmosphere: 4,000,000,000
- Land Plants: 3500
- Soils: 9500
- Oceans: 23,000,000
- Sediments and Rocks: 200,000,000,000

Did You Know?

Buried rocks and sediments are the largest pool of nitrogen, but this reservoir is a minor part of the cycle.

In their voluminous and authoritative text, *Wastewater Engineering*, Metcalf & Eddy (1991) devote several pages to describing the nitrogen cycle and its impact upon the wastewater treatment process. They point out that nitrogen is found in wastewater in the form of urea. During wastewater treatment, the urea is transformed into ammonia nitrogen. Since ammonia exerts a BOD and chlorine demand, high quantities of ammonia in wastewater effluents are undesirable. The process of nitrification is utilized to convert ammonia to nitrates. Nitrification is a biological process, which involves the addition of oxygen to the wastewater. If further treatment is necessary, another biological process called denitrification is used. In this process, nitrate is converted into nitrogen gas, which is lost to the atmosphere, as can be seen in Figure 34.

Did You Know?

Specialized bacteria and lightning are the only natural ways that nitrogen is fixed.

When attempting to address the important and complex factors that make up the topic of ecology, it is important to understand the impact that the nitrogen cycle can have on effluent that is dumped (outfalled) into the environment. At the same time, in regards to wastewater treatment, for example, one should remember that the nitrogen cycle that occurs in the wastewater stream is not the primary source of the nitrogen contamination of surface water bodies. As a case in point, Price (1984) points to the example of large inputs of nitrogen fertilizer from agricultural systems, which "may result in considerable leaching and unidirectional flow of nitrogen into aquatic systems which become polluted with excessive nitrogen."

Did You Know?

Lightning may have been necessary for life to begin:

No life → no bacteria → no bacterial fixation → no usable nitrogen

---

Phosphorus Cycle

Phosphorus is another element that is common in the structure of living organisms. However, of all the elements recycled in the biosphere, phosphorus is the scarcest and therefore the one most limiting in any given ecological system. Unlike many other biogeochemical cycles, the atmosphere does not play a significant role in the movements of phosphorus, because phosphorus and phosphorus-based compounds are usually soils at the typical ranges of temperature and pressure found on Earth. It is indispensable to life, being intimately involved in energy transfer and in the passage of genetic information in the DNA of all cells.

The ultimate source of phosphorus is rock, from which it is released by weathering, leaching, and mining. Phosphorus has no stable gas phase, so addition of P to land is slow. Phosphorus occurs as phosphate or other minerals formed in past geological ages. These massive deposits are gradually eroding to provide phosphorus to ecosystems. A large amount of eroded phosphorus ends up in deep sediments in the oceans and lesser amounts in shallow sediments. Part of the phosphorus comes to land when marine animals are brought out. Birds also play a role in the recovery of phosphorus. The great guano deposit, bird excreta, of the Peruvian coast is an example. Man has hastened the rate of loss of phosphorus through mining activities and the subsequent production of fertilizers, which are washed away and lost. Even with the increase in human activities, however, there is no immediate cause for concern, since the known reserves of phosphate are quite large.

Phosphorus has become very important in water quality studies, since it is often found to be a limiting factor (i.e., limiting plant nutrient). Control of phosphorus compounds that enter surface waters, and contribute to growth of algal blooms, is of much interest to stream ecologists. Phosphates, upon entering a stream, act as fertilizer, which promotes the growth of undesirable algae populations or algal blooms. As the organic matter decays, dissolved oxygen levels decrease, and fish and other aquatic species die. Figure 35 shows the phosphorus cycle.

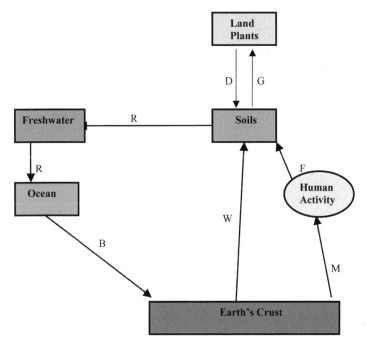

M = Mining
F = Fertilization
W = Weathering
R = Runoff
B = Burial
D = Decay
G = Growth

FIGURE 35. Phosphorus cycle

While it is true that phosphorus discharged into streams is a contributing factor to stream pollution (and causes eutrophication), it is also true that phosphorus is not the lone factor. Odum (1975) warns against what he calls the one-factor control hypothesis, i.e., the one-problem/one solution syndrome. He goes on to point out that environmentalists in the past have focused on one or two items, like phosphorus contamination, and "have failed to understand that the strategy for pollution control must involve reducing the input of all enriching and toxic materials."

Did You Know?

Because of its high reactivity, phosphorus exists in combined form with other elements. Microorganisms produce acids that form soluble phosphate from insoluble phosphorus compounds. The phosphates are utilized by algae and terrestrial green plants, which in turn pass into the bodies of animal consumers. Upon death and decay of organism, phosphates are released for recycling (Spellman 1996).

Sulfur Cycle

Sulfur, like nitrogen and carbon, is characteristic of organic compounds. However, an important distinction between cycling of sulfur and cycling of nitrogen and carbon is that sulfur is "already fixed." That is, plenty of sulfate anions are available for living organisms to utilize; the largest reservoir is Earth's crust. By contrast, the major biological reservoirs of nitrogen atoms (N_2) and carbon atoms (CO_2) are gases that must be pulled out of the atmosphere. Sulfur is rarely a limiting nutrient for ecosystems or organisms.

Did You Know?

In the sulfur cycle, elementary sulfur of the lithosphere is not available to plants and animals unless converted to sulfates.

The sulfur cycle (see Figure 36) is both sedimentary and gaseous. Tchobanoglous and Schroeder (1985) note that "the principal forms of sulfur that are of special significance in water quality management are organic sulfur, hydrogen sulfide, elemental sulfur and sulfate."

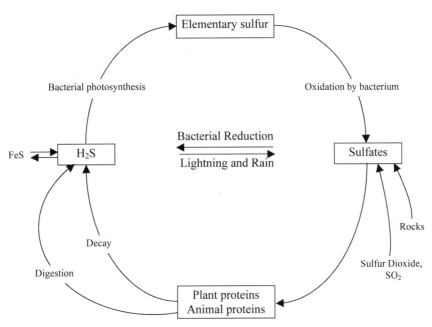

FIGURE 36. Sulfur cycle

Bacteria play a major role in the conversion of sulfur from one form to another. In an anaerobic environment, bacteria break down organic matter producing hydrogen sulfide with its characteristic rotten egg odor. A bacterium called *Beggiatoa* converts hydrogen sulfide into elemental sulfur. An aerobic sulfur bacterium, *Thiobacillus thiooxidans*, converts sulfur into sulfates. Other sulfates are contributed by the dissolving of rocks and some sulfur dioxide. Sulfur is incorporated by plants into proteins. Some of these plants are then consumed by organisms. Sulfur from proteins is liberated by many heterotrophic anaerobic bacteria, as hydrogen sulfide.

Energy Flow

The original source of all energy going into food is the sun. This is because plants that have chlorophyll are able to combine water and carbon dioxide in the presence of light energy and produce sugar. This sugar can be converted to energy as the plant needs it. Of course, some of the sugar is converted into other complex chemicals that permit growth, reproduction, and other life processes.

—A. N. Tomera, American ecologist, 1989

Three hundred trout are needed to support one man for a year. The trout, in turn, must consume 90,000 frogs, that must consume 27 million grasshoppers that live off of 1,000 tons of grass.

—G.T. Miller, Jr., American chemist, 1988

Nature provides an exception to every rule.

—Margaret Fuller, *The Dial*, 1843

Again, it is important to point out that energy "moves" through an ecosystem and is not cycled through it. If you can understand this, you are in good shape, because then you have an idea of how ecosystems are balanced, how they may be affected by human activities, and how pollutants will move through an ecosystem.

In an ecosystem, *energy flow*, also called the calorific flow, refers to the flow of energy through a food chain. Energy is the ability or capacity to do work. For an ecosystem to exist, it must have energy. All activities of living organisms involve work, which is the expenditure of energy. This means the degradation of a higher state of energy to a lower state. The flow of energy through an ecosystem is governed by two laws: The First and Second Laws of Thermodynamics.

The first law, sometimes called the conservation law, states that energy may not be created or destroyed. The second law states that no energy transformation is 100% efficient. That is, in every energy transformation, some energy is dissipated as heat. The term entropy is used as a measure of the nonavailability of energy to a system. Entropy increases with an increase in dissipation. Because of entropy, input of energy in any system is higher than the output or work done; thus, the resultant, efficiency, is less than 100%.

Odum (1975) explains that "the interaction of energy and materials in the ecosystem is of primary concern of ecologists." Earlier we discussed the biogeochemical nutrient cycles. It is important to remember that it is the flow of energy that drives these cycles. Again, it should be noted that energy does not cycle as nutrients do in biogeochemical cycles. For example, when food passes from one organism to another, energy contained in the food is reduced step by step until all the energy in the system is dissipated as heat. Price (1984) refers to this process as "a unidirectional flow of energy through the system, with no possibility for recycling of energy." When water or nutrients are recycled, energy is required. The energy expended in this recycling is not recyclable. And, as Odum (1975) points out, this is a "fact not understood by those who think that artificial recycling of man's resources is somehow an instant and free solution to shortages."

While there is a slight input of geothermal energy, as pointed out earlier, the principal source of energy for any ecosystem is sunlight. Green plants, through the process of photosynthesis, transform the sun's light energy into chemical energy: carbohydrates which are consumed by animals. This transfer of energy, as stated previously, is unidirectional—from producers to consumers. It is accomplished by cellular respiration, which is the process by which organisms (like mammals) break the glucose back down into its constituents, water and carbon dioxide, thus regaining the stored energy the sun originally gave to the plants. Often this transfer of energy to different organisms is called a food chain. It is safe to say that food energy passes through a community in various ways—each separate way is called a food chain. Figure 37 shows a simple aquatic food chain.

All organisms, alive or dead, are potential sources of food for other organisms. All organisms that share the same general type of food in a food chain are said to be at the same **trophic level** (nourishment or feeding level—each level of consumption in a food chain is called a trophic level). Since green plants use sunlight to produce food for animals, they are called the producers, or the first trophic level. The **herbivores**, which eat plants directly, are called the second trophic level or the primary consumers. The carnivores are flesh eating consumers; they include several trophic levels from the third on up. At each transfer, a large amount of energy (about 80 to 90%) is lost as heat and wastes. Thus, nature normally limits food chains to four or five links; however, in aquatic ecosystems, "food chains are commonly longer than those on land" (Dasmann 1984). The aquatic food chain is longer because several predatory fish may be feeding on the plant consumers. Even so, the built-in inefficiency of the energy transfer process prevents development of extremely long food chains.

Tomera (1989) describes a simple food chain that can be seen in a prairie dog community.

> The grass in the community manufactures food. The grass is called a food producer. The grass is eaten by a prairie dog. Because the prairie dog lives directly off the grass, it is termed a first-order consumer. A weasel [or other predator] may kill and eat the prairie dog. The weasel is, therefore, a predator and would be termed a second-order consumer. The second-order consumer is twice removed from the green grass. The weasel, in turn, may be eaten by a large hawk or eagle. The bird that kills and eats the weasel would therefore be a third-order

FIGURE 37. Aquatic food chain

consumer, three times removed from the grass. Of course, the hawk would give off waste materials and eventually die itself. Wastes and dead organisms are then acted on by decomposers. (p. 50)

Only a few simple food chains are found in nature. Thus, when attempting to identify the complex food relationships among many animals and plants within a community, it is useful to create a food web. Most simple food chains are interlocked; this interlocking of food chains forms a food web. A food web can be characterized as a map that shows what eats what (Miller 1988). Most ecosystems support a complex food web. A food web involves animals that do not feed on one trophic level. For example, humans feed on both plants and animals. The point is, an organism in a food web may occupy one or more trophic levels. Trophic level is determined by an organism's role in its particular community, not by its species. Food chains and webs help to explain how energy moves through an ecosystem.

An important trophic level of the food web that has not been discussed thus far is comprised of the decomposers (bacteria, mushrooms, etc.). The decomposers feed on dead plants or animals and play an important role in recycling nutrients in the ecosystem. As Miller (1988) points out, "there is no waste in ecosystems. All organisms, dead or alive, are potential sources of food for other organisms." An example of an aquatic food web is shown in Figure 38.

From the preceding discussion about food chains and food webs, the important point to be gained is that there is a distinct difference between the two. A food chain, for example, is a simple straight-line process going from

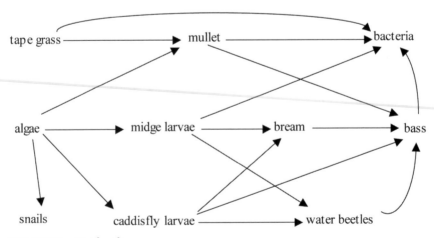

FIGURE 38. Food web

producer to first-, second-, and possibly third-order consumers, and ending with the decomposers. On the other hand, in a food web, there are a number of second- and third-order consumers.

Food Chain Efficiency

Earlier it was pointed out that energy from the sun is captured (via photosynthesis) by green plants and used to make food. Most of this energy is used to carry on the plant's life activities. The rest of the energy is passed on as food to the next level of the food chain.

It is important to note that nature limits the amount of energy that is accessible to organisms within each food chain. Not all food energy is transferred from one trophic level to the next. For ease of calculation, "ecologists often assume an ecological efficiency of 10% (10-percent rule) to estimate the amount of energy transferred through a food chain" (Moran et al. 1986). For example, if we apply the 10% rule to the diatoms-copepods-minnows-medium fish-large fish food chain shown in Figure 39, we can predict that 1000 grams of diatoms produce 100 grams of copepods, which will produce 10 grams of minnows, which will produce 1 gram of medium fish, which, in turn, will produce 0.1 gram of large fish. Thus, only about 10% of the chemical energy available at each trophic level is transferred and stored in usable form at the next level. What happens to the other 90%? The other 90% is lost to the environment as low-quality heat in accordance with the second law of thermodynamics.

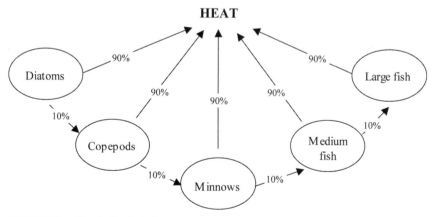

FIGURE 39. Simple food chain

Did You Know?

• A food chain is the path of food from a given final consumer back to a producer.
• The ratio of net production at one level to net production at the next higher level is called the *conversion efficiency*.
• When an organism loses heat, it represents one-way flow of energy out of the ecosystem. Plants only absorb a small part of energy from the sun. Plants store half of the energy and lose the other half. The energy plants lose is metabolic heat. Energy from a primary source will flow in one direction through two different types of food chains. In a grazing food chain, the energy will flow from plants (producers) to herbivores, and then through some carnivores. In a detritus-based food chain, energy will flow from plants through detrivores and decomposers. In terms of the weight (or biomass) of animals in many ecosystems, more of their body mass can be traced back to detritus than to living producers. Most of the time, the two food webs will intersect one another. For example, in the Chesapeake Bay bass fish of the grazing food web will eat a crab of the detrital food web (Spellman 1996).

Ecological Pyramids

As we proceed in the food chain from the producer to the final consumer it becomes clear that a particular community in nature often consists of several small organisms associated with a smaller and smaller number of larger organisms. A grassy field, for example, has a larger number of grass and other small plants, a smaller number of herbivores like rabbits, and an even smaller number of carnivores like fox. The practical significance of this is that we must have several more producers than consumers.

This pound-for-pound relationship, where it takes more producers than consumers, can be demonstrated graphically by building an **ecological pyramid**. In an ecological pyramid, the number of organisms at various trophic levels in a food chain is represented by separate levels or bars placed one above the other with a base formed by producers and the apex formed by the final consumer. The pyramid shape is formed due to a great amount of energy loss at each trophic level. The same is true if numbers are substituted by the corresponding biomass or energy. Ecologists generally use three types of ecological pyramids: pyramids of number, biomass, and energy. Obviously, there will be differences among them. Some generalizations:

1. **Energy pyramids** must always be larger at the base than at the top (because of the **2nd Law of Thermodynamics**, and have to do with dissipation of energy as it moves from one trophic level to another). Simply, energy pyramids depict the decrease in the total available energy at each higher trophic level.

2. Likewise, **biomass pyramids** (in which biomass is used as an indicator of production) are usually pyramid-shaped. This is particularly true of terrestrial systems and aquatic ones dominated by large plants (marshes), in which consumption by heterotroph is low and organic matter accumulates with time. A census of the population, multiplied by the weight of an average individual in it, gives an estimate of the weight of the population. This is called the biomass (or standing crop). However, it is important to point out that biomass pyramids can sometimes be inverted. This is especially common in aquatic ecosystems, in which the primary producers are microscopic **planktonic** organisms that multiply very rapidly, have very short **life spans**, and are heavily grazed by herbivores. At any single point in time, the amount of biomass in primary producers is less than that in larger, long-lived animals that consume primary producers.

3. **Numbers pyramids** can have various shapes (and not be pyramids at all, actually) depending on the sizes of the organisms that make up the trophic levels. In forests, the primary producers are large trees and the herbivore level usually consists of insects, so the base of the pyramid is smaller than the herbivore level above it; that is, the pyramid is inverted. In grasslands, the number of primary producers (grasses) is much larger than that of the herbivores above (large grazing animals) (Spellman 1996).

To get a better idea of how an ecological pyramid looks and how it provides its information, we need to look at an example of an energy pyramid. According to Odum (1983), the energy pyramid is a fitting example because among the "three types of ecological pyramids, the energy pyramid gives by far the best overall picture of the functional nature of communities."

In an experiment conducted in Silver Springs, Florida, Odum (1983) measured the energy for each trophic level in terms of kilocalories. A kilocalorie is the amount of energy needed to raise 1 cubic centimeter of water 1 degree centigrade. When an energy pyramid is constructed to show Odum's findings, it takes on the typical upright form (as it must because of the second law of thermodynamics) as shown in Figure 40.

Simply put, as reflected in Figure 40 and according to the second law of thermodynamics, no energy transformation process is 100% efficient. This fact is demonstrated, for example, when a horse eats hay. The horse cannot obtain, for his own body, 100% of the energy available in the hay. For this reason, the

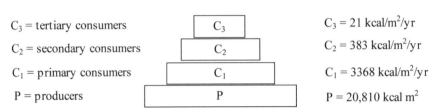

C_3 = tertiary consumers C_3 $C_3 = 21$ kcal/m²/yr

C_2 = secondary consumers C_2 $C_2 = 383$ kcal/m²/yr

C_1 = primary consumers C_1 $C_1 = 3368$ kcal/m²/yr

P = producers   P   P = 20,810 kcal m²

FIGURE 40.   Energy flow pyramid
Source: Adapted form Odum, 1971, Fundamentals of Ecology, p. 80

energy productivity of the producers must be greater than the energy production of the primary consumers. When human beings are substituted for the horse, it is interesting to note that according to the second law of thermodynamics only a small population could be supported. But this is not the case. Humans also feed on plant matter, which allows a larger population. Therefore, if meat supplies become scarce, we must eat more plant matter. This is the situation we see today in countries where meat is scarce. Consider this, if we all ate soybeans, there would be at least enough food for 10 times as many of us as compared to a world where we all eat beef (or pork, fish, chicken, etc.). Another way of looking at this: Every time we eat meat, we are taking food out of the mouths of 9 other people, who could be fed with the plant material that was fed to the animal we are eating (Spellman 1996). Food-energy relationships are often referred to as eater-eaten relationships. It's not quite that simple, of course, but you probably get the general idea.

Relationships in Living Communities

In addition to the pyramid-shaped relationships, there are other important relationships in living communities. Some of these involve food energy and some do not. In this section several of these relationships are described.

Symbiosis is a close (intimate) ecological relationship (organisms live together in close proximity) between the individuals of two (or more) different species. Sometimes a symbiotic relationship benefits both species, sometimes one species benefits at the other's expense, and in other cases neither species benefits. One thing is certain; the relationship is *obligate*, meaning at least one of the species must be involved in the relationship to survive.

Ecologists use a different term for each type of symbiotic relationship:

• Mutualism—both organisms benefit
• Commensalism—one organism benefits, the other is unaffected

- Parasitism—one organism benefits, the other is harmed
- Competition—neither organism benefits
- Neutralism—both organisms are unaffected

The effects of these interactions on population growth can be positive, negative, or neutral (see Table 14.2).

TABLE 14.2
Population Interactions, Two-Species System

	Response	
Type of Interaction	*A*	*B*
	A	B
Neutral	0	0
Mutualism	+	+
Commensalism	+	0
Parasitism	+	−
Competition	−	−

Productivity

As mentioned previously, the flow of energy through an ecosystem starts with the fixation of sunlight by plants through photosynthesis. In evaluating an ecosystem, the measurement of photosynthesis is important. Ecosystems may be classified into highly productive or less productive. Therefore, the study of ecosystems must involve some measure of the productivity of that ecosystem.

Smith (1974) defines production (or more specifically primary production, because it is the basic form of energy storage in an ecosystem) as being "the energy accumulated by plants." Stated differently, primary production is the rate at which the ecosystem's primary producers capture and store a given amount of energy, in a specified time interval. In even simpler terms, primary productivity is a measure of the rate at which photosynthesis occurs. Odum (1971) lists four successive steps in the production process as follows:

1. **Gross primary Productivity**—the total rate of photosynthesis in an ecosystem during a specified interval at a given trophic level.
2. **Net primary Productivity**—the rate of energy storage in plant tissues in excess of the rate of aerobic respiration by primary producers.
3. **Net community Productivity**—the rate of storage of organic matter not used.
4. **Secondary Productivity**—the rate of energy storage at consumer levels.

When attempting to comprehend the significance of the term productivity as it relates to ecosystems, it is wise to consider an example. Consider the productivity of an agricultural ecosystem such as a wheat field. Often its productivity is expressed as the number of bushels produced per acre. This is an example of the harvest method for measuring productivity. For a natural ecosystem, several one-square meter plots are marked off, and the entire area is harvested and weighed to give an estimate of productivity as grams of biomass per square meter per given time interval. From this method, a measure of net primary production (net yield) can be measured.

Productivity, both in the natural and cultured ecosystem, may vary considerably, not only between type of ecosystems, but also within the same ecosystem. Several factors influence year-to-year productivity within an ecosystem. Such factors as temperature, availability of nutrients, fire, animal grazing, and human cultivation activities are directly or indirectly related to the productivity of a particular ecosystem.

The following study of an aquatic ecosystem is used as an example of productivity. Productivity can be measured in several different ways in the aquatic ecosystem. For example, the production of oxygen may be used to determine productivity. Oxygen content may be measured in several ways. One way is to measure it in the water every few hours for a period of 24 hours. During daylight, when photosynthesis is occurring, the oxygen concentration should rise. At night the oxygen level should drop. The oxygen level can be measured by using a simple x-y graph. The oxygen level can be plotted on the y-axis with time plotted on the x-axis, as shown in Figure 41.

Another method of measuring oxygen production in aquatic ecosystems is to use light and dark bottles. **Biochemical oxygen demand** (BOD) bottles

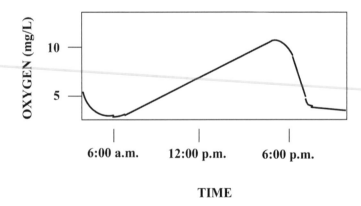

FIGURE 41.   The diurnal oxygen curve for an aquatic ecosystem

(300 ml) are filled with water to a particular height. One of the bottles is tested for the initial dissolved oxygen (DO), then the other two bottles (one clear, one dark) are suspended in the water at the depth they were taken from. After a twelve hour period, the bottles are collected and the DO values for each bottle are recorded. Once the oxygen production is known, the productivity in terms of grams/m/day can be calculated.

Table 14.3 shows representative values for the net productivity of a variety of ecosystems—both nature and managed. Keep in mind that these values are only approximations derived from Odum's (1971, 1983) work and are subject to marked fluctuations because of variations in temperature, fertility, and availability of water.

In the aquatic (and any other) ecosystem, pollution can have a profound impact upon the system's productivity. For example, certain kinds of pollution may increase the turbidity of the water. This increase in turbidity causes a decrease in energy delivered by photosynthesis to the ecosystem. Accordingly, this turbidity and its aggregate effects decrease net community productivity on a large scale (Laws 1993).

The ecological trends paint a clear picture. Wherever we look, ecological productivity is limping behind human consumption. Since 1984, the global fish harvest has been dropping, and so has the per capita yield of grain crops (Brown 1994). Moreover, stratospheric ozone is being depleted, the release of greenhouse gases has changed the atmospheric chemistry and might lead to climate change; erosion and desertification are reducing nature's biological productivity; irrigation water tables are falling; contamination of soil

TABLE 14.3
Estimated Net Productivity of Certain Ecosystems

Ecosystem	kilocalories/m²/year
Clear (oligotrophic) lake	800
Coastal marsh	12,000
Corn (maize) field, U.S.	4,500
Desert	500
Field of alfalfa (Lucerne)	15,000
Lake in advanced state of eutrophication	2,400
Lawn, Washington, D.C.	6,800
Ocean close to shore	2,500
Open ocean	800
Rice paddies, Japan	5,500
Silver Springs, Florida	8,800
Sugar cane, Hawaii	25,000
Tall-grass prairie	2,000
Temperate deciduous forest	5,000
Tropical rain forest	15,000

and water is jeopardizing the quality of food; other natural resources are being consumed faster than they can regenerate; and biological diversity is being lost—to reiterate only a small part of a long list. These trends indicate a decline in the quantity and productivity of nature's assets (Wachernagel 1997).

References and Recommended Reading

Abedon, S.T. 1997. *Ecosystems.* Accessed 02/15/07 @ http://www.mansfield.hio-state .edu.

Abrahamson, D.E., ed. 1988. *The Challenge of Global Warming.* Washington, DC: Island Press.

Asimov, I. 1989. *How Did We Find Out about Photosynthesis?* New York: Walker.

Barlocher, R., and Kendrick, L. 1975. Leaf conditioning by microorganisms. *Oecologia* 20: 359–362.

Benfield, E.F. 1996. Leaf breakdown in streams ecosystems. In *Methods in Stream Ecology.* Hauer, F.R., and Lambertic, G.A., eds. San Diego: Academic Press.

Benfield, E.F., Jones, D.R., and Patterson, M.F. 1977. Leaf pack processing in a pastureland stream. *Oikos* 29: 99–103.

Benjamin, C.L., Garman, G.R., Funston, J.H. 1997. *Human Biology.* New York: McGraw-Hill.

Brown, L.R. 1994. Facing food insecurity. In *State of the World.* Brown, L.R., et al., eds. New York: W.W. Norton.

Carson, R. 1962. *Silent Spring.* Boston: Houghton Mifflin.

Clements, E.S. 1960. *Adventures in Ecology.* New York: Pageant Press.

Crossley, D.A., Jr., House, G.J., Snider, R.M., Snider, R.J., and Stinner, B.R. 1984. The positive interactions in agroecosystems, In *Agricultural Ecosystems.* Lowrance, R., Stinner, B.R., and House, G.J., eds. New York: John Wiley & Sons.

Cummins, K.W. 1974. Structure and function of stream ecosystems. *Bioscience* 24: 631–641.

Cummins, K.W., and Klug, M.J. 1979. Feeding ecology of stream invertebrates. *Annual Review of Ecology and Systematics* 10: 631–641.

Darwin, C. 1998. *The Origin of Species.* Suriano, G. (ed.) New York: Grammercy.

Dasmann, R.F. 1984. *Environmental Conservation.* New York: John Wiley & Sons.

Dolloff, C.A., and Webster, J.R. 2000. Particulate organic contributions from forests to streams: Debris isn't so bad. In *Riparian Management in Forests of the Continental Eastern United States.* Verry, E.S., Hornbeck, J.W., and Dolloff, C.A. eds. Boca Raton, FL: Lewis Publishers.

Ecology. 2007. *Ecology.* Accessed 02/10/07 @ http://www.newworldencyclopedia.org/ preview/ecology.

Ecosystem. 2007. *Ecosystem.* Accessed 02/11/07 @ http://en.wikipedia.org/wiki/ Ecosystem.

Evans, F.C. 1956. Ecosystem as the basic unit in ecology. *Science* 23: 1127–1128.

ISWS. 2005. *Biogeochemical Cycles II: The Nitrogen Cycle.* Illinois State Water Survey. Accessed 02/15/07 @ http://www.sws.uiuc.edu.

Killpack, S.C., and Buchholz, D. 1993. *Nitrogen in the Environment.* Columbia: University of Missouri.

Krebs, C.H. 1972. *Ecology: The Experimental Analysis of Distribution and Abundance.* New York: Harper and Row.

Laws, E.A. 1993. *Environmental Science: An Introductory Text.* New York: John Wiley & Sons.

Lindeman, R.L. 1942. The trophic-dynamic aspect of ecology. *Ecology* 23: 399–418.

Lowdermilk, W.C. 1938. *Conquest of the Land through 7,000 Years.* Washington, DC: USDA.

Margulis, L., and Sagan, D. 1997. *Microcosmos: Four Billion Years of Evolution from our Microbial Ancestors.* Berkeley: University of California Press.

Marshall, P. 1950. *Mr. Jones, Meet the Master.* Grand Rapids, MI: Fleming H. Revell.

Metcalf & Eddy, Inc. 1991. *Wastewater Engineering: Treatment, Disposal, Reuse,* 3rd ed. New York: McGraw-Hill.

Miller, G.T. 1988. *Environmental Science: An Introduction.* Belmont, CA: Wadsworth.

Moran, J.M., Morgan, M.D., and Wiersma, J.H. 1986. *Introduction to Environmental Science.* New York: W.H. Freeman.

NSF. 2000. *Report of the Workshop on the Terrestrial Carbon Cycle.* National Science Foundation. Accessed 02/18/07 @ http://www.carboncyclescience.gov.

Odum, E.P. 1952. *Fundamentals of Ecology,* 1st ed. Philadelphia: W.B. Saunders Co.

Odum, E.P. 1971. *Fundamentals of Ecology,* 3rd ed. Philadelphia: Saunders College Publishing.

Odum, E.P. 1975. *Ecology: The Link between the Natural and the Social Sciences.* New York: Holt, Rinehart and Winston, Inc.

Odum, EP. 1983. *Basic Ecology.* Philadelphia: Saunders College Publishing.

Odum, E.P. 1984. Properties of agroecosystems. In *Agricultural Ecosystems.* Lowrance, R., Stinner, B.R., and House, G.J., eds. New York: John Wiley & Sons.

Odum, E.P., and Barrett, G.W. 2005. *Fundamentals of Ecology,* 5th ed. Belmont, CA: Thomson Brooks/Cole.

Paul, R.W., Jr., Benfield, E.F., and Cairns, J., Jr. 1978. Effects of thermal discharge on leaf decomposition in a river ecosystem. *Verhandlugen der Internationalen Vereinigung fur Thoeretsche and Angewandte Limnologie* 20: 1759–1766.

Peterson, R.C., and Cummins, K.W. 1974. Leaf processing in woodland streams. *Freshwater Biology* 4: 345–368.

Porteous, A. 1992. *Dictionary of Environmental Science and Technology.* New York: John Wiley & Sons.

Price, P.W. 1984. *Insect Ecology.* New York: John Wiley & Sons.

Ramalay, F. 1940. The growth of a science. *Univer. Colorado Stud.* 26: 3–14.

Smith, C.H. 2007. *Karl Ludwig Willdenow.* Accessed 02/09/07 @ http://www.wku.edu/~smithch/chronob/WILL1765.htm.

Smith, R.L. 1974. *Ecology and Field Biology.* New York: Harper & Row.

Smith, R.L. 1996. *Ecology and Field Biology.* New York: HarperCollins College Publishers.

Smith, T.M., and Smith, R.L. 2006. *Elements of Ecology*, 6th ed. San Francisco: Pearson, Benjamin Cummings.

Spellman, F.R. 1996. *Stream Ecology & Self-Purification.* Lancaster, PA: Technomic Publishing Company.

Suberkoop, K., Godshalk, G.L., and Klug, M.J. 1976. Changes in the chemical composition of leaves during processing in a woodland stream. *Ecology* 57: 720–727.

Tansley, A.G. 1935. The use and abuse of vegetational concepts and terms. *Ecology* 16: 284–307.

Tchobanoglous, G., and Schroeder, E.D. 1985. *Water Supply.* Reading, MA: Addison-Wesley Publishing Company, 1985.

Tomera, A.N. 1989. *Understanding Basic Ecological Concepts.* Portland, ME: J. Weston Walch, Publisher.

Townsend, C.R., Harper, J.L., and Begon, M. 2000. *Essentials of Ecology.* Oxford: Blackwell Science.

USDA. 1982. *Agricultural Statistics 1982.* Washington, D.C.: U.S. Government Printing Office.

USDA. 1999. *Autumn Colors—How Leaves Change Color.* Accessed 02/08/07 @http://www.na.fs.fed.us/spfo/pubs/misc/autumn/autumn_colors.htm.

USDA. 2007. *Agricultural Ecosystems and Agricultural Ecology.* Accessed 02/11/07 @ http://nrcs.usda.gov/technical/ECS/agecol/ecosystem.html.

USFWS. 2007. *Ecosystem Conservation.* U.S. Fish & Wildlife Service. Accessed 02/11/07 @ http://www.fws.gov/ecosystems/.

USGS. 2005. *The Water Cycle: Evapotranspiration.* Accessed 02/16/07 @ http://ga.water.usgs. Gov/edu/watercycleevapotranspiration.html.

Wachernagel, M. 1997. *Framing the Sustainability Crisis: Getting from Concerns to Action.* Accessed 02/26/07 @ http://www.sdri,ubc.ca/publications/wacherna.html.

Wessells, N.K., and Hopson, J.L. 1988. *Biology.* New York: Random House.

15

The Fundamental Unit of Life

Amoebas at the start
Were not complex;
They tore themselves apart
And started Sex.

—Arthur Guiterman

All flesh is grass.

—Isaiah

Introduction

CELLS ARE the fundamental units of life. The cell retains a dual existence as distinct entity and a building block in the construction of organisms. These conclusions about cells were observed and published by Schleiden (1838). Later, Rudolph Virchow added the powerful dictum, "Omnis cellula e cellula" . . . "All cells only arise from pre-existing cells." This important tenet, along with others, formed the basis of what we call Cell Theory. The modern tenets of the Cell Theory include:

- All known living things are made up of cells.
- The cell is a structural and functional unit of all living things.
- All cells come from pre-existing cells by division.

- Cells contain hereditary information which is passed from cell to cell during cell division.
- All cells are basically the same in chemical composition.
- All **energy** flow of life occurs within cells.

The modern tenets, of course, post-dated Robert Hooke's 1663 discovery of cells in a piece of cork, which he examined under his primitive microscope. Hooke drew the cell (actually it was the cell wall he observed) and coined the word CELL. The word cell is derived form the Latin word 'cellula' which means small compartment.

Thus, since the nineteenth century, we have known that all living things, whether animal or plant, are made up of cells. Again, the fundamental unit of all living things, no matter how complex, is the cell. A typical cell is an entity isolated from other cells by a membrane or cell wall. The cell membrane contains protoplasm (the living material found within them) and the nucleus.

In a typical mature plant cell (see Figure 42), the cell wall is rigid and is composed of nonliving material, while in the typical animal cell (see Figure 43)

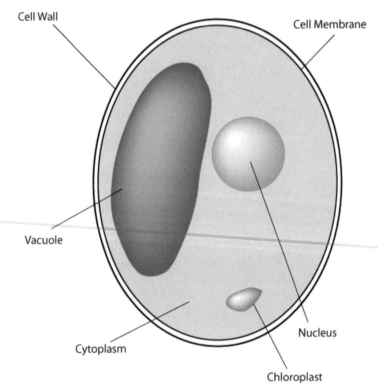

FIGURE 42.   Plant cell

the wall is an elastic living membrane. Cells exist in a very great variety of sizes and shapes, as well as functions. The cell is the smallest functioning unit of a living thing that still has the characteristics of the whole organism. Size ranges from bacteria too small to be seen with the light microscope to the largest single cell known, the ostrich egg. Microbial cells also have an extensive size range, some being larger than human cells.

Did You Know?

The small size of a cell is limited by the volume capable of holding genetic material, proteins, etc., which are necessary to carry out the basic cell functions and reproduction. The large size of a cell is limited by metabolism. A cell must take in adequate amounts of oxygen and nutrients and get rid of wastes.

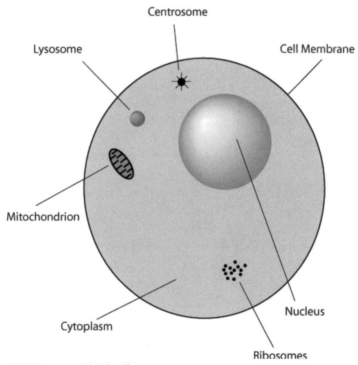

FIGURE 43.   Animal cell

Types of Cells

Cells are of two fundamental types, prokaryotic and eukaryotic. **Prokaryotic** (meaning before nucleus) cells are simpler in design than eukaryotic cells, possessing neither a nucleus nor the organelles (i.e., internal cell structures, each of which has a specific function within the cells) found in the cytoplasm of eukaryotic (meaning true nucleus) cells. Because prokaryotes do not have a nucleus, DNA is in a "nucleiod" region. With the exception of archaebacteria, proteins are not associated with bacterial DNA. Bacteria are the best known and most studied form of prokaryotic organisms (see Figure 44).

Did You Know?

Cells may exist as independent units, e.g., the Protozoa, or as parts of multicellular organisms in which the cells may develop specializations and form tissues and organs with specific purposes.

Prokaryotes are unicellular organisms that do not develop or differentiate into multicellular forms. Some bacteria grow in filaments, or masses of cells, but each cell in the colony is identical and capable of independent existence.

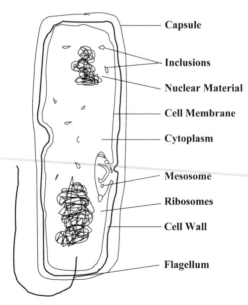

FIGURE 44.   Bacterial cell

Prokaryotes are capable of inhabiting almost every place on the earth, from the deep ocean, to the edges of hot springs, to just about every surface of our bodies.

As mentioned, prokaryotes are distinguished from eukaryotes on the basis of nuclear organization, specifically their lack of a nuclear membrane. Also, prokaryotes are smaller and simpler than eukaryotic cells. Again, prokaryotes also lack any of the intracellular organelles (i.e., internal cell structures, each of which has a specific function within the cell) and structures that are characteristic of eukaryotic cells. Most of the functions of organelles, such as mitochondria, chloroplasts, and the Golgi apparatus, are taken over by the prokaryotic plasma membrane. Prokaryotic cells have three architectural regions: **appendages** called flagella and pili—proteins attached to the cell surface; a cell envelope consisting of a capsule, a cell wall, and a plasma membrane; and a cytoplasmic region that contains the cell genome (DNA) and ribosomes and various sorts of inclusions.

Did You Know?

It is often stated that prokaryotic cells are among the most primitive forms of life on earth. However, it is important to point out that primitive does not mean they are outdated in the evolutionary sense, since primitive bacteria seem little changed, and thus may be viewed as well adapted.

Eukaryotic cells evolved about 1.5 billion years ago. Protists, fungi, plants, and animals have eukaryotic cells—all plants and animals are eukaryotes. They are larger, as much as 10 times the size of prokaryotic cells and most of their genetic material is found within a membrane-bound nucleus (a true nucleus), which is generally surrounded by several membrane-bound organelles. The

Did You Know?

An enormous gap exists between prokaryote cells and eukaryote type cells: "prokaryotes and eukaryotes are profoundly different from each other and clearly represent a marked dichotomy in the evolution of life . . . The organizational complexity of the eukaryotes is so much greater than that of prokaryotes that it is difficult to visualize how a eukaryote could have arisen from any known prokaryote" (Hickman et al. 1997).

presence of these membrane-bound organelles points to the significant difference between prokaryotes and eukaryotes. Although eukaryotes use the same genetic code and metabolic processes as prokaryotes, their higher level of organizational complexity has permitted the development of truly multicellular organisms.

Prokaryotic and eukaryotic cells also have their similarities. All cell types are bounded by a plasma membrane that encloses proteins and usually nucleic acids such as DNA and RNA. Table 15.1 shows a comparison of key features of both cell types.

TABLE 15.1
Comparison of Typical Prokaryotic and Eukaryotic Cells

Characteristic	Prokaryotic	Eukaryotic
Size	1-10 μm	10-100 μm
Nuclear envelope	Absent	Present
Cell wall	Usually	Present (plants)/Absent (animals)
Plasma membrane	Present	Present
Nucleolus	Absent	Present
DNA	Present (single loop)	Present
Mitochondria	Absent	Present
Chloroplasts	Absent	Present (plants only)
Endoplasmic reticulum	Absent	Present
Ribosomes	Present	Present
Vacuoles	Absent	Present
Golgi apparatus	Absent	Present
Lysosomes	Absent	Often Present
Cytoskeleton	Absent	Present

Did You Know?

Plant cells can *generally* be distinguished from animal cells by: (1) the *presence* of cell walls, chloroplasts, and central vacuoles in *plants* and their absence in animals, and (2) the *presence* of lysosomes and centrioles in *animals* and their absence in plants.

Cell Structure

As mentioned, cells contain structures called organelles. The **structure and function** of the major organelles in eukaryotic and prokaryotic cells are described below.

- **Plasma Membrane**—is located between the cell and its environment; it is a semi-permeable phospholipid bilayer. The membrane serves to separate and protect a cell from its surrounding environment and is made mostly from a double layer of proteins and **lipids**, fat-like molecules. The proteins within the membrane have a variety of functions. For example, some proteins are receptors which can detect the presence of certain kinds of molecules in the surrounding fluids. Also embedded within this membrane is a variety of other molecules that act as channels and pumps, moving different molecules into and out of the cell. Cells that are specialized for absorption (e.g., intestinal cells) have folds in the plasma membrane called *microvilli* that increase the surface area. Pseudopodia are temporary extensions of the plasma membrane used for movement or to engulf particles. A form of plasma membrane is also found in prokaryotes, but is usually referred to as the cell membrane.
- **Nucleus**—is the control center of the cell. The nucleus contains DNA and the place where RNA synthesis occurs; it is therefore the cell's control center because DNA contains instructions needed to produce proteins that control metabolism and other cell functions. *Chromatin* is the grainy threadlike DNA. During cell division, the nuclear membrane disintegrates and the DNA becomes coiled, producing visible structures called *chromosomes*. The material within the nucleus is referred to as the nucleoplasm. The nucleus is spheroid in shape and separated from the cytoplasm by a double membrane called the *nuclear envelope*. The nuclear envelope isolates and protects a cell's DNA from various molecules that could accidentally damage its structure or interfere with its processing. *Nuclear pores* allow materials to pass into and out of the nucleus. During processing, DNA is transcribed, or synthesized, into a special RNA, called mRNA. This mRNA is then transported out the nucleus, where it is translated into a specific protein molecule. In prokaryotes, DNA processing takes place in the cytoplasm.
- **Nucleolus**—is a structure made up of RNA and proteins within the nucleus where the ribosomal subunits are produced. The ribosomes exit pores in the nuclear envelope, and enter the cytoplasm, where they are involved in protein synthesis.
- **Endoplasmic Reticulum (ER)**—is continuous with the nuclear envelope and the plasma membrane. The ER is the transport network for molecules targeted for certain modifications and specific destinations, as compared to molecules that will float freely in the cytoplasm. ER has two forms that differ in structure and function: the *rough ER* and the *smooth ER*. In the rough ER, the rough appearance of rough endoplasmic reticulum is due to the presence of ribosomes on the membrane. The rough ER functions

in protein synthesis, especially proteins that are to be secreted to outside the cell. Proteins enter the interior (lumen) of the endoplasmic reticulum while being synthesized. The rough ER also functions in the modifications of newly formed proteins. For example, some **enzymes** may add carbohydrates chains forming glycoproteins. Other enzymes function to hold the newly-synthesized proteins into their proper shape. In regards to transport of molecules through the rough ER, the molecules go through the internal channel of the ER and fold into their own conformation. The ER surrounds the molecules with vesicles (sacs) that pinch off the ER or Golgi apparatus (discussed below) and transport the molecules (e.g., secretory proteins, such as insulin) to other parts of the cell.

The smooth ER has no ribosomes attached to it. As mentioned, it is continuous with rough ER. The smooth ER reticula have a variety of different functions but often function to produce lipid compounds such as phospholipids, steroids, and fatty acids. Certain kinds of cells have smooth endoplasmic reticulum with a specialized function, including producing steroid hormones in the adrenal cortex and testes; the smooth ER of liver cells helps detoxify drugs in the blood; and calcium ions needed for contraction are stored in the smooth ER of muscle cells. Vesicles pinch off the smooth ER and carry materials to other parts of the cell such as the Golgi apparatus or the plasma membrane.

- **Golgi Apparatus** (sometimes called a Golgi body or Golgi complex)—is a stack of 3 to 20 flattened, slightly curved sacs which appear like a stack of pancakes that is a continuation of the ER. The Golgi apparatus is where exportable vesicles of ER proteins are passed into the forming face for further processing, modifying, and packaging, then transported from the maturing face to a variety of other cellular locations.
- **Lysosomes** ("bodies that dissolve")—are formed in the Golgi apparatus and are often referred to as the garbage disposal system of the cell. Lysosomes are somewhat spherical, membrane-bounded vesicles of hydrolytic enzymes, enabling them to digest all major classes of macromolecules. Lysosomes can contain more than three dozen enzymes for degrading proteins, nucleic acids, and certain sugars called polysaccharides. Lysosomes also digest foreign bacteria that invade the cell. All of these enzymes work best at an optimal acidic environment (pH 5.0). Lysosomes point out the importance behind compartmentalization of the eukaryotic cell. The cell could not house such destructive enzymes if they were not contained in a membrane-bound system.
- **Mitochondria**—are commonly referred to as the powerhouse of the cell. Mitochondria are self-replicating organelles that contain their own DNA and occur in various numbers, shapes, and sizes in the cytoplasm of all

eukaryotic cells. Mitochondria have an external membrane and an inner membrane with numerous folds called *cristae*. Cristae project into the gel-like matrix. Enzymes involved in cellular respiration are found in the matrix and embedded in the membrane of the cristae. It is composed of three types of protein fibers: microtubules, actin filaments, and intermediate filaments. These protein fibers function to move materials within the cell, move the cell, and provide mechanical support. Mitochondria convert fuel to useable energy. It is semiautonomous, and replicates itself.

- **Peroxisomes**—are vesicles found in nearly all eukaryotes that function to rid the body of toxic substances, such as hydrogen peroxide, or other metabolites, and contain enzymes concerned with oxygen utilization. Peroxisomes often resemble a lysosome. However, peroxisomes are self-replicating, whereas lysosomes are formed in the Golgi apparatus. Peroxisomes also have membrane proteins that are critical for various functions, such as for importing proteins into their interiors and to proliferate and segregate into daughter cells. Peroxisome enzymes are also involved in the breakdown of fatty acids to acetyl CoA, which is transported to the mitochondria for fuel during cellular respiration. The liver, where toxic byproducts accumulate, contains high numbers of peroxisomes.

- **Vacuoles**—are membranous sacs similar to, but larger than vesicles. They store water and dissolved substances. They are more important in plant cells. Most of the center of a plant cell is occupied by a long *central vacuole*, which gives support because pressure within the vacuole makes the cell turgid (rigid). The cell wall prevents the cell from bursting. Some organisms (e.g., some fresh-water protozoa) have specialized *contractile vacuoles* for eliminating excess water and *food vacuoles* (formed by phagocytosis, the process in which white blood cells consume particles) that contain food within the cell.

- **Ribosomes**—are cytoplasmic granules composed of RNA and protein, at which protein synthesis takes place. Ribosomes are large complexes composed of many molecules found in both prokaryotes and eukaryotes. Protein synthesis is the process by which proteins are made from individual amino acids. Ribosomes read the code in messenger RNA (mRNA) and synthesize protein accordingly. Several ribosomes may be attached to a strand of mRNA forming a unit called a *polysome*. The process of converting mRNA's genetic code into the exact sequence of amino acids that make up a protein is called *translation*. A ribosome is composed of two subunits, one large and one small, each having a different function during protein synthesis. In eukaryotic cells, the subunits are synthesized in the *nucleolus* and move into the cytoplasm. During the process of protein synthesis, two subunits will come together along with mRNA. Ribosomes

are composed of both RNA (called ribosomal RNA and tRNA) and protein. Ribosomes in eukaryotes are about 33% larger than those in prokaryotes.

- **Cell Wall**—functions to give shape, support, and protect the cell; they are
  found in almost all plant cells. The cell wall surrounds the cell on the outside and is the secretory product of the cytoplasm. The cell wall is permeable to water. Plants have cell walls composed of cellulose; fungi have walls
  composed of chitin.
- **Cytoskeleton** (the cell's scaffold)—is an important, complex, and dynamic cell component. It acts to organize and maintain the cell's shape;
  anchors organelles in place; helps during endocytosis, the uptake of external materials by a cell; and moves parts of the cell in processes of growth
  and motility. There is a great number of proteins associated with the cytoskeleton, each controlling a cell's structure by directing, bundling, and
  aligning filaments. There are three types of cytoskeleton filaments or fibers: microtubules, microfilaments, and intermediate filaments.

  - **Microtubules**—are in all eukaryotic cells. They are straight, hollow fi-
    bers made from globular proteins called *tubulin*. They have many func-
    tions including cellular support, organelle movement, cell motility (they
    move the cilia and flagella), and the separation of chromosomes during
    cell division. The assembly of microtubules is controlled by an area near
    the nucleus called the *centrosome* or microtubule organizing area.
  - **Microfilaments**—are made up of two intertwined strands of the pro-
    tein actin. They provide cellular support when they combine with other
    proteins just inside of the plasma membrane and play a role in cell
    shape. They participate in muscle contraction and localized contrac-
    tion of cells (i.e., they move the cell).
  - **Intermediate filaments**—are composed of keratin subunits and are
    more permanent than either microtubules or microfilaments. They
    may be the framework of the cytoskeleton; they provide mechanical
    support. They reinforce cell shape and probably fix an organelle's posi-
    tion in the cell. These fibers line the interior of the nuclear envelope.

- **Cytoplasm** (cell's inner space)—is the material that lies within the cytoplasmic membrane, or the membrane that surrounds a cell. It contains
  none of a cell's genetic material, because this is contained in the nucleus.
  It does, however, contain a lot of water, and the other organelles of the
  cells. It provides a platform upon which they can operate within the cell.
  It is made up of proteins, vitamins, ions, nucleic acids, amino acids, sugars, carbohydrates, and fatty acids. All of the functions for cell expansion,
  growth, and replication are carried out in the cytoplasm of a cell. The

cytoplasm also contains many salts and is an excellent conductor of electricity, creating the perfect environment for the mechanics of the cell. The function of the cytoplasm, and the organelles which reside in it, are critical for the cell's survival.

- **Chloroplast**—is a large, complex double membraned organelle that performs the function of photosynthesis within plant cells and contains the substance chlorophyll that is essential for the process. All reactions of photosynthesis occur in the chloroplast, and in addition, the chloroplasts also create sugar from the sun for the cell and make all the food for other organelles. The chloroplasts use photosynthetic chlorophyll pigment and take in sunlight, water, and carbon dioxide to produce glucose and oxygen. This is the process of photosynthesis.
- **Cilia and Flagella**—are hairlike structures projecting from the cell that function to move the cell by their movements (e.g., sperm are motile; they use flagella to move). They contain cytoplasm and are enclosed by the plasma membrane. Cells that contain cilia are *ciliated*. Cilia are shorter than flagella but are similar in construction. Cilia are shorter than flagella but are similar in construction. Cilia and flagella are formed from a core of nine outer microtubules and two inner single microtubules ensheathed in an extension of the plasma membrane.

Intercellular Junctions (Animal Cells)

The way cells interact with neighbor cells varies, but they all are virtually in contact with each other. Contact is maintained via *cell junctions*. These junctions function to anchor cells to one another or to provide a passageway for cellular exchange. There are at least three different types of cell junctions (contacts): desmosomes, tight junctions, and gap junctions.

- **Desmosomes**—are strong protein attachments between adjacent *animal* cells. They are interacting complementary folds of membrane. Desmosomes act like spot welds or interlinking fingers to hold together tissues, such as skin or heart muscle tissues that undergo considerable stress.
- **Tight Junctions**—are tight bans of proteins that prevent fluids and small molecules from crossing the membrane. The junction completely encircles each cell, preventing the movement of material between the cells. Tight junctions in the stomach lining protect the stomach cells from hydrochloric acid and are characteristic of *animal* cells lining the digestive tract where materials are required to pass through cells to penetrate the bloodstream.

- **Gap Junctions**—are narrow tunnels (doorways) between *animal* cells that consist of proteins called *connexions*. The proteins prevent the cytoplasm of each cell from mixing, but allow the passage of ions and small molecules. In this manner, they allow the flow of materials and electrical charge.

References and Recommended Reading

Cibas, E.S., and Ducatman, B.S. 2003. *Cytology: Diagnostic Principles and Clinical Correlates*. London: Saunders Ltd.

DeDuve, C. 1984. *A Guided Tour of the Living Cell*. New York: Scientific Library. W.H. Freeman.

Finean, J.B. 1984. *Membranes and Their Cellular Functions*. Oxford, Boston: Blackwell Scientific Publications.

Frank, J., et al. 1995. A model of synthesis based on cryo-electron microscopy of the E. coli ribosome. *Nature* 376: 440–444.

Garrett, R., et al. 2000. *The Ribosome: Structure, Function, Antibiotics, and Cellular Interactions*. American Society Microbiology.

Hickman, C.P., et al. 1997. *The Biology of Animals*, 7th ed. New York: William C. Brown/McGraw Hill.

Martin, S. 1981. *Understanding Cell Structure*. Cambridge, New York: Cambridge University Press.

Murray, A.W. 1993. *The Cell Cycle: An Introduction*. New York: W.H. Freeman.

Schleiden, M. 1838. *Concept of Photogenesis*.

Serafini, A. 1993. *The Epic History of Biology*. New York: Plenum.

Spirin, A. 1986. *Ribosomes Structure and Protein Biosynthesis*. Menlo Park, CA: The Benjamin/Ammins Publishing Co. Inc.

Thomas, L. 1995. *The Life of a Cell: Notes of a Biology Watcher*. New York: Penguin Books.

16

Biological Diversity

Introduction

NATIONAL INSTITUTES OF HEALTH (NIH 2006) makes the point that perhaps the best way to illustrate the importance of biodiversity is by analogy to the diversity of human knowledge stored in books (an argument made by Tom Lovejoy and others). When the library in Alexandria was consumed by fire in 391 AD, when Constantinople was sacked in 1453, or when Maya codices were burned in the sixteenth century, thousands of works of literature were destroyed. Hundreds of works of genius are now known to us only by their titles, or from quoted fragments. Thousands more will never be known; several millennia of collective human memory have been irretrievably lost.

Like books, living species represent a kind of memory, the cumulative record of several *million* millennia of evolution. Every species has encountered and survived countless biological problems in its evolutionary history; molecules, cells, and tissues record their solutions. Because we are biological beings ourselves, nature offers a vast library of solutions to many of our current health, environmental, and economic problems. Unfortunately that precious and irretrievable information is now being destroyed at an unprecedented rate.

Fast-forward to the present. Biological diversity, or "biodiversity," refers to the variety of genes, species, and ecosystems found on Earth. Population pressures and demographic changes threaten biodiversity worldwide. Our cities produce a growing stream of industrial poisons and human wastes. In the countryside, overproduction depletes the soil, and pesticides contaminate water supplies; deforestation for farming, pasture, and building material leads to erosion and heavy flooding. The resulting contraction of natural habitats—including the destruction of species-rich tropical forests—will have profound consequences for the future.

We have identified more than 2 million species of organisms on earth, but estimate 40 million species inhabit the earth. Some estimate that there may be millions of species in the tropical rain forest and an unspecified number living in the oceans yet undiscovered. Then there is this: Of the millions of species presently identified how many do we really understand, know, or can accurately explain? When studying the various species, it is important to remember the words of Ralph Waldo Emerson: "What is a weed? A plant whose virtues have not yet been discovered."

Classification

For centuries, scientists classified the forms of life visible to the naked eye as either animal or plant. The Swedish naturalist Caroulus Linnaeus organized much of the current knowledge about living things in 1735.

The importance of classifying organisms cannot be overstated. Without a classification scheme, how could we establish criteria for identifying organisms and arranging similar organisms into groups? The most important reason for classification is that a standardized system allows us to handle information efficiently—it makes the vastly diverse and abundant natural world less confusing.

Linnaeus's classification system was extraordinarily innovative. His binomial system of nomenclature is still with us today. Under the binomial system, all organisms are generally described by a two-word scientific name, the genus and species. Genus and species are groups that are part of a hierarchy of groups of increasing size, based on their nomenclature (taxonomy). This hierarchy is:

Kingdom
    Phylum
        Class
            Order
                Family
                    Genus
                        Species

Using this hierarchy and Linnaeus's binomial system of nomenclature, the scientific name of any organism (as stated previously) includes both the genus and the species name. The genus name is always capitalized, while the species name begins with a lowercase letter. On occasion, when little chance or confusion is present, the genus name is abbreviated to a single capital letter. The names are always in Latin, so they are usually printed in italics or underlined. Some organisms also have English common names. Some microbe names of interest, for example, are listed as follows:

- *Salmonella typhi*—the typhoid bacillus
- *Escherichia coli*—a coliform bacterium
- *Giardia lamblia*—a protozoan

Kingdoms of Life

Linnaeus classified all then-known (1700s) organisms into two large groups: the kingdoms Plantae and Animalia. In 1969, Robert Whittaker proposed five kingdoms: Monera, Protista, Fungi, Plantae, and Animalia. Other schemes involving an even greater number of kingdoms have lately been proposed, however this text employs Whittaker's five kingdoms. Moreover, recent studies suggest that three domains (super-kingdoms) be employed: Archaea, Bacteria, and Eukarya.

The basic characteristics of each kingdom are summarized in the following:

1. **Kingdom Monera** (10,000 species)—unicellular and colonial—including *archaebacteria* (from the Greek meaning "ancient") and *eubacteria* (meaning "true"). Archaebacteria include methanogens (producers of methane), halophiles (live in bodies of concentrated salt water) and thermocidophiles (live in the hot acidic waters of sulfur springs). Eubacteria include heterotrophs (decomposers), autotrophs (make food from photosynthesis), and proteobacteria (one of largest phyla of bacteria). All prokaryotic cells (without nuclei and membrane-bound organelles) are in this kingdom. Both reproduce by binary fission, but they do have some ways to recombine genes, allowing change (evolution) to occur.

2. **Kingdom Protista** (250,000 species)—unicellular protozoans and unicellular and multicellular (macroscopic) algae with cilia and flagella. Kingdom Protista contains all eukaryotes that are not plants, animal, or fungi. Includes amoebas and *euglena*.

3. **Kingdom Fungi** (100,000 species)—eukaryotes, multicellular, and heterotrophic, having multinucleated cells enclosed in cells with cell walls. Fungi

act either as decomposers or as parasites in nature. Includes molds, mildews, mushrooms, and yeast.

4. **Kingdom Plantae** (250,000 species)—immobile, eukaryotes, multicellular and carry out photosynthesis (autotrophs) and have cells encased in cellulose cell walls. Plants are important sources of oxygen, food, and clothing/construction materials, as well as pigments, spices, drugs, and dyes.

5. **Kingdom Animalia** (1,000,000 species)—multicellular, eukaryotes, heterotrophic, without photosynthetic pigment, and mostly move from place to place. Animal cells have no cell walls.

Bacteria

Of all organisms, bacteria are the most widely distributed, the smallest in size, the simplest in morphology (i.e., structure; see Figure 44), the most difficult to classify, and the hardest to identify. Because of their considerable diversity, even providing a descriptive definition of what a bacterial organism is can be difficult. About the only generalization that can be made for the entire group is that they are single-celled, prokaryotic, seldom photosynthetic, and reproduce by binary fission.

Bacteria cells are usually measured in microns, or micrometers, m; 1 m = 0.001 or 1/1000 of a millimeter, mm. A typical coliform bacterial cell that is rod-shaped is about 2 m long and about 0.7 microns wide. The size of each cell changes with time during growth and death.

Did You Know?

Note that recent practice is to place archaebacteria in a separate kingdom, Kingdom Archaebacteria. This is the case because data from DNA and RNA comparisons indicate that archaebacteria are so different that they should not even be classified with bacteria. Thus, a separate and distinct classification scheme higher than kingdom has been devised to accommodate the archaebacteria, called *domain*. In this new system, these organisms are now placed in the domain *Archaea*— the chemosynthetic bacteria. Other prokaryotes, including eubacteria are placed in the domain *Bacteria*—the disease-causing bacteria. All the kingdoms of eukaryotes, including Protista, Fungi, Plantae, and Animalia, are placed in the domain *Eukarya*.

The arrangement of bacterial cells, viewed under the microscope, may be seen as separate (individual) cells or as cells in groupings. Within their species, cells may appear in pairs (diplo), chains, groups of four (tetrads), cubes (Sarcinae), and in clumps. Long chains of cocci result when cells adhere after repeated divisions in one plane; this pattern is seen in the genera *Enterococcus* and *Lactococcus*. In the genus *Sarcina*, cocci divide in three planes, producing cubical packets of eight cells. The shape of rod-shaped cells varies, especially the rod's end, which may be flat, cigar-shaped, rounded, or bifurcated. While many rods do occur singly, they may remain together after division to form pairs or chains (see Figure 45). These characteristic arrangements are frequently useful in bacterial identification.

Bacteria are found everywhere in our environment. They are present in soil, in water, and in the air. Bacteria are also present in and on the bodies of all living creatures—including people. Most bacteria do not cause disease; they

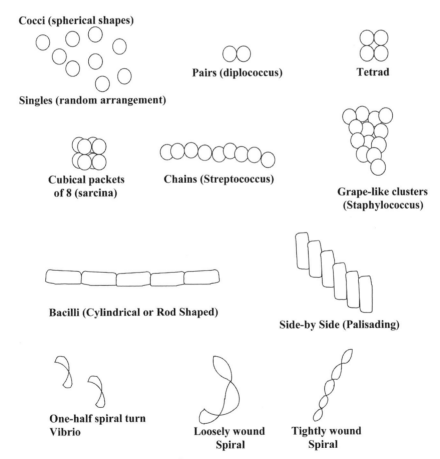

FIGURE 45. Bacterial shapes and arrangements

are not pathogenic. Many bacteria carry on useful and necessary functions, related to the life of larger organisms.

However, when we think about bacteria in general terms, we usually think of the damage they cause. In water, for example, the form of water pollution that poses the most direct menace to human health is bacteriological contamination, part of the reason that bacteria are of great significance to water and wastewater specialists. For water treatment personnel tasked with providing the public with safe, potable water, disease-causing bacteria pose a constant challenge (see Table 16.1).

The conquest of disease has placed bacteria high on the list of microorganisms of great interest to the scientific community. There is more to this interest and accompanying large research effort than just an incessant search for understanding and the eventual conquest of disease-causing bacteria. Not all bacteria are harmful to man. Some, for example, produce substances (antibiotics) that help in the fight against disease. Others are used to control insects that attack crops. Bacteria also have an impact on the natural cycle of matter. Bacteria work to increase soil fertility, which increases the potential for more food production. With the burgeoning world population, increasing future food productivity is no small matter.

We have a lot to learn about bacteria, because we are still principally engaged in making observations and collecting facts, trying wherever possible to relate one set of facts to another, but lacking much of a basis for grand unifying theories. Like most learning processes, gaining knowledge about bacteria is a slow and deliberate process. With more knowledge about bacteria, we can minimize their harmful potential and exploit their useful activities.

Bacteria come in three shapes: elongated rods called *bacilli*, rounded or spherical cells called *cocci*, and spirals (helical and curved) called *spirilla* (for the less rigid form) and *spirochaete* (for those which are flexible). Elongated rod-shaped bacteria may vary considerably in length; have square, round, or pointed ends; and may be motile (possess the ability to move) or nonmotile. The spherical-shaped bacteria may occur singly, in pairs, in tetrads, in chains, and in

TABLE 16.1
Disease-causing Bacterial Organisms Found in Polluted Water

Microorganism	Disease
Campylobacter jejuni	campylobacter enteritis
Escherichia coli	E. coli
Salmonella sp.	salmonellosis
Salmonella Typhi	typhoid fever
Shigella sp.	shigellosis
Yersinia entercolitice	yersiniosis

irregular masses. The helical and curved spiral-shaped bacteria exist as slender spirochaetes, spirillum, and bent rods (see Figure 45).

Viruses

Viruses are parasitic intracellular particles that are the smallest living infectious agents known. They are not cellular—they have no nucleus, cell membrane, or cell wall. Viruses, like cells, carry genetic information encoded in their nucleic acid, and can undergo mutations and reproduce; however they cannot carry out metabolism, and thus are not considered alive. They multiply only within living cells (hosts) and are totally inert outside of living cells, but can survive in the environment. Just a single virus cell can infect a host. As far as measurable size goes, viruses range from 20–200 millimicrons in diameter, about 1–2 of magnitude smaller than bacteria. More that 100 virus types excreted from humans through the enteric tract could find their way in to sources of drinking water. In sewage, these average between 100–500 enteric infectious units/1000 ml. If the viruses are not killed in various treatment processes and become diluted by a receiving stream, for example, to 0.1–1 viral infectious units/100 ml, the low concentrations make it very difficult to determine virus levels in water supplies. Since tests are usually run on samples of less than 1 ml, at least 1,000 samples would have to be analyzed to detect a single virus unit in a liter of water.

Viruses differ from living cells in at least three ways: (1) they are unable to reproduce independently of cells and carry out cell division; (2) they possess only one type of nucleic acid, either DNA or RNA; and (3) they have a simple cellular organization. Some viruses that may be transmitted by water include hepatitis A, adeno virus (a DNA virus that causes colds and "pink eye"), polio, Coxsacki's, echo, and Norwalk agent. A virus that infects a bacterium is called a bacteriophage.

Lewis Thomas, in *The Lives of a Cell*, points out that when humans "catch diphtheria it is a virus infection, but not of us." That is, when humans are infected by the virus causing diphtheria, it is the bacterium that it really infects—humans simply "blundered into someone else's accident" (1974, 76). The toxin of diphtheria bacilli is produced when the organism has been infected by a bacteriophage.

Did You Know?

The Papillomavirus is a DNA virus that causes warts. These infectious particles are small, about 15 nm in diameter.

A bacteriophage (phage) is any viral organism whose host is a bacterium. Most of the bacteriophage research that has been carried out has been on the bacterial Escherichia coli, one of the gram-negative bacteria that environmental specialists such as water and wastewater operators are concerned about because it is a dangerous typical coliform.

A virus does not have a cell-type structure from which it is able to metabolize or reproduce. However, when the *genome* (a complete **haploid** set of chromosomes) of a virus is able to enter into a viable living cell (a bacterium) it may "take charge" and direct the operation of the cell's internal processes. When this occurs, the genome, through the host's synthesizing process, is able to reproduce copies of itself, move on, and then infect other hosts. Hosts of a phage may involve a single bacterial species or several bacteria genera.

The most important properties used in classifying bacteriophages are nucleic acid properties and phage morphology. That is, viruses are classified by the type of nucleic acid they contain, and the shape of their protein capsule (*capsid*) bacterial viruses may contain either DNA or RNA; most phages have double-stranded DNA.

Did You Know?

The influenza virus causes the flu. It has RNA as its genetic material instead of DNA.

Many different basic structures have been recognized among phages. Phages appear to show greater variation in form than any other viral group. (Basic morphological structures of various viruses are shown in Figure 46.) The T-2 phage virus has two prominent structural characteristics: The head (a polyhedral capsid) and the tail.

The effect of phage infection depends on the phage and host and to a lesser extent on conditions. Some phages multiply with and *lyse* (destroy) their hosts. When the host lyses (dies and breaks open), phage progeny are released.

Viruses cause a variety of diseases among all groups of living organisms. Viral diseases include the flu, common cold, herpes, measles, chicken pox, small pox, and encephalitis. Antibiotics are not effective against viruses. Vaccination offers protection for uninfected individuals.

Protists

Protists are unicellular and multicellular eukaryotes, which exhibit a great deal of variation in their life cycles. The Protists include heterotrophs, autotrophs,

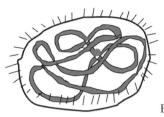

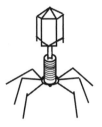

Elaborate Irregular

Long Slender Rod   Geometric Polyhedral

FIGURE 46.   Virus shapes.

and some organisms that can vary their nutritional mode depending on environmental conditions. Protists occur in freshwater, saltwater, soil, and as symbionts within other organisms; they include protozoa, algae, and slime molds. Because of this tremendous diversity, classification of the Protista is difficult.

Did You Know?

Protists are not plants, animals, or fungi, but they act enough like them that scientists believe protists pave the way for the evolution of early plants, animals, and fungi.

Protozoa

The *protozoa* ("first animals" or "little animals") are a large group of eukaryotic organisms (65,000 species, of which more than 50,000 known species have adapted a form or cell to serve as the entire body). All protozoans are single-celled organisms. Typically they lack cell walls, but have a plasma membrane that is used to take in food and discharge waste. They can exist as solitary (colonies are rare) or independent organisms (the stalked ciliates such as *Vorticella* sp., for example) or they can colonize like the sedentary *Carchesium* sp. As the principal hunters and grazers of the microbial world, protozoa play a key role in maintaining the balance of bacterial, algal, and other microbial life. Protozoa are microscopic and get their name because they employ the same type of feeding strategy as animals. The animal-like protozoans differ from animals in that they are unicellular and do not have specialized tissues, organs, or organ systems that carry out life functions. They vary in shape. Most are harmless, but some are parasitic and can be spread by insect vectors. Some forms have two life stages: active *trophozoites* (vegetative state, capable of feeding and growing) and dormant *cysts* (survival state; form cyst when conditions are bad or need to move from one host to the next; they convert to trophozoite in favorable conditions).

Did You Know?

Although they are efficient hunters and grazers, feed on bacteria, and also eat other protozoa and bits of material that has come off of other living things (organic matter), protozoans also are themselves an important food source for larger creatures and the basis of many food chains.

As mentioned, as unicellular eukaryotes, protozoa cannot be easily defined because they are diverse and, in most cases, only distantly related to each other. Also, again, protozoa are distinguished from bacteria by their eukaryotic nature and by their usually larger size. Protozoa are distinguished from algae because protozoa obtain energy and nutrients by taking in organic molecules, detritus, or other protists rather than from photosynthesis. Each protozoan is a complete organism and contains the facilities for performing all the body functions for which vertebrates have many organ systems.

Like bacteria, protozoa depend upon environmental conditions (the protozoan quickly responds to changing physical and chemical characteristics of the environment), reproduction, and availability of food for their existence. Relatively large microorganisms, protozoans range in size from 4 microns to about 500 microns. They can both consume bacteria (limit growth) and feed on organic matter (degrade waste).

Interest in types of protozoa is high among water treatment practitioners because certain types of protozoans can cause disease. In the United States, the most important of the pathogenic parasitic protozoans is *Giardia lamblia*, which causes a disease known as *giardiasis*. Two other parasitic protozoans that carry waterborne disease are *Entamoeba histolytica* (amoebic dysentery) and *Cryptosporidia* (cryptosporidosis).

Protozoa are divided into four groups based on their method of motility and reproduction as shown in Table 16.2.

TABLE 16.2
Classification of Protozoans

Group	Common name	Movement	Reproduction
Ciliophora	Ciliates	Cili	Asexual by transverse fission
			Sexual by conjugation
Mastigophora	Flagellates	Flagella	Asexual
Sarcodina	Amoebas	Pseudophodia	Asexual & Sexual
Sporozoa	Sporozoans	nonmotile	Asexual & Sexual

- **Mastigophora** (flagellates)—are mostly unicellular, lack specific shape (have an extremely flexible plasma membrane that allows for the flowing movement of cytoplasm), and possess whip-like structures called flagella. The flagella, which can move in whip-like motion (to move in a relatively straight path, or create currents that spin them through fluids), are used for locomotion, as sense receptors, and to attract food. These organisms are common in both fresh and marine waters. The group is subdivided into the *phytomastigophrea*, most of which contain chlorophyll and are thus plant-like. A characteristic species of *phytomastigophrea* is the *Euglena* sp., often associated with high or increasing levels of nitrogen and phosphate in the wastewater treatment process.

Did You Know?

The mastigophora trypanosomes require two hosts, one a mammal, to complete their life cycle, and cause the diseases African sleeping sickness, Chagas disease, and leishmaniasis. Trichonymphs are symbionts inside the intestines of termites.

- **Ciliophora** (ciliates)—are the most advanced and structurally complex of all protozoans. They are heterotrophic and use multiple small cilia for locomotion. The *Paramecium* is probably the most commonly studied ciliate in basic biology classes. Movement and food-getting is accomplished with short hairlike structures called cilia that are present in at least one stage of the organism's life cycle. Three groups of ciliates exist: free-swimmers, crawlers, and stalked. The majority are free-living. They are usually solitary, but some are colonial and others are sessile. They are unique among protozoa in having two kinds of nuclei: a micronucleus

and a macronucleus. The micronucleus is concerned with sexual reproduction. The macronucleus is involved with metabolism and the production of RNA for cell growth and function.

The ciliate **pellicle** may also act as thick armor. In other species, the
pellicle may be very thin. The cilia are short and usually arranged in rows.
Their structure is comparable to flagella except that cilia are shorter. Cilia
may cover the surface of the animal or may be restricted to banded regions. Like many freshwater protozoans, ciliates are hypotonic; however,
removal of water crossing the cell membrane by osmosis is a problem.
Therefore, one commonly employed mechanism is a contractile vacuole.
Water is collected into the central ring of the vacuole and actively transported from the cell.

- **Sarcodina** (pseudopods, "false feet")—Members of this group have fewer
organelles and are simpler in structure than the ciliates and flagellates.
Sarcodina move about by the formation of flowing protoplasmic projections called *pseudopodia*. The formation of pseudopodia is commonly
referred to as *amoeboid movement*. The *Amoebae* are well known for this
mode of action (see Figure 47). The pseudopodia not only provide a
means of locomotion, but also serve as a means of feeding; this is accomplished when the organism puts out the pseudopodium to enclose the
food. Most amoebas feed on algae, bacteria, protozoa, and rotifers. Several species in the Sarcodina group, including some species of amoebas,
cover themselves with protective shell-like coverings called *tests*. These
tests (made of silica) are stippled with many small and large openings
through which water can flow in and out, and through which the pseudopods protrude.

Did You Know?

Pseudopodia are used by many cells, and are not fixed structures like
flagella but rather are associated with action near the moving edge of
the cytoplasm.

- **Sporozoans**—are protozoans that are obligatory intracellular parasites:
they must spend at least part if not all of their life cycle in a host animal.
They have no special structures used for locomotion. The life cycle often
involves more than one host, such as when plasmodium infects both
mosquitoes and humans, causing human malaria.

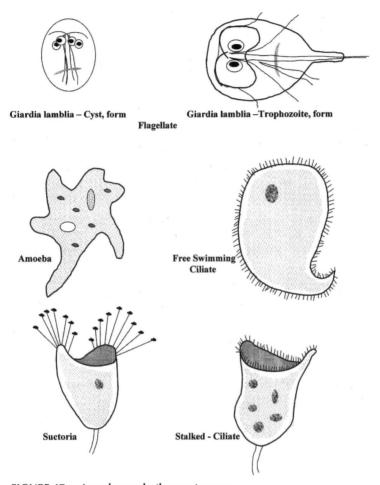

Giardia lamblia – Cyst, form
Flagellate

Giardia lamblia –Trophozoite, form

Amoeba

Free Swimming Ciliate

Suctoria

Stalked - Ciliate

FIGURE 47. *Amoebae* and other protozoans

Did You Know?

The plasmodium includes the malaria parasites transmitted by Anopheles mosquitoes.

Algae

The plant-like protists that perform photosynthesis are called **algae**. Algae can be both a nuisance and an ally. Many ponds, lakes, rivers, streams, and

bays (e.g., Chesapeake Bay) in the United States (and elsewhere) are undergoing *eutrophication*, the enrichment of an environment with inorganic substances (phosphorous and nitrogen). When eutrophication occurs, when filamentous algae like *Caldophora* break loose in a pond, lake, stream, or river and
wash ashore, algae makes its stinking, noxious presence known. Algae provides
basis of food in most aquatic habitats. Algae are allies in many wastewater
treatment operations. They can be valuable in long-term oxidation ponds
where they aid in the purification process by producing oxygen. Algae are also
used for cosmetics, food, and medical products.

Algae are photoautotrophic, contain the green pigment chlorophyll, require
high moisture, and are a form of aquatic plant. Algae differ from bacteria and
fungi in their ability to carry out photosynthesis—the biochemical process
requiring sunlight, carbon dioxide, and raw mineral nutrients. Photosynthesis
takes place in the **chloroplasts**. The chloroplasts are usually distinct and visible. They vary in size, shape, distribution, and number. In some algal types the
chloroplast may occupy most of the cell space. They usually grow near the
surface of water because light cannot penetrate very far through water. Although in mass (multicellular forms like marine kelp) the unaided eye easily
sees them, many of them are microscopic. Algal cells may be nonmotile, motile
by one or more flagella, or exhibit gliding *motility* as in diatoms. They occur
most commonly in water (fresh and polluted water, as well as in salt water), in
which they may be suspended (planktonic) phytoplanktons or attached and
living on the bottom (benthic). A few algae live at the water-atmosphere interface and are termed **Neustonic**. Within the fresh and saltwater environments,
they are important primary producers (the start of the food chain for other
organisms). During their growth phase, they are important oxygen-generating
organisms and constitute a significant portion of the plankton in water.

According to the five kingdom system of Whittaker, the algae belong to seven
divisions distributed between two different kingdoms. Although seven divisions
of algae occur, only five divisions are discussed in this text:

- **Chlorophyta**—Green algae
- **Euglenophyta**—Euglenids
- **Chrysophyta**—golden-brown algae, diatoms
- **Phaeophyta**—Brown algae
- **Pyrrophyta**—Dinoflagellates

The primary classification of algae is based on cellular properties. Several
characteristics are used to classify algae, including: (1) cellular organization
and cell wall structure; (2) the nature of chlorophyll(s) present; (3) the type
of motility, if any; (4) the carbon polymers that are produced and stored; and

(5) the reproductive structures and methods. Table 16.3 summarizes the properties of the five divisions discussed in this text.

Algae show considerable diversity in the chemistry and structure of their cells. Some algal cell walls are thin, rigid structures usually composed of cellulose modified by the addition of other polysaccharides. In other algae, the cell wall is strengthened by the deposition of calcium carbonate. Other forms have chitin present in the cell wall. Complicating the classification of algal organisms are the Euglenids, which lack cell walls. In diatoms the cell wall is composed of silica. The **frustules** (shells) of diatoms have extreme resistance to decay and remain intact for long periods of time, as the fossil records indicate.

The principal feature used to distinguish algae form other microorganisms (for example, fungi) is the presence of chlorophyll and other photosynthetic pigments in algae. All algae contain chlorophyll a. Some, however, contain other types of chlorophylls. The presence of these additional chlorophylls is characteristic of a particular algal group. In addition to chlorophyll, other pigments encountered in algae include fucoxanthin (brown), xanthophylls (yellow), carotenes (orange), phycocyanin (blue), and phycoerythrin (red).

TABLE 16.3
Comparative Summary of Algal Characteristics

Algal Group	Common Name	Structure	Pigments	Carbon Reserve	Motility	Reproduction
Chlorophyta	Green algae	Unicellular to Multicellular	Chlorophylls a and b, carotenes, xanthophylls	Starch, oils	Most are nonmotile	Asexual and sexual
Euglenophyta	Euglenoids	Unicellular	Chlorophylls a and b, carotenes, xanthophylls	Fats	Motile	Asexual
Chrysophyta	Golden brown algae, diatoms	Multicellular	Chlorophylls a and b, special carotenoids, xanthophylls	Oils	Gliding by diatoms; others by flagella	Asexual and sexual
Phaeophyta	Brown algae	Unicellular	Chlorophylls a and b, carotenoids xanthophylls	Fats	Motile	Asexual and sexual
Pyrrophyta	Dinoflagellated	Unicellular	Chlorophylls A and b, Carotenes, xanthophylls	Starch	Motile	Asexual; sexual rare

Many algae have flagella (a threadlike appendage). As mentioned, the flagella are locomotor organelles that may be the single polar or multiple polar types. The *Euglena* is a simple flagellate form with a single polar flagellum. Chlorophyta have either two or four polar flagella. Dinoflagellates have two flagella of different lengths. In some cases, algae are nonmotile until they form motile gametes (a haploid cell or nucleus) during sexual reproduction. Diatoms do not have flagella, but have gliding motility.

Algae can be either autotrophic or heterotrophic. Most are photoautotrophic; they require only carbon dioxide and light as their principal source of energy and carbon. In the presence of light, algae carry out oxygen-evolving photosynthesis; in the absence of light, algae use oxygen. Chlorophyll and other pigments are used to absorb light energy for photosynthetic cell maintenance and reproduction. One of the key characteristics used in the classification of algal groups is the nature of the reserve polymer synthesized as a result of utilizing carbon dioxide present in water.

Algae may reproduce either asexually or sexually. Three types of **asexual reproduction** occur: binary fission, **spores**, and **fragmentation**. In some unicellular algae, binary fission occurs where the division of the cytoplasm forms new individuals like the parent cell following nuclear division. Some algae reproduce through spores. These spores are unicellular and germinate without fusing with other cells. In fragmentation, the **thallus** breaks up and each fragment grows to form a new thallus.

Sexual reproduction can involve union of cells where eggs are formed within vegetative cells called **Oogonia** (which function as female structures) and sperm are produced in a male reproductive organ called antheridia. Algal reproduction can also occur through a reduction of chromosome number and/or the union of nuclei.

Characteristics of Algal Divisions

- **Chlorophyta** (Green Algae)—The majority of algae found in ponds belong to this group; they also can be found in salt water and soil. Several thousand species of green algae are known today. Many are unicellular; others are multicellular filaments or aggregated colonies. The green algae have chlorophylls a and b, along with specific carotenoids, and they store carbohydrates (food) as starch. Few green algae are found at depths greater than 7–10 meters, largely because sunlight does not penetrate to that depth. Some species have a holdfast structure that anchors them to the bottom of the pond and to other submerged inanimate objects. Green algae reproduce by both sexual and asexual means. Multicellular green algae have some division of labor, producing various reproductive cells and structures.

- **Euglenophyta** (**Euglenoids**)—are a small group of unicellular microorganisms that have a combination of animal and plant properties. Euglenoids lack a cell wall, possess a gullet, have the ability to ingest food, have the ability to assimilate organic substances, and, in some species, are absent of chloroplasts. They occur in fresh, brackish, and salt waters, and on moist soils. A typical *Euglena* cell is elongated and bound by a plasma membrane; the absence of a cell wall makes them very flexible in movement. Inside the plasma membrane is a structure called the pellicle that gives the organisms a definite form and allows the cell to turn and flex. Euglenoids that are photosynthetic contain chlorophylls *a* and *b*, and they always have a red eyespot (stigma) that is sensitive to light (photoreceptive). Some Euglenoids move about by means of flagellum; others move about by means of contracting and expanding motions. The characteristic food supply for Euglenoids is a lipopolysaccharide. Reproduction in Euglenoids is by simple cell division.

Did You Know?

Some autotrophic species of *Euglena* become heterotrophic when light levels are low.

- **Chrysophycophyta** (Golden Brown Algae)—The Chrysophycophyta group is quite large—several thousand diversified members. They differ from green algae and Euglenoids in that: (1) chlorophylls *a* and *c* are present, (2) fucoxanthin, a brownish pigment, is present; and (3) they store food in the form of oils and leucosin, a polysaccharide. The combination of yellow pigments, fucoxanthin, and chlorophylls causes most of these algae to appear golden-brown. The Chrysophycophyta is also diversified in cell wall chemistry and flagellation. The division is divided into three major classes: golden-brown, yellow-brown algae, and diatom.

Some Chrysophyta lack cell walls; others have intricately patterned coverings external to the plasma membrane, such as walls, plates, and scales. The diatoms are the only group that has hard cell walls of pectin, cellulose, or silicon, constructed in two halves (the **epitheca** and the hypotheca) called a frustule. Two anteriorly attached flagella are common among Chrysophyta; others do not have flagella.

Most Chrysophyta are unicellular or colonial. Asexual cell division is the usual method of reproduction in diatoms; other forms of Chrysophyta can reproduce sexually. Diatoms have direct significance for humans.

Because they make up most of the phytoplankton of the cooler ocean parts, they are the ultimate source of food for fish.

Did You Know?

Diatoms secrete a silicon dioxide shell (frustule) that forms the fossil deposits known as diatomaceous earth, which is used in filters and as abrasives in polishing compounds.

- **Phaeophyta** (Brown Algae)—With the exception of a few freshwater species, all algal species of this division exist in marine environments as seaweed. They are a highly specialized group, consisting of multicellular organisms that are sessile (attached and not free-moving). These algae contain essentially the same pigments seen in the golden-brown algae, but they appear brown because of the predominance of and the masking effect of a greater amount of fucoxanthin. Brown algal cells store food as the carbohydrate laminarin, mannitol, and some lipids. Brown algae reproduce asexually. Brown algae are used in foods, animal feeds, and fertilizers and as a source for alginate, a chemical emulsifier added to ice cream, salad dressing, and candy.
- **Pyrrophyta** (**Dinoflagellates**)—The principal members of this division are the dinoflagellates. The dinoflagellates comprise a diverse group of biflagellated and nonflagellated unicellular, eukaryotic organisms. The dinoflagellates occupy a variety of aquatic environments with the majority living in marine habitats. Most of these organisms have a heavy cell wall composed of cellulose-containing plates. They store food as starch, fats, and oils. These algae have chlorophylls a and c and several xanthophylls. The most common form of reproduction in dinoflagellates is by cell division, but sexual reproduction has also been observed.
    Cell division in dinoflagellates differs from most protistans, with chromosomes attaching to the nuclear envelope and being pulled apart as the nuclear envelope stretches. During cell division in most other eukaryotes, the nuclear envelope dissolves.

Slime and Water Molds

Slime mold, most commonly found in a forest, is a heterotrophic organism that was once regarded as a fungus but later classified with the Protista. They have very complex life cycles involving multiple forms and stages. In their visible, aggregate states, they look like blobs. They may be red, yellow, brown, bright orange, black, white, or blue. They spend part of their life as single-celled forms, but can aggregate to form multicellular forms. Slime molds eat decaying vegetation, bacteria, fungi, and even other slime molds.

Water molds (Oomycota) are decomposers found in freshwater aquatic environments. They are known as downy mildews and white rusts. The downy mildew caused the great Potato Famine in Ireland in 1846–1847.

Did You Know?

Slime molds act like giant amoebas, creeping slowly along and engulfing food particles along the way.

Fungi

Fungi (singular fungus) constitute an extremely important and interesting group of eukaryotic, aerobic microbes ranging from the unicellular yeasts to the extensively mycelial molds. Fungi first evolved in water but made the transition to land through the development of specialized structures that prevented their drying out. Not considered plants, they are a distinctive life-form of great practical and ecological importance. Fungi are important because, like bacteria, they metabolize dissolved organic matter; they are the principal organisms responsible for the decomposition of carbon in the biosphere. Fungi, unlike bacteria, can grow in low moisture areas and in low pH solutions, which aid them in the breakdown of organic matter.

Fungi comprise a large group of organisms that include such diverse forms as molds, mushrooms, puffballs, and yeasts. Because they lack chlorophyll (and thus are not considered plants) they must get nutrition from organic substances. They are either *parasites*, existing in or on animals or plants, or more commonly are *saprophytes*, obtaining their food from dead organic matter. Fungi also are important crop parasites, causing loss of food plants, spoilage of food, and some infectious diseases. Fungi are classified in their own kingdom but the main groups are called divisions rather than phyla. The study of fungi is called *mycology*.

Did You Know?

Fungi range in size from the single-celled organism we know as yeast to the largest known living organism on Earth—a 3.5 mile mushroom dubbed "the humongous fungus," which covers some 2000+ acres in Oregon's Malheur National Forest.

McKinney, in *Microbiology for Sanitary Engineers*, complains that the study of mycology has been directed solely toward classification of fungi and not toward the actual biochemistry involved with fungi. McKinney goes on to point out that for those involved in the sanitary field it is important to recognize the "sanitary importance of fungi . . . and other steps will follow" (1962, p. 40). For students of environmental science, understanding the role of fungi as it relates to the water purification process is important. Environmental practitioners need knowledge and understanding of the organism's ability to function and exist under extreme conditions, which make them important elements in biological waste-stream treatment processes and in the degradation that takes place during waste-composting processes.

Fungi may be unicellular or filamentous. They are large, 5–10 microns wide, and can be identified by a microscope. The distinguishing characteristics of the group, as a whole, include: (1) they are non-photosynthetic, (2) lack tissue differentiation, (3) have cell walls of polysaccharides (chitin), and (4) propagate by spores (sexual or asexual).

Fungi are divided into five classes:

- Myxomycetes, or slime fungi
- Phycomycetes, or aquatic fungi (algae)
- Ascomycetes, or sac fungi
- Basidiomycetes, or rusts, smuts, and mushrooms
- Fungi imperfecti, or miscellaneous fungi

Did You Know?

Although fungi are limited to only five classes, more than 80,000 known species exist.

Fungi differ from bacteria in several ways, including in their size, structural development, methods of reproduction, and cellular organization. They differ from bacteria in another significant way as well: their biochemical reactions (unlike the bacteria) are not important for classification; instead, their structure is used to identify them. Fungi can be examined directly, or suspended in liquid, stained, dried, and observed under microscopic examination where they can be identified by their appearance (color, texture, and diffusion of pigment) or their mycelia.

One of the tools available to environmental science students and specialists for use in the fungal identification process is the distinctive terminology used in mycology. Fungi go through several phases in their life cycle; their structural characteristics change with each new phase.

Fungi can be grown and studied by cultural methods. However, when culturing fungi, use culture media that limit the growth of other microbial types—controlling bacterial growth is of particular importance. This can be accomplished by using special agar (culture media) that depresses pH of the culture medium (usually Sabouraud glucose or maltose agar) to prevent the growth of bacteria. Antibiotics can also be added to the agar that will prevent bacterial growth.

Did You Know?

Fungi can be found in rising bread, moldy bread, and old food in the refrigerator, and on forest floors.

As part of their reproductive cycle, fungi produce very small spores that are easily suspended in air and widely dispersed by the wind. Insects and other animals also spread fungal spores. The color, shape, and size of spores are useful in the identification of fungal species.

Reproduction in fungi can be either sexual or asexual. The union of compatible nuclei accomplishes sexual reproduction. Most fungi form specialized asexual and/or **sexual spore**-bearing structures (fruiting bodies). Some fungal species are self-fertilizing and other species require outcrossing between different but compatible vegetative thalluses (mycelia).

Most fungi are asexual. Asexual spores are often brightly pigmented and give their colony a characteristic color (green, red, brown, black, blue—the blue spores of *Penicillium roquefort* are found in blue or Roquefort cheese).

Did You Know?

Fungi usually reproduce without sex. Single-celled yeasts reproduce asexually by **budding**. A single yeast cell can produce up to 24 offspring.

Asexual reproduction is accomplished in several ways:

• Vegetative cells may bud to produce new organisms. This is very common in the yeasts.
• A parent cell can divide into two daughter cells.
• The most common method of asexual reproduction is the production of spores.

Several types of asexual spores are common:

* A **hypha** may separate to form cells (*arthrospores*) that behave as spores.
* If a thick wall before separation encloses the cells, they are called *chlamydospores*.
* If budding produces the spore, they are called **blastospores**.
* If the spores develop within sporangia (sac), they are called **sporangiospores**.
* If the spores are produced at the sides or tips of the hypha, they are called *conidiospores*.

Fungi are found wherever organic material is available. They prefer moist habitats and grow best in the dark. Most fungi can best be described as grazers, but a few are active hunters. Most fungi are saprophytes, acquiring their nutrients from dead organic matter, gained when the fungi secrete hydrolytic enzymes, which digest external substrates. They are able to use dead organic matter as a source of carbon and energy. Most fungi use glucose and maltose (carbohydrates) and nitrogenous compounds to synthesize their own proteins and other needed materials. Knowing from what materials fungi synthesize their own protein and other needed materials in comparison to what bacteria are able to synthesize is important to those who work in the environmental disciplines for understanding the growth requirements of the different microorganisms.

Did You Know?

Some fungi produce a sticky substance on their hyphae, which then act like flypaper, trapping passing prey.

Plants: The Great Starch Producers

So far as we know, there is no starch made which is not made in plants. Simply, plants are starch factories. Actually, plants are much like us; they have to have food to make them grow. Okay, you might ask: where is the food and how do they find it? Every green leaf is a factory to make food for the plant; the green pulp in the leaf is the internal mechanism; the leaves get the raw materials from the sap and from the air, and the internal mechanism unites them and makes them into plant food. In photosynthesis, plants use light energy to produce glucose from carbon dioxide. The glucose is stored mainly in the form of starch granules (the chief food of starch), in plastids (site of manufacture and storage) such as chloroplasts, and especially amyloplasts (synthesizes and stores starch granules). The internal mechanism is run by sunshine power, so the leaf-factory can make nothing without the aid of light. As the sun rises, the leaf-factories begin to work and stop working when it sets. Starch production is only part of the operation. It has to be changed to sugar before the plant can use it for nourishment and growth. After the leaves make the starch from the sap and the air, it is digested, changing the starch to sugar. This sweet, sugary sap feeds the growing parts of the plant. Again, the starch factory in the leaves can work only in the daytime, but the leaves can change the starch to sugar during the night. Leaves are the workhorses of plants.

The Plant Kingdom

The plant kingdom ranks second in importance only to the animal kingdom (at least from the human point of view). The importance of plants and plant communities to humans and all life on earth and their environment cannot be overstated. Some of the important things plants provide are listed below.

- **Aesthetics**—Plants add to the beauty of the places we live.
- **Medicine**—80% of all medicinal drugs originate in wild plants.
- **Food**—90% of the world's food comes from only 20 plant species.
- **Industrial Products**—Plants are very important for goods they provide (e.g., plant fibers provide clothing); wood is used to build homes; some important fuel chemicals come from plants, such as ethanol from corn and soy diesel from soybeans.
- **Recreation**—Plants form the basis for many important recreational activities, including fishing, nature observation, hiking, and hunting.
- **Air Quality**—The oxygen in the air we breathe comes form the photosynthesis of plants.
- **Water Quality**—Plants aid in maintaining healthy watersheds, streams, and lakes by holding soil in place, controlling stream flows, and filtering sediments from water.
- **Erosion Control**—Plant cover helps to prevent wind or water erosion of the top layer of soil that we depend on.
- **Climate**—Regional climates are impacted by the amount and type of plant cover.
- **Fish and Wildlife Habitat**—Plants provide the necessary habitat for wildlife and fish populations.

- **Ecosystem**—Every plant species serves an important role or purpose in their community.

John Muir (1868) in his work *The Yosemite* pointed out that there was a time before the pioneers settled the western parts of what would eventually become the United States of America when the western landscape was all one variegated plant garden before ploughs and scythes and trampling, biting horses came to make its wide open spaces look like farmers' pasture fields. Nevertheless, countless plants of all varieties still sprout and bloom every year in glorious profusion on the grand talus slopes, endless prairies, wall benches and tablets, and in all the fine, cool side canyons up to the base of the great western mountains, and beyond, higher and higher, to the summits of the western peaks. Even on the open floor and in the easily reached side-nooks many common flowering plants have survived and still make a brave show in the spring and early summer.

TABLE 16.4
Important Differences between Plants and Animals

Plants	Animals
Plants contain chlorophyll and can make their own food.	Animals cannot make their own food and are dependent on plants and other animals for food.
Plants give off oxygen and take in carbon dioxide given off by animals.	Animals give off carbon dioxide which plants need to make food and take in oxygen which they need to breathe.
Plants generally are rooted in one place and do not move on their own.	Most animals have the ability to move fairly freely.
Plants have either no or very basic ability to sense.	Animals have a much more highly developed sensory and nervous system.

Though both are important kingdoms of living things, plants and animals differ in many important aspects. Some of these differences are summarized in Table 16.4.

Although not typically acknowledged, plants are as intricate and complicated as animals. Plants evolved from photosynthetic protists and are characterized by photosynthetic nutrition, cell walls made from cellulose and other polysaccharides, lack of mobility and a characteristic life cycle involving an alternation of generations. The phyla/division of plants and examples is listed in table 16.5.

TABLE 16.5
The Main Phyla/Divisions of Plants

Phylum/Division	Examples
Anthophyta	flowering plants including oak, corn, maize, and herbs
Bryophyta	mosses, liverworts, and hornworts
Coniferophyta	conifers such as redwoods, pines, and firs
Cycadophyta	cycads, sago palms
Ginkophyta	Ginkgo is the only genus
Gnetophyta	shrub trees and vines
Lycophyta	lycopods (look like mosses)
Pterophyta	ferns and tree-ferns

The Plant Cell

The cell was covered earlier, but a brief summary of the plant cell is provided here.

- *Plants have all the organelles animal cells have* (i.e., nucleus, ribosomes, mitochondria, endoplasmic reticulum, Golgi apparatus, etc.).
- *Plants have chloroplasts.* Chloroplasts are special organelles that contain chlorophyll and allow plants to carry out photosynthesis.
- *Plant cells can sometimes have large vacuoles for storage.*
- *Plant cells are surrounded by a rigid cell wall made of cellulose*, in addition to the cell membrane that surrounds animal cells. Those walls provide support.

Vascular Plants

Vascular plants, also called Tracheophytes, have special vascular tissue for transport of necessary liquids and minerals over long distances. **Vascular tissues** are composed of specialized cells that create "tubes" through which materials can flow throughout the plant body. These vessels are continuous throughout the plant, allowing for the efficient and controlled distribution of water and nutrients. In addition to this transport function, vascular tissues also support the plant. The two types of vascular tissue are xylem and **phloem**.

- **Xylem** consists of a tube or a tunnel (pipeline) in which water and minerals are transported throughout the plant to leaves for photosynthesis. In addition to distributing nutrients, xylem (wood) provides structural support. After a time, the xylem at the center of older trees ceases to function in transport and takes on a supportive role only.

- **Phloem** tissue consists of cells called *sieve tubes* and **companion cells**. Phloem tissue moves dissolved sugars (carbohydrates), amino acids, and other producers of photosynthesis from the leaves to other regions of the plant.

The two most important Tracheophytes are gymnosperms (gymno = naked; sperma = seed) and angiosperms (angio = vessel, receptacle, container).

- **Gymnosperms**—The plants we recognize as gymnosperms represent the sporophyte generation (i.e., the spore-producing phase in the life cycle of a plant that exhibits alternation of generation). Gymnosperms were the first tracheophytes to use seeds for reproduction. The seeds develop in protective structures called cones. A gymnosperm contains some cones that are female and some that are male. Female cones produce spores that, after fertilization, become eggs enclosed in seeds that fall to the ground. Male cones produce pollen, which is taken by the wind and fertilizes female eggs by that means. Unlike flowering plants, the gymnosperm does not form true flowers or fruits. Coniferous trees such as firs and pines are good examples of gymnosperms.
- **Angiosperms**—The flowering plants are the most highly evolved plants and the most dominant in present times. They have stems, roots, and leaves. Unlike gymnosperms such as conifers and cycads, angiosperm's seeds are found in a flower. Angiosperm eggs are fertilized and develop into a seed in an ovary that is usually in a flower.

   There are two types of angiosperms: **monocots** and **dicots**.

- **Monocots**—These angiosperms start with one seed-leaf (**cotyledon**); thus, their name, which is derived from the presence of a single cotyledon during embryonic development. Monocots include grasses, grains, and other narrow-leaved angiosperms. The main veins of their leaves are usually parallel and unbranched, the flower parts occur in multiples of three, and a fibrous root system is present. Monocots include orchids, lilies, irises, palms, grasses, and wheat, corn, and oats.
- **Dicots**—Angiosperms in this group grow two seed-leaves (two cotyledons). Most plants are dicots and include maples, oaks, elms, sunflowers, and roses. Their leaves usually have a single main vein or three of more branched veins that spread out from the base of the leaf.

Leaves

The principal function of leaves is to absorb sunlight for the manufacturing of plant sugars in photosynthesis. The leaves' broad, flattened surfaces gather energy from sunlight, while apertures on their undersides bring in carbon dioxide and release oxygen. Leaves develop as a flattened surface in order to present a large area for efficient absorption of light energy. On its two exteriors, the leaf has layers of epidermal cells that secrete a waxy, nearly impermeable cuticle (chitin) to protect against water loss (dehydration) and fungal or bacterial attack. Gases diffuse in or out of the leaf through **stomata**, small openings on the underside of the leaf. The opening or closing of the stomata occurs through the swelling or relaxing of **guard cells**. If the plant wants to limit the diffusion of gases and the transpiration of water, the guard cells swell together and close the stomata. Leaf thickness is kept to a minimum so that gases that enter the leaf can diffuse easily throughout the leaf cells.

Chlorophyll/Chloroplast

The green pigment in leaves is chlorophyll. Chlorophyll absorbs red and blue light from the sunlight that falls on leaves. Therefore, the light reflected by the leaves is diminished in red and blue and appears green. The molecules of chlorophyll are large. They are not soluble in the aqueous solution that fills plant cells. Instead, they are attached to the membranes of disc-like structures, called **chloroplasts**, inside the cells. Chloroplasts are the site of photosynthesis, the process in which light energy is converted to chemical energy. In chloroplasts, the light absorbed by chlorophyll supplies the energy used by plants to transform carbon dioxide and water into oxygen and carbohydrates.

Chlorophyll is not a very stable compound; bright sunlight causes it to decompose. To maintain the amount of chlorophyll in their leaves, plants continuously synthesize it. The synthesis of chlorophyll in plants requires sunlight and warm temperatures. Therefore, during summer chlorophyll is continuously broken down and regenerated in the leaves of trees.

Photosynthesis

Because our quality of life, and indeed our very existence, depends on photosynthesis, it is essential to understand it. In photosynthesis, plants (and other photosynthetic autotrophs) use the energy from sunlight to create the carbohydrates necessary for cell respiration. More specifically, plants take water and carbon dioxide and transform them into glucose and oxygen:

$$6CO_2 + 6H_2O + \text{light energy} \rightarrow C_6H_{12}O_6 + 6O_2$$

This general equation of photosynthesis represents the combined effects of two different stages. The first stage is called the light reaction and the second stage is called the dark reaction. The light reaction is the photosynthesis process in which solar energy is harvested and transferred into the chemical bonds of ATP; it can only occur in light. The dark reaction is the process in which food (sugar) molecules are formed from carbon dioxide from the atmosphere with the use of ATP; it can occur in the dark as long as ATP is present.

Did You Know?

Charles Darwin was the first to discuss how plants respond to light. He found that the new shoot of grasses bend toward the light because the cells on the dark side grow faster than the lighted side.

Roots

Roots absorb nutrients and water, anchor the plant in the soil, provide support for the stem, and store food. They are usually below ground and lack nodes, shoots, and leaves. There are two major types of root systems in plants. Taproot systems have a stout main root with a limited number of side-branching roots. Examples of taproot system plants are nut trees, carrots, radishes, parsnips, and dandelions. Taproots make transplanting difficult. The second type of root system, fibrous, has many branched roots. Examples of fibrous root plants are most grasses, marigolds, and beans. Radiating from the roots is a system of root hairs, which vastly increase the absorptive surface area of the roots. Roots also anchor the plant in the soil.

Growth in Vascular Plants

Vascular plants undergo two kinds of growth (growth is primarily restricted to **meristems**), primary and secondary. Primary growth occurs relatively close to the tips of roots and stems. It is initiated by **apical meristems** and is primarily involved in the extension of the plant body. The tissues that arise during primary growth are called primary tissues and the plant body composed of these tissues is called the primary plant body. Most primitive vascular plants are entirely made up of primary tissues. Secondary growth occurs in some plants; secondary growth thickens the stems and roots. Secondary growth results from the activity of lateral meristems. Lateral meristems are called cambia (**cambium**) and there are two types:

1. **Vascular cambium**—gives rise to secondary vascular tissues (secondary xylem and phloem). The vascular cambium gives rise to xylem to the inside and phloem to the outside.

2. **Cork cambium**—forms the **periderm** (bark). The periderm replaces the epidermis in woody plants.

Plant Hormones

Plant growth is controlled by plant hormones, which influence cell differentiation, elongation, and division. Some plant hormones also affect the timing of reproduction and germination.

- **Auxins**—affect cell elongation (tropism), apical dominance, and fruit drop or retention. Auxins are also responsible for root development, secondary growth in the vascular cambium, inhibition of lateral branching, and fruit development. Auxin is involved in absorption of vital minerals and fall color. As a leaf reaches its maximum growth, auxin production declines. In deciduous plants this triggers a series of metabolic steps which cause the reabsorption of valuable materials (such as chlorophyll) and their transport into the branch or stem for storage during the winter months. Once chlorophyll is gone the other pigments typical of fall color become visible.
- **Kinins**—promote cell division and tissue growth in leaf, stem, and root. Kinins are also involved in the development of chloroplasts, fruits, and flowers. In addition, they have been shown to delay senescence (aging), especially in leaves, which is one reason that florists use cytokinins on freshly cut flowers—when treated with cytokinins they remain green, protein synthesis continues, and carbohydrates do not break down.
- **Gibberellins**—produced in the root growing tips and act as a messenger to stimulate growth, especially elongation of the stem, and can also end the dormancy period of seeds and buds by encouraging germination. Additionally, gibberellins play a role in root growth and differentiation.
- **Ethylene**—controls the ripening of fruits. Ethylene may insure that flows are carpelate (female) while gibberellin confers maleness on flowers. It also contributes to the senescence of plants by promoting leaf loss and other changes.
- **Inhibitors**—restrain growth and maintain the period of dormancy in seeds and buds.

Tropisms: Plant Behavior

Tropism is the movement (and growth in plants) of an organism in response to an external **stimulus**. For example, tropisms, controlled by hormones, are a unique characteristic of sessile organisms such as plants that enable them to adapt to different features of their environment—gravity, light, water, and touch—so that they can flourish. There are three main tropisms:

- **Phototropism**—the tendency of plants growing or bending (moving) in response to light. Phototropism results from the rapid elongation of cells on the dark side of the plant, which causes the plant to bend in the opposite direction. For example, the stems and leaves of a geranium plant growing on the windowsill always turn toward the light.
- **Gravitropism**—refers to a plant's tendency to grow toward or against gravity. A plant that displays positive gravitropism (plant roots) will grow downward, toward the center of earth. That is, gravity causes the roots of plants to grow down so that the plant is anchored in the ground and has enough water to grow and thrive. Plants that display negative gravitropism (plant stems) will grow upward, away from the earth. Most plants are negatively gravitropic. Gravitropism is also controlled by auxin. In a horizontal root or stem, auxin is concentrated in the lower half, pulled by gravity. In a positively gravitropic plant, this auxin concentration will inhibit cell growth on the lower side, causing the stem to bend downward. In a negatively gravitropic plant, this auxin concentration will inspire cell growth on that lower side, causing the stem to bend upward.
- **Thigmotropism**—some people notice that their houseplants respond to thigmotropism (i.e., growing or bending in response to touch), growing better when they touch them and pay attention to them. Touch causes parts of the plant to thicken or coil as they touch or are touched by environmental entities. For instance, tree trunks grow thicker when exposed to strong winds and vines tend to grow straight until they encounter a substrate to wrap around.

Photoperiodism

Photoperiodism is the response of an organism (e.g., plants) to naturally occurring changes in light during a 24-hour period. The site of perception of photoperiod in plants is leaves. For instance, sunflowers are known for their photoperiodism, or their ability to open and close in response to the changing position of the sun throughout the day.

All flowering plants have been placed in one of three categories with respect to photoperiodism: short-day plants, long-day plants, and day-neutral plants.

- **Short-day Plants**—flowering promoted by day lengths shorter than a certain critical day length—includes poinsettias, chrysanthemums, goldenrod, and asters

- **Long-day Plants**—flowering promoted by day lengths longer than a certain critical day length—includes spinach, lettuce, and most grains
- **Day-neutral Plants**—flowering response insensitive to day length—includes tomatoes, sunflowers, dandelions, rice, and corn

Plant Reproduction

Plants can reproduce both sexually and asexually. Each type of reproduction has its benefits and disadvantages. A comparison of sexual and asexual plant reproduction is provided in the following.

- **Sexual Reproduction:**

  - Sexual reproduction occurs when a sperm nucleus from the pollen grain fuses with egg cell from ovary of pistil (pistil defined: the female reproductive structures in flowers, consisting of the stigma, style, and ovary).
  - Each brings a complete set of genes and produces genetically unique organisms.
  - The resulting plant embryo develops inside the seed and grows when seed is germinated.

- **Asexual reproduction:**

  - Occurs when a vegetative part of a plant, root, stem, or leaf, gives rise to a new offspring plant whose genetic content is identical to the "parent plant." An example would be a plant reproducing by root suckers, shoots that come from the root system. The breadfruit tree is an example.
  - Asexual reproduction is also called vegetative propagation. It is an important way for plant growers to get many identical plants from one very quickly.
  - By sexual reproduction plants can spread and colonize an area quickly (e.g., crab grass).

Animals

All animals are members of the Kingdom Animalia. With over 2 million species, Kingdom Animalia is the largest of the kingdoms in terms of its species diversity. Not surprisingly, in regards to diversity among different animal species, it's difficult to imagine what they all might have in common. First, animals are composed of many cells—they are "multicellular." In most animals, these cells are organized into tissues that make up different organs and organ systems. Second, animals must get their food by eating other organisms, such as plants, fungi, and other animals—they are "heterotrophs." Third, animals are eukaryotic. Fourth, animals develop (or not) an internal cavity called a coelom. In addition, all animals require oxygen for their metabolism, can sense and respond to their environment, many animals have tissues specialized for specific functions (nerve tissue, muscle), and have the capacity to reproduce sexually (though many reproduce asexually as well). During their development from a fertilized egg to adult, all animals pass through a series of embryonic stages as part of their normal life cycle.

There are two main types of animals, invertebrates and vertebrates. These types are discussed in the sections below.

Invertebrates

Invertebrates—creatures without backbones—are the most abundant creatures on earth (more than 98% of the known animal species), crawling, flying, floating, or swimming in virtually all of Earth's habitats. Many invertebrates have a fluid-filled, hydrostatic skeleton, like the jellyfish or worm. Others have a hard outer shell, like insect crustaceans. There are many types of invertebrates. The most common invertebrates include the sponges, arachnids, insects, crustaceans, mollusks, and echinoderms.

Mollusks

Mollusks are an amazingly diverse group of animals that live in a wide variety of environments. They can be found inhabiting trees, gardens, freshwater ponds and streams, estuaries, tidal pools, beaches, the continental shelf, and the deep ocean. Some mollusks are excellent swimmers, others crawl or burrow in mud and sand. Others remain stationary by attaching themselves to rocks, other shells, or plants, or by boring into hard surfaces, such as wood or rocks. Adult mollusks can range in size from a few mm (0.1 in.) to over 10m (~33 ft.) in length as documented for some giant deep-sea squids. Their weight can vary from a few mg (a fraction of an ounce) to over 227 kg (500 lb.) as recorded for the giant south Pacific *Tridacna* clams.

The number of living species of mollusks has been estimated to range from 50,000 to 130,000. Everyone is probably familiar with some type of mollusk. They are the slugs and shelled pests in your backyard garden; the scallops, clams, mussels, or oysters on your dinner plant; the pretty shells you see washed up on the beach; the pearls or other treasures in many jewelry boxes; the octopus or squid at an aquarium.

The word Mollusca is translated from Latin as soft-bodied, but few physical characteristics are unique to all mollusks. The mollusks are invertebrates and therefore lack a backbone; they are unsegmented and most exhibit bilateral symmetry. Most mollusks can be described as free-living, multicellular animals that possess a true heart, and that have a calcareous exterior skeleton that covers at least the back or upper surfaces of the body. This exterior skeleton provides support for a muscular foot and the internal body organs, including the stomach mass. A thin flap of tissue called the mantle surrounds the internal organs of most mollusks, and it is this mantle that secretes the animal's shell. The nervous system of mollusks varies greatly from group to group; the clams and tusk shells have very simple nervous systems, while the squids, octopi, and some other mollusks have concentrated complex nerve centers and eyes equivalent to vertebrates.

Did You Know?

Because of the many movies in which octopi and squids attack people, boats, etc., there is a misconception that they are aggressive and dumb creatures. In fact, there are only two species of octopi that are aggressive (they are located in Australia), and they are highly intelligent. They are probably the most intelligent of all the invertebrates.

Annelids

Annelids are earthworms, leeches, a large number of mostly marine worms known as polychaetes (meaning "many bristles"), and other worm-like animals whose bodies are segmented. Segments each contain elements of such body systems as circulatory, nervous, and excretory tracts. Besides being segmented, the body wall of annelids is characterized by being made up of both circular and longitudinal muscle fibers surrounded by a moist, acellular cuticle that is secreted by an epidermal epithelium. All annelids except leeches also have chitonous hair-like structures, called setae, projecting from their cuticle. They can reproduce asexually by regeneration, but they usually reproduce sexually. There are about 9000 species of annelid known today.

Did You Know?

Ecologically, annelids range from passive filter feeders to voracious and active predators.

Arthropods

Insects and spiders belong to the group of animals known as *arthropods*. By nearly any measure, they are the most successful animals on the planet. They have conquered land, sea, and air, and make up over three-fourths of all currently known living organisms, or over one million species in all.

Arthropods have segmented bodies with jointed appendages and a chitonous exoskeleton, which must be molted and shed for growth to continue. Insect bodies are divided into three parts: the head, the thorax, and the abdomen. Nearly all insects have wings, and they are the only invertebrate group that can fly. Spiders and their relatives have bodies that are divided into two parts. The head and thorax together are called the cephalothorax, and then comes the abdomen. Most have four pairs of legs.

Did You Know?

There are 200 million insects for every person on Earth.

Echinoderms

Echinoderms (from the Greek for *spiny skin*) are a phylum of marine animals found at all depths. Along with spiny skin they are characterized by an endoskeleton, radial symmetry, and a water vascular system. Echinoderms include starfish, sea stars, asteroids, sea daisies, crinoids, feather stars or sea lilies, sand dollars, sea urchins, echinoids, sea cucumbers, brittle stars, and basket stars.

Did You Know?

Echinodermata is the largest animal phylum to lack any freshwater or terrestrial representatives.

Chordata

We are most intimately familiar with the *Chordata*, because it includes humans and other vertebrates. However, not all chordates are vertebrates. Chordates are defined as organisms that posses a notochord; a structure that is present at least during some part of their development. The notochord is a rod that extends most of the length of the body when it is fully developed. Other characteristics shared by chordates include the following (Hickman and Roberts 1994):

- bilateral symmetry
- segmented body, including segmented muscles
- three germ layers and a well-developed coelom
- single, dorsal, hollow nerve cord, usually with an enlarged anterior end (brain)
- tail projecting beyond (posterior to) the anus at some stage of development
- pharyngeal pouches present at some stage of development
- ventral heart, with dorsal and ventral blood vessels and a closed blood system
- complete digestive system
- bony or cartilaginous endoskeleton usually present.

The invertebrate chordates, which do not have a backbone, include the tunicates and lancelets. The adult form of most *tunicates* shows no resemblance to vertebrate animals, but such a resemblance is evident in the larva. The most familiar tunicates are the sea squirts. *Lancelets* are filter feeders with their tails buried in the sand and only their anterior end protruding.

Vertebrates

Although *vertebrates* represent only a very small percentage of all animals, their size and mobility often allow them to dominate their environments. Vertebrates include: primates, such as humans and monkeys; amphibians; reptiles; birds; and, fish. Vertebrates consist of about 43,000+ species of animals with backbones. Vertebrates exhibit all of the chordate characteristics at some point during their lives. The embryonic notochord is replaced by a vertebral column in the adult. The vertebral column is made of individual hard segments (vertebrae) surrounding the dorsal hollow nerve cord. The nerve cord is the one chordate feature present in the adult phase of all vertebrates. The vertebral column, part of a flexible but strong endoskeleton, is evidence that vertebrates are segmented. The vertebrate skeleton is living tissue (either cartilage or bone) that grows as the animal grows. The post-anal tail is the only characteristic of chordates that most vertebrates keep throughout their lives.

Human Evolution

Human evolution is the biological and cultural development and change of our hominid/hominin (hominid term is old system; hominin term is new system; in this text, the new term hominin is used) ancestors to modern humans. Hominins evolved between 5 to 8 million years ago. To date, fossil records provide evidence of this development and date from about 4.5 million years ago. There were about nine different hominin species. Evidence indicates that Homo sapiens make their appearance as early as 300,000 years ago.

References and Recommended Reading

Hickman, C.P., Jr., Roberts, L.S., and Hickman, F.M. 1994. *The Biology of Animals*. St. Louis: Mosby College Publishing.

Kendrick, B. 2001. *The Fifth Kingdom*, 3rd ed. Newburyport, MA: Focus Publishing.

McKinney, R.E. 1962. *Microbiology for Sanitary Engineers*. New York: McGraw-Hill.

NIH. 2006. *NIH Report on Biodiversity*. National Institutes of Health. Accessed 7/06 @ http://www.easi.org/nape/senrep.html.

Sleigh, M. 1975. *Biology of Protozoa*. London: Edward Arnold.

Spellman, F.R. 2001. *Microbiology for Water/Wastewater Operators*. Boca Raton, FL: CRC Press.

Thomas, L. 1974. *The Lives of a Cell*. New York: Viking.

17

Animal Groups

The fall of snowflakes in a still air, reserving to each crystal is perfect form;
The blowing of sleet over a wide sheet of water, and over plains, the waving
rye-field, the mimic waving of acres of houstonia, whose innumerable
floret whiten and ripple before the eye; the reflections of trees and flowers
in glass lakes; the musical steaming odorous south wind, which converts all
trees to windharps; the crackling and spurting of hemlock in the flames; or
of pine logs, which yield glory to the walls and faces in the sittingroom,—
these are the music and pictures of the most ancient religion.

—Ralph Waldo Emerson (1844)

We have forgotten how to be good guests, how to walk lightly on the earth
as its other creatures do.

—Barbara Ward, *Only One Earth*, 1972

Birds

A RISTOTLE, arguably the Father of Science, divided all living things between plants, which generally do not move fast enough for humans to
notice, and animals. For the novice, the word animal is often thought of as

---

Most of the material in this chapter is from, adapted from, and/or based on information from F.R.
Spellman's (2008) *Ecology for Non-Ecologists* and United States Geological Survey's (USGS) *Our Living
Resources* (1995).

just mammals. As a matter of fact, the fish, the insect, the snake, and the bird have as much right to be called animals as the raccoon or the bear. While in this study it is not important to precisely classify or memorize each individual animal organism by specific class, it is important to know the difference between one and the other. Well, this is rather easy to do, you say. And you would be correct, of course. For instance, we easily see that the fish differs in many ways from the dog and that the cat differs from the snake; and it is easy for us to grasp the fact that the mammals differ from all other animals in that their young are nourished by milk from the breasts of the mother; when we learn to appreciate this fundamental fact, we will understand that such diverse forms as the whale, the horse, the cow, the bat, and man are members of one great class of animals.

Did You Know?

Each bird feather consists of three parts, the shaft or quill, which is the central stiff stem of the feather, giving it strength. From this quill protrude the barbs which, toward the outer end, join together in a smooth web, making the thin, fanlike portion of the feather; at the base is the fluff, which is soft and downy and near to the body of the fowl.

Birds are animals. A Bird is studied to ascertain what it is and what it does. However, there are those practitioners in the field who feel that it is necessary to only identify the bird—all the needed knowledge is thus attained. The professional ecologist knows better. As Comstock (1986) puts it, "the identification of birds is simply the alphabet to the real study, the alphabet by means of which we may spell out the life habits of the bird." Knowledge of birds adds a valuable tool to our personal tool kit. Several of the most common have been selected as subjects for discussion in this book.

According to USGS (1995), migratory bird populations are an international resource for which there is special federal responsibility. Moreover, birds are valued and highly visible components of natural ecosystems that may be indicators of environmental quality. Consequently, many efforts have been directed toward measuring and monitoring the condition of North America's migratory bird fauna. The task is not an easy one because the more than 700 U.S. species of migratory birds are highly mobile and may occur in the United States during only part of their annual cycle. One often cannot tell whether a bird observed at a given moment is a resident, a migrant, a visitor from another locality, or the same individual seen 10 minutes earlier.

Determining status and trends is further complicated by the fact that each of these species has its own patterns of distribution and abundance, and each species has populations that respond to different combinations of environmental factors. Finally, the sheer abundance of birds, estimated at 20 billion individuals in North America at its annual late-summer peak (Robbins et al. 1966) may make it difficult to obtain accurate counts of common species, and the absolute abundance of some may mask important changes in their status.

> "How beautiful your feathers be!"
> The Redbird sang to the tulip-tree
> New garbled in autumn gold.
> "Alas!" the bending branches sighed,
> "They cannot like your leaves abide
> To keep us from the cold!"
>
> —John B. Tabb (1845–1909)

Results from the nationwide Breeding Bird Survey (BBS; Peterjohn et al. 1993) and a portion of the large-scale Christmas Bird Count (CBC; Root and McDaniel 1995) show that some populations are declining, others increasing, and many show what appears to be normal fluctuations around a more or less stable average.

Of the 245 species considered in the BBS, 130 have negative trend estimates, 57 of which exhibit significant declines. Species with negative trend estimates are found in all families, but they are especially prevalent among the mimids (mockingbirds and thrashers) and sparrows. A total of 115 species exhibits positive trends, 44 of which are significant increases. Flycatchers and warblers have the largest proportion of species with increasing populations (Peterjohn et al. 1993). Some wading birds, such as the American White Ibis, are holding their own with increases and offsetting decreases.

Songbirds

For the American farmer, songbirds provide a double measure of pleasure with their winged melodies and voracious appetite for insects, which helps farmers (and others), as songbirds annually consume millions of insects that, if unchecked, could damage crops and trees. Some birds eat as many as 300 insects a day during the summer months. For the rest of us, the pleasure and uplifting of a walk in the woods or local park in the spring is accentuated by the melodies of these winged singers.

These beneficial species can be found in virtually every habitat in the U.S. Forests, prairies, wetlands, deserts, and many other kinds of habitats are home to songbirds. Warblers, tanagers, orioles, finches, and hundreds of other species make up this diverse group of birds. The names often denote their colorful plumage: Indigo Bunting, Yellow Warbler, Ruby-crowned Kinglet, Purple Martin, and many more (USFSW 2002).

Of the 50 songbirds examined in the CBC, 27 (54%) exhibited a statistically and biologically significant trend in at least one state. Of these 27 species, 16 (59%) had populations declining in more states than states in which they were increasing; 12 exhibited only declines and 4 had a population increase in at least one state. Ten (37%) of the 27 species had populations increasing in more states than states exhibiting declines, with 7 exhibiting only population increases. One (4%) species had populations increasing and decreasing in the same number of states (Root and McDaniel 1995).

Did You Know?

Among all the vocalists in the bird world, the mockingbird is seldom rivaled in the variety and richness of his repertoire.

Soft and low the song began: I scarcely
   caught it as it ran
Through the melancholy trill of the plain-
   tive whip-poor-will,
Through the ringdove's gentle wail, chat-
   tering jay and whistling quail,
Sparrow's twitter, catbird's cry, redbird's
   whistle, robin's sigh;
Blackbird, bluebird, swallow, lark, each
   his native note might mark.

—Joseph Rodman Drake (1795–1820)

Did You Know?

Birds do most of their singing in the early morning and during the spring and early summer months.

Overall, approximately equal numbers of species appear to be increasing and decreasing over the past two to three decades. Groups of species with the most consistent declines are those characteristic of grassland habitats, apparently reflecting conversion of these habitats to other types of vegetative cover.

Did You Know?

Population health is a measure of a population's ability to sustain itself over time as determined by the balance between birth and death rates. Indices of population size do not always provide an accurate measure of population health because population size can be maintained in unhealthy populations by immigration of recruits from healthy populations (Pulliam 1988). Poor population health across many populations in a species eventually results in the decline of that species. Early detection of population declines allows managers to correct problems before they are critical and widespread.

Waterfowl Populations

North American waterfowl are a resource of economic, recreational, and aesthetic value. They are appreciated as game birds by millions of hunters, and their colorful plumage, elaborate displays, and observability attract even large numbers of nonconsumptive users. Each year in the U.S., more than 20 million people spend millions of dollars on waterfowl-related recreational activities (Teisl and Southwick 1995). Waterfowl populations are monitored closely as a basis for regulating annual harvests at levels consistent with maintenance of populations.

Did You Know?

Forty-two native species of waterfowl occur commonly in North America, representing 7 taxonomic tribes. Waterfowl present complex management challenges because of the diversity, widespread distribution, and seasonal migration. Among waterfowl species there is wide variation in reproductive potential and mortality rates, as well as predation rates. Swans, geese, and sea ducks tend to have lesser reproductive potential than other waterfowl groups, but they also have lesser annual mortality rates for eggs and juveniles (Bellrose 1980). In contrast, dabbling ducks and bay (diving) ducks have greater reproductive potential but also greater rates of mortality.

Geese

Goose populations (Rusch et al. 1995; Hestback 1995a; Hupp et al. 2007) have shown some impressive gains over the past decades, but most gains have been registered by large-bodied geese, with several smaller species and smaller subspecies of the highly variable Canada goose having depressed populations.

> There is a sound, that, to the weather-wise farmer, means cold and snow, even though it is heard through the hazy atmosphere of an Indian summer day; and that is the honking of wild geese as they pass on their southward journey. And there is not a more interesting sight anywhere in the autumn landscape than the wedge-shaped flock of these long-necked birds with their leader at the front apex. (Comstock 1986)

According to Rusch et al. (1995), Canada geese are probably more abundant now than at any time in history. They rank first among wildlife watchers and second among harvests of waterfowl species in North America. Canada geese are also the most widely distributed and phenotypically (visible characteristics of the birds) variable species of bird in North America.

Breeding populations now exist in every province and territory of Canada and in 49 of the 50 United States. The sizes of the 12 recognized subspecies range from the 1.4-kg (3-lb) cackling Canada goose to the 5.0-kg (11-lb) giant Canada goose (Delacour 1954; Bellrose 1976).

Did You Know?

Market hunting and poor stewardship led to record low numbers of geese in the early 1900's, but regulated seasons including closures, refuges, and law enforcement led to restoration of most populations.

Large changes have occurred in the geographic wintering distribution and sub-species composition of the Atlantic Flyway population of Canada geese over the last 40 years. The Atlantic Flyway can be thought of as being partitioned into four regions: South, Chesapeake, mid-Atlantic, and New England. Wintering numbers have declined in the southern states, increased then decreased in the Chesapeake region, and increased markedly in the mid-Atlantic region. In the New England region, wintering numbers increased from around 6,000 during 1948–1950 to between 20,000 and 30,000 today (Serie 1993).

North American populations of most goose species have remained stable or have increased in recent decades (USFWS and Canadian Wildlife Service 1986). Some populations, however, have declined or historically have had small numbers of individuals, and thus are of special concern. Individual populations of geese should be maintained to ensure that they provide aesthetic, recreational, and ecological benefits to the nation. Monitoring and management efforts for geese should focus on individual populations to ensure that genetic diversity is maintained (Anderson et al. 1992).

Ducks

Duck—sometimes confused with several types of unrelated water birds with similar forms, such as loons or divers, gallinules, grebes, and coots—is the common name for a number of species in the family of birds. Ducks are mostly aquatic birds, mostly smaller than the swans and geese, and may be found in both freshwater and seawater.

Did You Know?

The word "duck" comes from Old English *duce* "diver," a derivative of the verb "ducan" to duck, bend down low as if to get under something, or dive.

Even though some species are stable or even increasing, many duck populations have declined in the past decade. Biologists attribute these declines to losses of breeding and wintering habitats and a long period of drought in breeding areas. Among species receiving special emphasis, canvasbacks showed a complex pattern with regional changes in distribution and abundance, and pintails showed a widespread and nearly consistent pattern of decline (USGS 1995).

Increased predation and habitat degradation and destruction coupled with drought, especially on breeding grounds, have caused the decline of some such populations. More than 30 species of ducks breed in North America, in areas as diverse as the arctic tundra and the subtropics of Florida and Mexico. For many of these species, however, the Prairie Pothole region of the north-central United States and south-central Canada is the most important breeding area, although migratory behavior and the life histories of different species lead them to use many wetland habitats (Caithamer and Smith 1995).

Duck population changes occur on breeding, staging, and wintering habitats, with the changes on breeding habitats having the greatest effect on populations. Degradation and destruction of wetlands over the last 200 years have diminished duck populations; wetland alteration and degradation continue. The rate of wetland loss has been greatest in prime agricultural areas such as the Prairie Pothole region and lowest in northern boreal forests and tundra. Thus, species such as dabbling ducks that mostly nest in the severely altered Prairie Potholes have been harmed more than species like sea ducks and mergansers that nest farther north (Bellrose 1980; Johnson and Grier 1988).

Did You Know?

A duck has the same number of toes as a hen, but there is a membrane, called the web, which joins the second, third, and fourth toes, making a fan-shaped foot; the first toe or hind toe has a web of its own. A webbed foot is first of all a paddle for propelling its owner through the water; it is also a very useful foot on the shores of ponds and streams, since its breath and flatness prevent it from sinking into the soft mud (Comstock 1986).

Because most dabbling ducks need grassy cover for nesting (Kaminski and Weller 1992), conversion of native grasslands to agricultural production, including pastures, has reduced available nesting cover and contributed to a reduced nesting success for dabblers. This condition is especially true in the Prairie Pothole region of the United States and Canada. In addition, highly variable precipitation in the Prairie Potholes has changed the number of wetlands available for nesting. For example, in 1979 there were 6.3 million wetlands in the surveyed portion of the Prairie Pothole region, but by the next spring, wetlands in the same area had decreased 55% to 2.9 million. Two years later they increased more than 100% to 4.2 million. These annual changes can temporarily mask the long-term declining trend in wetland abundance across the Prairie Pothole region.

Did You Know?

Dabbling ducks are so-named because they feed mainly on vegetable matter by upending on the water surface, or grazing, and only rarely dive (Avianweb 2007).

The changing availability of wetland habitats in the Prairie potholes region causes substantial fluctuations in some duck populations. During periods of high precipitation, larger wetland basins are full or overflowing, and shallow wetlands are abundant. Species such as the northern pintail, which tend to use shallow or ephemeral wetlands for feeding, produce more young when wetland numbers increase (Smith 1970; Hochbaum and Bossenmaier 1972). Consequently, population numbers increase as they did during the 1970's.

Stewart and Kantrud (1973) report that during the direst periods, however, such as those in the 1980's, only the deepest and most permanent wetlands retain water, causing population declines in species such as pintails that rely primarily on shallow wetlands. Population numbers are more stable for species like the canvasback, which rely on deeper marshes, and are therefore less affected by annual changes in wetland numbers because deeper marshes consistently retain water, providing ample habitat in most years.

Nest success in the Prairie Pothole region has declined in recent years largely because of increased nest predation caused by the range expansion of some predators and by reduced nesting habitat (Sargeant and Raveling 1992). Fewer and smaller areas of nesting habitat concentrate duck nests, enhancing the ability of predators to find nests. Predators such as raccoons have expanded their range northward, probably because they can den in buildings, rock piles, and other human-made sites during winter.

Although wetland drainage, urbanization, and other human-caused changes have resulted in wintering habitat losses, these loses have been offset, at least for dabbling ducks, by increased fall and winter food from waste grain left in stubble fields. In addition, the national wildlife refuge system has protected and managed many staging and wintering areas for the benefit of water (USGS 1995).

Modern duck-hunting regulations are believed to keep recreational harvest at levels compatible with the long-term welfare of duck populations. The proportion of ducks harvested varies regionally and by species, age, and sex. In 1992, 2%–12% of the adult mallards from the Prairie Pothole region were killed by hunters. Harvest rates of other species were generally lower. These conservative harvest rates are unlikely to cause population declines (Blohm 1989).

The size of the continental breeding population of northern pintail has greatly varied since 1955, with numbers in surveyed areas ranging from a high of 9.9 million in 1956 to a low of 1.8 million in 1991. This variation results primarily from differences in the numbers of breeding pintails in the prairie region of Canada and the United States; these numbers ranged from 8.6 million in 1956 to 9.5 million in 1991; numbers in the northern regions from Alaska to northern Alberta and northern Manitoba varied primarily between 1 and 2 million (Hestback 1995b).

Breeding pintails prefer seasonal shallow-water habitats without tall emergent aquatic vegetation (Smith 1968). The proportions and distribution of breeding pintails on the prairies vary annually depending on the amount of annual precipitation and the resulting increase or decrease in the availability of suitable breeding habitat (Smith 1970; Johnson and Grier 1988).

Did You Know?

USGS (2007) points out that the northern pintail is one of several species of dabbling ducks belonging to the tribe Anatini in the family Anatidae. Pintails are medium-size ducks with slender, elegant lines and conservative plumage coloration. Males are larger than females. The male in breeding plumage is readily distinguished from other dabbling ducks by a combination of chocolate brown head, white neck stripes, black/gray body feathers, and very long black central tail feathers, which give the species its name; the male has a bright green/purple iridescent speculum. The female is colored similarly to females of other dabbling ducks, basically a dull brown with black markings. Pintails are graceful, acrobatic flyers capable of darting and wheeling routines especially during pursuit or courtship flights, which are fast and vigorous with sudden rapid dives from great heights to low ground-level flight. Common vocalizations by females include the traditional decrescendo "quack" call that consists of one loud quack followed by a softer quack; males issue a series of "chirps" and whistles. Pintails employ an elaborate series of displays during courtship, including "head-up-tail-up," "grunt-whistles," and "chin lifts." Pintails are very wary, long-lived ducks that tend to survive at higher rates than other ducks. The maximum longevity in the wild was recorded at 21 years 4 months based on a California-banded adult male recorded by a hunter in Idaho. Average life spans would be considerably shorter than this, but relative to other duck species, pintails are long-lived in the wild (Austin and Miller 1995; Bellrose 1980).

Changes in the size of the continental pintail population result from changes in reproduction, survival, or both. Consequently, understanding population changes involves detecting variation in survival and reproduction over time and relating that variation to changes in population size. Once the cause of the decline is determined, appropriate management strategies can be developed to reverse it.

Did You Know?

The canvasback duck is a large diving duck that breeds in prairie potholes and winters on ocean bays. Its sloping profile distinguishes it from other ducks.

Canvasbacks are unique to North America and are one of our most widely recognized waterfowl species. Unlike other ducks that nest and feed in uplands, diving ducks such as canvasbacks are totally dependent on aquatic habitats throughout their life cycle. Canvasbacks nest in prairie, parkland, subarctic, and Great Basin wetlands; stage during spring and fall on prairie marshes, northern lakes, and rivers; and winter in Atlantic, Pacific, and Gulf of Mexico bays, estuaries, and some inland lakes. They feed on plant and animal foods in wetland sediments. Availability of preferred foods, especially energy-rich subterranean plant parts, is probably the most important factor influencing geographic distribution and habitat use by canvasbacks (Hohman et al. 1995).

In spite of management efforts that have included restrictive harvest regulations and frequent hunting closures in all or some of the flyways (Anderson 1989), canvasback numbers declined from 1955 to 1993 and remain below the population goal (540,000) of the North American Waterfowl Management Plan (USFWS and Canadian Wildlife Service 1994). Causes for this apparent decline are not well understood, but habitat loss and degradation, low rates of recruitment, a highly skewed sex ratio favoring males, and reduced survival of canvasbacks during their first year are considered important constraints on population growth.

Did You Know?

In regard to one of the factors involved with population control, ducks have many predators. Ducklings are particularly vulnerable, since their inability to fly makes them easy prey not only for predatory birds but also large fish like pike, crocodilians, and other aquatic hunters, including fish-eating birds such as herons.

Shorebirds

Shorebirds are usually small bodied with long thin bills. The differences in their bill lengths and shape allow the different shorebird species to forage for food within their habitat either on dry soil, mud, or in shallow water. They are highly migratory, and status and trends of their populations are largely determined from observations made during periods in their life cycles in which birds congregate in limited breeding, staging, or migratory stopover areas. Populations of eastern and western species show general patterns of decline, although some species, including those using inland areas, are too poorly studied to detect trends. Apparent dependence on critical breeding and staging areas suggest that populations of many species are vulnerable to habitat loss and disturbance (USGS 1995).

Shorebirds are a diverse group that includes oystercatchers, stilts, avocets, plovers, and sandpipers. They are familiar birds of seashores, mudflats, tundra, and other wetlands, and they also occur in deserts, high mountains, forests, and agricultural fields. Widespread loss and alteration of these habitats, coupled with unregulated shooting at the turn of the century, resulted in population declines and range contraction of several species throughout North America (Gill et al. 1995). Most populations recovered after passage of the Migratory Bird Treaty Act of 1918, although some species never recovered and others have declined again. In the western portion of the continent, efforts to monitor the status and trends of shorebirds have been in effect for only the past 15–25 years and for only a few species (Harrington 1995).

Seabirds

> Ah! well a-day! what evils looks
> Had I from old and young!
> Instead of the cross, the Albatross
> About my neck was hung.
>
> —S.T. Coleridge,
> *The Rime of the Ancient Mariner* (1798)

> . . . but I shall kill no albatross . . .
>
> —Mary Shelley, *Frankenstein* (1818)

While the albatross can be both an omen of good or bad luck (they are often regarded as the souls of lost sailors), as well as a metaphor for a burden to be carried as penance, for our purpose, the albatross is just one of many seabirds, including Common Murre, Brandt's Cormorant, Double-crested Cormorant, Pelagic Cormorant, Pigeon Guillemot, Tufted Puffin, Rhinoceros Auklet, Cassin's Auklet, Marbled Murrelet, Western Gull, Caspian Tern, and Leach's Storm Petrel.

Seabirds are birds that spend almost all their time on or near the sea (USFWS 2007d). They are medium-size to large birds; most are between the size of a robin and a crow. They get all their food from the water. Some spend the winter at sea, several hundred miles from land. Seabirds come to land to raise the young birds each summer. They nest on protected cliffs or islands, often in dense groups called colonies.

> **Did You Know?**
>
> Seabirds have special adaptations that allow them to live at sea and get all their food there. Some eat small fish or shrimp-like invertebrates called zooplankton, which they catch from the sea. Seabirds such as kittiwakes pick their prey from the water's surface. Others, such as auks and cormorants, dive for their prey and chase it underwater (USFWS 2007a).

Seabirds in the Pacific region include many diverse species that respond differently to factors such as human proximity to nesting areas, oil spills, introduction of predators, depletion of fishery stocks, and availability of human refuse as food. Some species, including certain gulls, brown pelicans, and double-crested cormorants, have responded positively to recent changes in some areas, whereas others, including murrelets and murres and kittiwakes, have shown declining trends. Populations of other species appear to fluctuate widely, and information for many species is insufficient to determine long-term trends (Carter et al. 1995; Hatch and Platt, 1995).

More than two million seabirds of 29 species nest along the west coasts of California, Oregon, and Washington, including three species listed on the federal list of threatened and endangered species: the brown pelican, least tern, and marbled murrelet. The size and diversity of the breeding seabird community in this region reflects excellent nearshore prey conditions; subtropical waters within the southern California Bight area; complex tidal waters of the Strait of Juan de Fuca and Puget Sound in Washington; large estuaries at San Francisco Bay, the Columbia River, and Grays Harbor-Willapa bays; and the variety of nesting habitats used by seabirds throughout the region, including islands, mainland cliffs, old-growth forests, and artificial structures (USGS 1995).

Breeding seabird populations along the west coast have declined since European settlement began in the later 1700's because of human occupation of, commercial use of, and introduction of mammalian predators to seabird nesting islands. In the 1900's, further declines occurred in association with rapid human population growth and intensive commercial use of natural resources in the Pacific region. In particular, severe adverse impacts have occurred from partial or complete nesting habitat destruction on islands or the mainland, human disturbance of nesting islands or areas, marine pollution, fisheries, and logging of old-growth forests (Ainley and Lewis 1974; Bartonek and Nettleship 1979; Hunt et al. 1979; Sowls et al. 1980; Nettleship et al. 1984;

Speich and Wahl 1989; Ainley and Boekelheide 1990; Sealy 1990; Ainley and Hunt 1991; Carter and Morrison 1992; Carter et al. 1992; Vermeer et al. 1993).

> Paddling as quietly as I could, I slipped through the fog and stopped abruptly as a brown, robin-sized seabird—a marbled murrelet dove almost immediately, reappearing again with a tiny fish in its bill. Loggers working along the Northwest coast in the early years gave this enigmatic little seabird the name "fog lark."

> —Audrey Benedict (2007)

About 100 million seabirds reside in marine waters of Alaska during some part of the year. Perhaps half this population is composed of 50 species of nonbreeding residents, visitors, and breeding species that use marine habitats only seasonally (Gould et al. 1982). Another 30 species include 40–60 million individuals that breed in Alaska and spend most of their lives in U.S. territorial waters (Sowls et al. 1978). Alaskan populations account for more than 95% of the breeding seabirds in the continental U.S., and eight species nest nowhere else in North America (USFWS 1992).

Seabird nest sites include rock ledges, open ground, underground burrows and crevices in cliffs or talus. Seabirds take a variety of prey from the ocean, including krill, small fish, and squid. Suitable nest sites and oceanic prey are the most important factors controlling the natural distribution and abundance of seabirds (Hatch and Platt 1995).

The impetus for seabird monitoring is based partly on public concern for the welfare of these birds, which are affected by a variety of human activities like oil pollution and commercial fishing. Equally important is the role seabirds serve as indicators of ecological change in the marine environment. Seabirds are long-lived and slow to mature, so parameters such as breeding success, diet, or survival rates often give earlier signals of changing environmental conditions than population size itself. Seabird survival data are of interest because they reflect conditions affecting seabirds in the nonbreeding season, when most annual mortality occurs (Hatch et al. 1993b).

Techniques for monitoring seabird populations vary according to habitat types and the breeding behavior of individual species (Hatch and Hatch 1978, 1989; Byrd et al. 1983). An affordable monitoring program can include but a few of the 1,300 seabird colonies identified in Alaska, and since the mid-1970's, monitoring efforts have emphasized a small selection of surface-feeding and diving species, primarily kittiwakes and murres. Little or no information on trends is available for other seabirds (Hatch 1993a). The existing monitoring program occurs largely on sites within the Alaska Maritime National Wildlife Refuge, which was established primarily for the conservation of marine birds.

Data are collected by refuge staff, other state and federal agencies, private organizations, university faculty, and students.

Colonial-nesting Waterbirds

"Colonial-nesting waterbird" is a tongue-twister of a collective term used by bird biologists to refer to a large variety of different species that share common characteristics: (1) they tend to gather in large assemblages, called colonies, during the nesting season, and (2) they obtain all or most of their food (fish and aquatic invertebrates) from the water. Colonial-nesting waterbirds can be further divided into two major groups depending on where they feed (USFWS 2002).

Seabirds (also called marine birds, oceanic birds, or pelagic birds) feed primarily in saltwater. Some seabirds are so marvelously adapted to marine environments that they spend virtually their entire lives at sea, returning to land only to nest; others (especially the gulls and terns) are confined to the narrow coastal interface between land and sea, feeding during the day and loafing and roosting on land. Included among the seabirds are such groups as the albatrosses, shearwaters, storm-petrels, tropic birds, boobies, pelicans, cormorants, frigatebirds, gulls, terns, murres, guillemots, murrelets, auklets, and puffins. A few species of cormorants, gulls, and terns also occupy freshwater habitats.

Wading Birds seek their prey in fresh or brackish waters. As the name implies, these birds feed principally by wading or standing still in the water, patently waiting for fish or other prey to swim within striking distance. The wading birds include the bitterns, herons, egrets, night-herons, ibises, spoonbills, and storks.

Did You Know?

Colonial-nesting waterbirds have attracted the attention of scientists, conservationists, and the public since the early 1900s when plume hunters nearly drove many species to **extinction** (Erwin 1995).

Colonial-nesting waterbirds of the continental and east coast regions of the United States show trends related to many of the same factors operating in the Pacific region, with some species recovering from past losses from pesticides while some other species that exploit human refuse are increasing dramatically. Populations of other species, especially certain terns, are declining, probably as a result of habitat loss and degradation or other kinds of human disturbance. Special efforts have been made to determine status and trends of the piping plover, a species listed as endangered in certain parts of its range and as threatened in others (Robbins et al. 1966).

Raptors

Possessing several unique anatomical characteristics—binocular vision, acute hearing, large powerful grasping feet with razor-sharp talons and generally large, hooked bills—raptors are superior hunters, who, in superior fashion, swiftly capture, efficiently kill, and consume their vertebrate prey. Raptors, or birds of prey, include the hawks, falcons, eagles, vultures, and owls; they occur throughout North American ecosystems. Again, as predators, most of them kill other vertebrates for their food. Compared to most other animal groups, birds of prey naturally exist at relatively low population levels and are widely dispersed within their habitats. The natural scarcity of raptors, combined with their ability to move quickly, the secretive behavior of many species, and the difficulties of detecting them in rugged terrain or vegetation, all make determining their population status difficult.

As top predators, raptors are key species for our understanding and conservation of ecosystems. Changes in raptor status can reflect changes in the availability of their prey species, including population declines of mammals, birds, reptiles, amphibians, and insects. Changes in raptor status also can be indicators of more subtle detrimental environmental changes such as chemical contamination and the occurrence of toxic levels of heavy metals (e.g., mercury, lead). Consequently, determining and monitoring the population status of raptors are necessary steps in the wise management of our natural resources (Fuller et al. 1995).

The *California condor* is a member of the vulture family. With a wingspan of about 9 ft and weighing about 20 lb, it spends much of its time in soaring flight visually seeking dead animals as food. The California condor has always been rare (Wilbur 1978; Pattee and Wilbur 1989). Although probably numbering in the thousands during the Pleistocene epoch in North America, its numbers likely declined dramatically with the extinction of most of North America's large mammals 10,000 years ago. Condors probably numbered in the hundreds and were nesting residents in British Columbia, Washington, Oregon, California, and Baja California around 1800. In 1939 the condor population was estimated at 60–100 birds, and its home range was reduced to the mountains and foothills of California, south of San Francisco and north of Los Angeles (Pattee and Mesta 1995).

Franson et al. (1995) report that the U.S. Department of the Interior has investigated the deaths of more than 4,300 *bald* and *gold eagles* since the early 1960's as part of an ongoing effort to monitor causes of wildlife mortality. The availability of dead eagles for study depends on finding carcasses in fair to good condition and transporting them to the laboratory. Such opportunistic collection and the fact that recent technological advances have enhanced our diagnostic capabilities, particularly for certain toxins, means that results reported here do not necessarily reflect actual proportional causes of death for all eagles in the U.S. throughout the 30-year period. This type of sampling does, however, identify major or frequent causes of death.

Most diagnosed deaths of eagles in our study (Franson et al. 1995) resulted from accidental trauma, gunshot, electrocution, and poisoning. Accidental trauma, such as impacts with vehicles, power lines, or other structures, was the most frequent cause of death in both eagle species (23% of bald and 27% of golden). Gunshot killed about 15% of each species. Electrocution was twice as frequent in golden (255) as in bald eagles (12%), probably because of the preference of golden eagles for prairie habitats and their use of utility poles as perches.

Lead poisoning was diagnosed in 338 eagles from 34 states. Eagles become poisoned by lead after consuming lead shot and, occasionally, bullet fragments present in food items. Agricultural pesticides accounted for most remaining poisonings, organophosphorus and carbamate compounds killed 139 eagles in 25 states. Eagles are exposed to these chemicals in a variety of ways, often by consuming other animals that died of direct poisoning or from baits placed to deliberately kill wildlife.

Overall, poisonings were more frequent in bald eagles (16%) than golden eagles (6%). The reasons for this are unclear, but may be related to factors that influence submission of carcasses for examination or differences in species' preferences for agricultural, rangeland, and wetland habitats (Franson et al. 1995).

Wild Turkeys

The wild turkey is a large gallinaceous (domestic fowl) bird characterized by strong feet and legs adapted for walking and scratching, short wings adapted for short rapid flight, a well-developed tail, and a stout beak useful for pecking (Figure 48). These birds probably originated some 2 to 3 million years ago in the Pliocene epoch (5.3 to 1.8 mya). Molecular data suggest this genetic line descended from pheasant-like birds about 11 million years ago. There are two species in the genus, the wild turkey of the United States, portions of southern Canada, and northern Mexico; and the ocellated (eyelike spots on the tail) turkey in the Yucatan region of southern Mexico, Belize, and northern Guatemala. The wild turkey has shown dramatic increases in distribution and abundance in recent decades because of translocations, habitat restoration, and harvest control (Dickson 1995).

FIGURE 48.   Wild Turkey
Photo by Frank R. Spellman at Zion National Park, Utah, 2010

Mourning Doves

The mourning dove is one of the most widely distributed and abundant birds in urban and rural areas of North America (Droege and Sauer 1990). It is also the most important U.S. game bird in terms of numbers harvested. The U.S. fall population of mourning doves has been estimated to be about 475 million (Tomlinson et al. 1988; Tomlinson and Dunks 1993).

The breeding range of the morning dove extends from the southern portions of the Canadian Provinces throughout the continental United States into Mexico, the islands near Florida and Cuba, and scattered areas in Central America (Aldrich 1993). Although some mourning doves are nonmigratory, most migrate south to winter in the United States from northern California and Connecticut, south throughout most of Mexico and Central America to western Panama.

Within the United States, three areas contain breeding, migrating, and wintering mourning dove populations that are largely independent of each other (Kiel 1959). In 1960 three areas were established as separate management units: the Eastern, Central, and Western (Dolton 1995).

Did You Know?

Mourning Doves feed their nestlings crop milk or "pigeon milk," which is secreted by the crop lining. This is an extremely nutritious food with more protein and fat than is found in either cow or human milk. Crop milk, which is regurgitated by both adults, is the exclusive food of hatchlings for three days, after which it is gradually replaced by a diet of seeds (Cornell U. 1999).

Common Ravens

... Then this ebony bird beguiling my sad fancy into smiling,
By the grave and stern decorum of the countenance it wore.
"Though thy crest be shorn and shaven, thou," I said, "art sure no craven.
Ghastly grim and ancient raven wandering from the Nightly shore—
Tell me what thy lordly name is on the Night's Plutonian shore!"
    Quoth the Raven, "Nevermore."
Much I marveled this ungainly fowl to hear discourse so plainly,
Though its answer little meaning—little relevancy bore;
For we cannot help agreeing that no living human being
Ever yet was blest with seeing bird above his chamber door—
Bird or beast upon the sculptured bust above his chamber door,
    With such name as "Nevermore."
But the raven, sitting lonely on the placid bust, spoke only
That one word, as if his soul in that one word he did outpour.
Nothing further then he uttered—not a feather then he fluttered—
Till I scarcely more than muttered, "other friends have flown before—
On the morrow he will leave me, as my hopes have flown before."
    Then the bird said, "Nevermore."

—Edgar Allen Poe (1845)

The common raven is a large black passerine (perching bird) found throughout the Northern Hemisphere including western and northern North America. Long recognized as one of the most intelligent birds, the raven also has a less than savory image throughout history as a scavenger that does not discriminate between humans and animals (Nature 2007). Ravens are scavengers that frequently feed on road-killed animals, large dead mammals, and human refuse. They kill and eat prey including rodents, lambs (Larsen and Dietrich 1970), birds, frogs, scorpions, beetles, lizards, and snakes. They also feed on nuts, grains, fruits, and other plant matter (Knight and Call 1980; Heinrich 1989). Their recent population increase is of concern because ravens eat agricultural crops and animals whose populations may be depleted.

Ravens are closely associated with human activities, frequently visiting solid-waste landfills and garbage containers at parks and food establishments, being pests of agricultural crops, and nesting on many human-made structures. In two recent surveys in the deserts of California (FaunaWest Wildlife Consultants 1989; Knight and Kawashima 1993), ravens were more numerous in areas with more human influences, and were often indicators of the degree to which humans affect an area (Boarman and Berry 1995).

Mississippi Sandhill Cranes

Cranes look superficially like herons and their relatives. Both are tall, thin, and have long legs, necks, and beaks. Despite appearance, though, cranes are not closely related to herons, and their biology and way of living are quite different. Cranes fly with necks outstretched and issue loud, rattling bugle calls. They are long-lived, monogamous, nest on ground, and solitary. Cranes are territorial nesters. They do not even allow the offspring of the previous year in the nesting territory.

Resident Sandhill Cranes formed a continuous population in Georgia and Florida and widely separated populations along the Gulf coastal plain of Texas, Louisiana, Mississippi, and Alabama. The Mississippi Sandhill Crane was one of the widely separated populations on the coastal plain that bred in pine savannas in southeastern Mississippi, just east of the Pascagoula River to areas just west of the Jackson County line, south to Simmons Bayou, and north to an east-west line (5–10 mi) north of VanCleave (Gee and Hereford 1995).

Did You Know?

Cranes are unique and are among the most spectacular of the bird families. In fact, they have captured the human imagination as few other birds have. Learned naturalist and pioneering wildlife biologist Aldo Leopold called them, "nobility in the midst of mediocrity." The Mississippi Sandhill Crane was described as a distinct subspecies in 1972 and there are physiological, morphological, behavioral, and other differences between them and other Sandhill cranes. The Mississippi Sandhill Crane is a noticeably different darker shade of gray resulting in a more distinct cheek patch (NWR 2007).

Mississippi Sandhill Cranes are a critically endangered subspecies found only on and adjacent to the Mississippi Sandhill Crane National Wildlife Refuge. There are only about 100 individuals remaining, including about 20 breeding pairs. Without intensive management from the U.S. Fish and Wildlife Service and its cooperators and partners, this unique bird may disappear from the wild (Gee and Hereford 1995).

Piping Plovers

The piping plover is a small, stocky, sandy-colored bird resembling a sandpiper. It is a wide-ranging, beach-nesting shorebird whose population viability continues to decline as a result of habitat loss from development and other human disturbance (Haig 1992). In 1985 the species was listed as endangered in the Great Lakes basin and Canada and threatened in the northern Great Plains and along the U.S. Atlantic coast (Haig and Plissner 1995).

Piping plovers are approximately seven inches long with sand-colored plumage on their backs and crown and white underparts. Breeding birds have a single black breastband, a black bar across the forehead, bright orange legs and bill, and a black tip on the bill. During winter, the birds lose their black bands, the legs fade to pale yellow, and the bill becomes mostly black (USFWS 2007b).

Red-cockaded Woodpeckers

The red-cockaded woodpecker, *Picoides borealis*, or RWC, is a territorial, nonmigratory, cooperative breed species (Lennartz et al. 1987). It is a small black-and-white bird about the size of a cardinal. The red patches, or "cockades," on either side of the head on males are rarely seen as they usually conceal the red until excited or agitated. This woodpecker usually does not frequent urban settings and is not a familiar backyard species. It is not likely to be observed at a bird feeder, unlike the Downy and Red-bellied woodpeckers, for example (TPWD 2007).

In the world of North American woodpeckers, red-cockaded woodpeckers stand out as an exception to the usual rules. They are the only woodpeckers to excavate nest and roost sites in living trees. Living in small family groups, red-cockaded woodpeckers are a social species, unlike others. These groups chatter and call throughout the day, using a great variety of vocalizations (USFWS 2008).

Historically, the southern pine ecosystem, contiguous across large areas and kept open with recurring fire (Christensen 1981), provided ideal conditions for a nearly continuous distribution of RCWs throughout the South. Within this extensive ecosystem red-cockaded woodpeckers were the only species to excavate cavities in living pine trees, thereby providing essential cavities for other cavity-nesting birds and mammals, as well as some reptiles, amphibians, and invertebrates (Kappes 1993). The loss of open pine habitat since European settlement precipitated dramatic declines in the bird's population and led to its being listed as endangered in 1970 (Federal Register 35:16047).

Southwestern Willow Flycatchers

Sogge (1995) points out the southwestern willow flycatcher occurs, as its name implies, throughout most of the southwestern United States. It is a Neotropical migrant songbird, i.e., one of many birds that return to the United States and Canada to breed each spring after migrating south to the Neotropics (Mexico and Central America) to winter in milder climates. It is an olive-gray bird with a white throat and yellow-gray rump that measures about 5-3/4 inches in length. In recent years, there has been strong evidence of decline in many Neotropical migrant songbirds (e.g., Finch and Stangel 1993), including the southwestern willow flycatcher (Federal Register 1993). The flycatcher appears to have suffered significant declines throughout its range, including total loss from some areas where it historically occurred. These declines, as well as the potential for continued and additional threats, prompted the U.S. Fish and Wildlife Service (USFWS) to propose listing the southwestern willow flycatcher as an endangered species (Federal Register 1993).

Did You Know?

The flycatcher reminds observers of a sentinel constantly at attention, whose flitty wing movements resemble salutes and constant tail motions signal a readiness for action. It feeds on insects in lush, multilayered riparian zones by snatching them on the wing or harvesting them from dense vegetation. Its mission to control insects in riparian areas is an essential function benefiting people as well as plant life (USFWS 2004).

Bottom Line on Birds

Even though the preceding information is representative of only a small sample of bird species, if any overall conclusion is possible on status and trends of bird populations it is this: apparent stability for many species; increase in some species, many of which are generalists adaptable to altered habitats; and decreases in other species, many of which are specialists most vulnerable to habitat loss and degradation (USGS 1995).

Mammals

Mammals could be categorized as a group of animals with backbones, and bodies insulated by hair, which nurse their infants with milk and share a unique jaw articulation. Yet this fails to convey their truly astonishing features—their intricate adaptations, thrilling behavior, and highly complex societies.

—D.W. Macdonald (2006)

Mammals, like birds, in contrast to amphibians, fishes, and reptiles, are **warm-blooded animals**. In contrast to feather-covered birds, scale-covered fish, and skin-covered reptiles, the skin of mammals is more or less hairy. The young of most mammals are born alive, whereas the young of fish, birds, amphibians, and some species of reptiles hatch from eggs. After birth young mammals breathe by lungs rather than by gills as do fish; for a time they are nourished with milk produced by the mother (Comstock 1986). The small representative sample discussed in the following demonstrates that mammals are exceedingly diverse in size, shape, form, and function.

Tuggle (1995) reports many mammalian population studies have been initiated to determine a species' biological and/or ecological status. These studies were conducted because of their perceived economic importance, their abundance, their threatened or endangered state, or because they are viewed as our competitor. As a result, data on mammalian populations in North America have been amassed by researchers, naturalists, trappers, farmers, and land managers for years.

Inventory and monitoring programs that produce data about the status and trends of mammalian populations are significant for many reasons. One of the most important reasons, however, is that as fellow members of the most advanced class of organisms in the animal kingdom, the condition of mammal populations most closely reflects our condition. In essence, mammalian species are significant biological indicators for assessing the overall health of advanced organisms in an ecosystem.

Habitat changes, particularly those initiated by humans, have profoundly affected wildlife populations in North America. Though Native Americans used many wildlife species for food, clothing, and trade, their agricultural and land-use practices usually had minimal adverse effects on mammal populations during the pre-European settlement era. In general, during the post-Columbian era, most North American mammalian populations significantly declined, primarily because of their inability to adapt and compete with early European land-use practices and pressures.

Habitat modification and destruction during the settlement of North America occurred very slowly initially. Advances in agriculture and engineering accelerated the loss or modification of habitats that were critical to many species in climax communities. These landscapes transformations often occurred before we had any knowledge of how these environmental changes would affect native flora and fauna. Habitat alterations were almost always economically driven and in the absence of land-use regulations and conservation measures many species were extirpated.

In addition to rapid and sustained habitat and landscape changes from agricultural practices, other factors such as unregulated hunting and trapping, indiscriminate predator and pest control, and urbanization also contributed significantly to the decline of once-bountiful mammalian populations. These practices, individually and collectively, have been directly correlated with the decline or extinction of many sensitive species.

The turn of the century brought a new focus on conservation efforts in this country. Populations of some species, such as the white-tailed deer, showed marked recovery after regulatory and conservation strategies began. Ancient wildlife management and conservation programs, started primarily for game species, have increased our knowledge and understanding of species and habitat interactions. Conservation programs have also positively affected many species that share habitat with the target species the programs are designed to aid. To complement these efforts, however, integrated regulatory legislation and conservation policies that specifically help sustain nontarget species and their habitats are still imperative.

The increased emphasis on the importance of managing for biological diversity and adopting an ecosystem approach to management has enhanced our efforts to move from resource-management practices that are oriented to single species to strategies that focus on the long-term conservation of native populations and their natural habitats. Thus, an integrated and comprehensive inventory and monitoring program that coordinates data on the status and trends of our natural resources is critical to successfully manage habitats that support a diverse array of plant and animal species.

Marine Mammals

At least 35 species of marine mammals are found along the U.S. Atlantic coast and in the Gulf of Mexico: 2 seal species, 1 manatee, and 32 species of whales, dolphins, and porpoises. Seven of these species are listed as endangered under the Endangered Species Act (ESA). At least 50 species of marine mammals are found in U.S. Pacific waters: 11 species of seals and sea lions; walrus; polar bear; sea otter; and 36 species of whales, dolphins, and porpoises; 11 species are listed as endangered or threatened under the ESA (Kinsinger 1995).

Information on the size, distribution, and productivity of the California *sea otter* population is broadly relevant to two federally mandated goals: removing the population's listing as threatened under the Endangered Species Act (ESA) and obtaining an "optimal sustainable population" under the Marine Mammal Protection Act. Except for the population in central California, sea otters were hunted to extinction between Prince William Sound, Alaska, and Baja California (Kenyon 1969). Wilson et al. (1991), based on variations in cranial morphology, recently assigned sub-specific status to the California sea otter. Furthermore, mitochondrial DNA analysis has revealed genetic differences among populations in the California, Alaska, and Asia sea otter (NBS, Unpublished data).

Did You Know?

Sea otters have no blubber layer, unlike Pinnipeds and cetaceans, and rely entirely on their thick fur's ability to trap air for insulation.

In 1977, the California sea otter was listed as threatened under the ESA, largely because of its small population size and perceived risks from such factors as human disturbance, competition with fisheries, and pollution. Because of unique threats and growth characteristics, the California population is treated separately from sea otter populations elsewhere in the north Pacific (Estes et al. 1995).

Whales, dolphins, and *porpoises* all belong to the same taxonomic order called cetaceans. Cetaceans spend their whole lives in water and some live in family groups called "pods." Cetaceans are known for their seemingly playful behavior. Pinnipeds are carnivorous aquatic mammals that use flippers for movement on land and in the water. Seals, sea lions, and walruses all belong to the suborder called Pinnipedia or the "fin-footed." Pinnipeds spend the majority of their lives swimming and eating in water and have adapted their bodies to move easily through their aquatic habitat (NOAA 2007).

Manatees and *dugongs,* order Sirenia, are so ugly that they are really cute (NOAA 2007)! Sirenians spend their whole lives in water. The word "Sirenia" came from the world "siren." "Sirens" are legendary Greek sea beauties that lured sailors into the sea. It is thought that old-time mermaid sightings were actually sirenians rather than mythical half women, half fish.

Indiana Bats

His small umbrella, quaintly halved,
Describing in the air an arc alike inscrutable,—
Elate philosopher!

—Emily Dickinson

The Indiana bat is a medium-sized, dull gray, black, or chestnut bat listed as an endangered species, and occurs throughout much of the eastern United States. Drobney and Clawson (1995) write that although bats are sometimes viewed with disdain, they are of considerable ecological and economic importance. Bats consume a diet consisting largely of nocturnal insects and thereby are a natural control for both agricultural pests and insects that are annoying to humans. Furthermore, many forms of cave life depend upon nutrients brought into caves by bats in the form of guano or feces (Missouri Department of Conservation 1991).

Indiana bats use distinctly different habitats during summer and winter. In winter, bats congregate in a few large caves and mines for hibernation and have a more restricted distribution than at other times of the year. Nearly 85% of the known population winters in only seven caves and mines in Missouri, Indiana, and Kentucky, and approximately one-half of the population uses only two of three hibernacula (i.e., hibernation location).

Did You Know?

Indiana bats live an average of 7.5 years, but some have reached 14 years of age.

In spring, females migrate north from their hibernacula and form maternity colonies in predominately agricultural areas of Missouri, Iowa, Illinois, Indiana, and Michigan. Three colonies, consisting of 50 to 150 adults and their young, normally roost under the loose bark of dead, large-diameter trees throughout summer; however, living shagbark hickories and tree cavities are also used occasionally (Humphrey et al. 1977; Gardner et al. 1991; Callahan 1993; Kurta et al. 1993).

As a consequence of their limited distribution, specific summer and winter habitat requirements, and tendency to congregate in large numbers during winter, Indiana bats are particularly vulnerable to rapid population reductions resulting from habitat change, environmental contaminants, and other human disturbances (Brady et al. 1983). Additionally, because females produce only one young per year, recovery following a population reduction occurs slowly. Concerns arising from the high potential vulnerability and slow recovery rate have led to a long-term population monitoring effort for this species.

Gray Wolves

Second only to humans in adapting to climate extremes, gray wolves (*Canis lupus*) once ranged from coast to coast and from Alaska to Mexico in North

America. USFWS (1998) points out that, historically, most Native Americans revered gray wolves, trying to emulate their cunning and hunting abilities. However, by 1960 the wolf was exterminated by federal and state governments from all of the United States except Alaska and northern Minnesota. Until recently, 24 sub-species of the gray wolf were recognized for North America, including 8 in the contiguous 48 states. After the gray wolf was listed as an endangered species in 1967, recovery plans were developed for the eastern timber wolf, the northern Rocky Mountain wolf, and the Mexican wolf. The other subspecies in the contiguous United States were considered extinct (Mech et al. 1995).

The Eastern Timber Wolf Recovery Plan (USFWS 1992) set as criteria for recovery the following conditions: a viable wolf population in Minnesota consisting of at least 200 animals, and either a population of at least 100 wolves in the United States within 100 mi of the Minnesota population, or a population of at least 200 wolves if farther than 100 mi from the Minnesota population. The Northern Rocky Mountain Wolf Recovery Plan (USFWS 1987) defines discovery as when at least 10 breeding pairs of wolves inhabit each of their specified areas in the northern Rockies for 3 successive years. The Mexican Wolf Recovery Plan (USFWS 1982) called for a self-sustaining population of at least 100 Mexican wolves in a 4,941 square mile range.

A recent revision of wolf subspecies in North America (Novak 1994), however, reduced the number of subspecies originally occupying the contiguous 48 states from eight to four. It classified the wolf currently inhabiting northern Montana as being *C.I. occidentalis*, primarily a Canadian and Alaskan Wolf. It considered *C.I. nubilus* to be the wolf remaining in most of the range of the former northern Rocky Mountain wolf and the present range of the eastern timber wolf; this leaves the eastern timber wolf extinct in its former U.S. range, surviving now only in southeastern Canada. The new classification may have implications for the recovery criteria propounded by the Eastern Timber Wolf and Northern Rocky Mountain Wolf recovery plans. The reclassification did not change the status of the Mexican wolf.

Did You Know?

Gray wolves have a longevity span of 8–16 years in the wild and up to 20 years in captivity.

North American Black Bears

Perhaps no other animals have so excited the human imagination as bears. References of bears are found in ancient and modern literature, folk songs, legends, mythology, children stories, and cartoons. Bears are among the first animals that children learn to recognize. Bear folklore is confusing because it is based on caricature, with Teddy Bears and the kindly Smoky on one hand and ferocious magazine cover drawings on the other. Dominant themes of our folklore are fear of the unknown and man against nature, and bears have traditionally been portrayed as the villains to support those themes, unfairly demonizing them to the public. A problem for black bears is that literature about bears often does not separate black bears from grizzly bears.

—Lynn L. Rogers, 2002

Habitat loss, habitat fragmentation, and unrestricted harvest have significantly changed the distribution and abundance of black bears in North America since colonial settlement. Even though bears have been more carefully managed in the last 50 years and harvest levels are limited, threats from habitat alteration and fragmentation still exist and are particularly acute in the southeastern U.S. In addition, the increased efficiency in hunting techniques and the illegal trade in bear parts, especially gall bladders, have raised concerns about the effect of poaching on some bear populations. Because bears have lower reproductive rates, their populations recover more slowly from losses than do those of most other North American mammals (Vaughan and Pelton 1995).

Black bear populations are difficult to inventory and monitor because the animals occur in relatively low densities and are secretive by nature. Black bears are an important game species in many states and Canada and are an important component of their ecosystems. It is important that they be continuously and carefully monitored to ensure their continued existence.

Did You Know?

Not all black bears are black. In the East, they are nearly black; in the West, black to cinnamon, with white blaze on chest. A "blue" phase occurs near Yakutat Bay, Alaska, and a nearly white population on Gribble Island, British Columbia, and the neighboring mainland.

Grizzly Bears

Mattson et al. (1995) point out that the grizzly bear, sometimes called the silvertip bear, is a powerful brownish-yellow bear that once roamed over most of the U.S. from the high plains to the Pacific coast. In the Great Plains, they seem to have favored areas near rivers and streams, where conflict with humans was also likely. These grassland grizzlies also probably spent considerable time searching out and consuming bison that died from drowning, birthing, or winter starvation, and so were undoubtedly affected by the elimination of bison from most of the Great Plains in the late 1800's. They are potential competitors for most foods valued by humans, including domesticated livestock and agricultural crops, and under certain limited conditions are also a potential threat to human safety. For these and other reasons, grizzly bears in the U. S. were vigorously sought out and killed by European settlers in the 1800's and early 1900's.

Did You Know?

Grizzly bears reach weights of 400–1,500 pounds; the male is on average 1.8 times as heavy as the female, an example of sexual dimorphism.

Between 1850 and 1920 grizzlies were eliminated from 95% of their original range, with extirpation occurring earliest on the Great Plains and later in remote mountainous areas. Unregulated killing of grizzlies continued in most places through the 1950's and resulted in a further 52% decline in their range between 1920 and 1970. Grizzlies survived this last period of slaughter only in remote wilderness areas larger than 10,000 square mi. Altogether, grizzly bears were eliminated from 98% of their original range in the contiguous U.S. during a 100-year period.

Did You Know?

It is a common misconception that grizzly bears cannot climb trees. They will climb trees if they have a food incentive.

Because of the dramatic decline and the uncertain status of grizzlies in areas where they had survived, their populations in the contiguous U.S were listed as threatened under the Endangered Species Act in 1975. High levels of grizzly bear mortality in the Yellowstone area during the early 1970's were also a major impetus for this listing.

Did You Know?

The grizzly has the second slowest reproduction rate of all North American mammals (the musk ox has the slowest), making it harder for it to rebound from threats to its survival. The grizzly can be distinguished from the black bear by its concave face, high-humped shoulders, and long, curved claws.

Black-footed Ferrets

The black-footed ferret is a medium-sized mustelid (from Latin *mustela*, weasel) typically weighing 1.4 to 2.5 pounds (lbs) and measuring 19 to 24 inches in total length. Upper body parts are yellowish buff, occasionally whitish; feet and tail tip are black; and a black "mask" occurs across the eyes.

Biggins and Godbey (1995) report that the black-footed ferret was a charter member of endangered species lists for North America, recognized as rare long before the passage of the Endangered Species Act of 1973. This member of the weasel family is closely associated with prairie dogs of three species, a specialization that contributed to its downfall. Prairie dogs make up 90% of the ferret diet; in addition, ferrets dwell in prairie dog burrows during daylight, venturing out mostly during darkness. Trappers captured black-footed ferrets during their quests for other species of furbearers. Although the species received increased attention as it became increasingly rare, the number of documented ferrets fell steadily after 1940, and little was learned about the animals before large habitat declines made studies of them difficult. These declines were brought about mainly by prairie dog control campaigns begun before 1900 and reaching high intensity by the 1920's and 1930's.

Much of what is known about black-footed ferret biology was learned from research during 1964–1974 on a remnant population in South Dakota (Linder et al. 1972; Hillman and Linder 1973), and from 1981 to the present on a population found at Meeteetse, Wyoming, and later transferred to captivity (Biggins et al. 1985; Forrest et al. 1988; Williams et al. 1988). Nine ferrets from the sparse South Dakota population (only 11 ferret litters were located during 1964–1972) were taken into captivity from 1971 to 1973, and captive breeding was undertaken at the U.S. Fish and Wildlife Service's Patuxent Wildlife Research Center in Maryland (Carpenter and Hillman 1978). Although litters were born there, no young were successfully raised. The last of the Patuxent captive ferrets died in 1978, and no animals were located in South Dakota after 1979.

In 1981, at Meeteetse, black-footed ferrets were "rediscovered" in prairie dog complexes. This rediscovery gave conservationists what seemed a last chance to learn about the species and possibly save it from extinction. That population remained healthy (70 ferret litters were counted from 1982 to 1986) through 1984, a period when much was learned about ferret life history and behavior. In 1985, sylvatic plague, a disease deadly to prairie dogs, was confirmed in the prairie dogs at Meeteetse, creating fear that the prairie dog habitat vital for ferrets would be lost. In addition, field biologists were reporting a substantial decrease in the number of ferrets detected. The fear of plague was quickly overshadowed by the discovery of canine distemper in the ferrets themselves. It is a disease lethal to ferrets.

In 1985 six ferrets were captured to begin captive breeding, but two brought the distemper virus into captivity, and all six died (Williams et al. 1988). A plan was formulated to place more animals from Meeteetse into captivity to protect them from distemper and to start the breeding program. By December 1985, only 10 ferrets were known to exist, 6 in captivity and 4 in Meeteetse. The following year, the surviving free-ranging ferrets at Meeteetse produced only two litters, a number thought too small to sustain the wild population. Because both the Meeteetse and captive populations were too small to sustain themselves, all remaining ferrets were removed from the wild, resulting in a captive population of 18 individuals by early 1987.

Did You Know?

The ferret's large ears and eyes suggest it has acute hearing and sight, but smell is probably its most important sense for hunting prey underground in the dark (BFFRP 2005).

Captive breeding of ferrets eventually became successful. Although the captive population is growing, researchers fear the consequences of low genetic diversity (already documented by O'Brien et al. 1989) and of inbreeding depression (i.e., reduced fitness as a result of breeding of related individuals). A goal of the breeding program is to retain as much genetic diversity as possible, but the only practical way to increase diversity is to find more wild ferrets. In spite of intensive searches of the remaining good ferret habitat and investigations of sighting reports, no wild ferrets have been found.

Biggins and Godbey (1995) point out that the captive breeding program now is producing sufficient surplus ferrets for reintroduction into the wild; 187 ferrets were released into prairie dog colonies in Shirley Basin, Wyoming, during 1991–1993. Challenges facing the black-footed ferret reintroduction include low survivorship of released ferrets due to high dispersal and losses to other predators; unknown influence of low genetic diversity; canine distemper hazard; indirect effect of plague on prairie dogs and possible direct effect on ferrets; and low availability of suitable habitat for reintroduction. The scarcity of habitat reflects a much larger problem with the prairie dog ecosystem and needs increased attention.

At the turn of the twentieth century, prairie dogs reportedly occupied more than 100 million acres of grasslands, but by 1960 that area had been reduced to about 1.5 million acres (Marsh 1984). Much reduction was attributed to prairie dog control programs, which continue. For example, in South Dakota in the late 1980's, millions were spent to apply toxicants to prairie dog colonies on Pine Ridge Indian Reservation (Sharp 1988). At least two states (Nebraska and South Dakota) have laws prohibiting landowners from allowing prairie dogs to flourish on their properties; if the land manager does not "control" the "infestation," the state can do so and bill expenses to the owner (Clarke 1988).

Several prairie dog complexes have been evaluated as sites for reintroduction of black-footed ferrets. The evaluation involves grouping clusters of colonies separated by fewer than 4.3 mi into complexes, based on movement capabilities of ferrets (Biggins et al. 1993); these areas include some of the best prairie dog complexes remaining in the states. Nevertheless, other extensive prairie dog complexes were not considered for ferret reintroduction.

Ramifications of a healthy prairie dog ecosystem extend well beyond black-footed ferrets. The prairie dog is a keystone species of the North American prairies. It is an important primary consumer, converting plants to animal biomass at a higher rate than other vertebrate herbivores of the short-grass prairies, and its burrowing provides homes for many other species of animals and increases nutrients in surface soil. This animal also provides food for many predators. We estimated it takes 700–800 prairie dogs to annually support a reproducing pair of black-footed ferrets and a similar biomass of associated predators (Biggins et al. 1993), suggesting that large complexes of prairie dog colonies are necessary to support self-sustaining populations of these second-order consumers.

The black-footed ferret cannot be reestablished on the grasslands of North America in viable self-sustaining populations without large complexes of prairie dog colonies. The importance of this system to other species is not completely understood, but large declines in some of its species should serve as a warning. The case of the black-footed ferret provides ample evidence that timely preventive action would be preferable to the inefficient "salvage" operations. Furthermore, there is considerable risk of irreversible damage (e.g., genetic impoverishment) with such rescue efforts (Biggins and Godbey 1995).

American Badgers

Primarily because of their burrows, which are dangerous for livestock and horsemen, the American badger has been considered "a malignant creature that had to be destroyed with all means possible, including trapping, shooting and other ways" (A-B.com 2007). In addition to the livestock problem with their burrows, the badger burrows also present a problem for farmers because the holes damage crops. Moreover, badgers prey on livestock. All of this led to significant species destruction and reduced the number of badgers in some regions to the point of extinction.

Steeg and Warner (1995) describe the American badger as a medium-sized carnivore found in treeless areas across North America, such as the tall-grass prairie (Lindzey 1982). Badgers rely primarily on small burrowing mammals as a prey source: availability of badger prey may be affected by changes in land-use practices that alter prey habitat. In the midwestern United States most native prairie was plowed for agricultural use in the mid-1800's (Burger 1978). In the past 100 years, Midwest agriculture has shifted from a diverse system of small farms with row crops, small grains, hay, and livestock pasture to larger agricultural operations employing a mechanized and chemical approach to cropping. The result is a more uniform agricultural landscape dominated by two primary row crops, corn and soybeans. The effects of such land-use alterations on badgers are unknown. In addition, other human activities such as hunting and trapping have no doubt had an impact on native vertebrates such as the badger.

Trends in carnivore abundance are difficult to evaluate because most species are secretive or visually cryptic. Trapping records, one of the earliest historical data sources for furbearers, are virtually nonexistent for badgers in the 1800's (Obbard et al. 1987).

Did You Know?

Most research on badgers has been limited to the western United States. Although results have varied somewhat among these studies, average densities (estimated subjectively from mark-recapture and home range data) have ranged from 0.98–12.95 badgers/square mile.

Northeastern White-tailed Deer

Storm and Palmer (1995) point out that the populations of white-tailed deer have changed significantly during the past 100 years in the eastern United States (Halls 1984). After near extirpation in the eastern states by 1900, deer numbers increased during the first quarter of the century. The effects of growing deer populations on forest regeneration and farm crops have been a concern to foresters and farmers for the past 50 years.

Did You Know?

The fur of white-tailed deer is a grayish color in the winter; the red comes out during the summer. It has a band of white fur behind its nose, in circles around the eyes, and inside the ears. More white fur goes down the throat, on the upper insides of the legs and under the tail.

In recent years, deer management plans have been designed to maintain deer populations at levels compatible with all land uses. Conflicts, however, between deer and forest management or agriculture still exist in the Northeast. Areas that were once exclusively forests are now a mixture of forest, farm, and urban environments that create increased interactions and conflicts between humans and deer, including deer-vehicle collisions. Management of deer near urban environments presents a unique challenge for local resource managers (Porter 1991).

North American Elk

North American elk, or wapiti (the name "elk" was given by early explorers because they resembled the elk or moose of Europe . . . the American Indian term "Wapiti" is sometimes used to identify the animal), represent how a wildlife species can recover even after heavy exploitation of populations and habitats around the early 1900's. Elk are highly prized by wildlife enthusiasts and by the hunting public, which has provided the various state wildlife agencies with ample support to restore elk populations to previous occupied habitats and to manage elk populations effectively. Additionally, the Rocky Mountain Elk Foundation, founded in 1984, has promoted habitat management, acquisition, and proper hunting ethics among many segments of the hunting public (Peek 1995).

Current population size is estimated at 782,500 animals for the entire elk range (Rocky Mountain Elk Foundation 1989). Projections of population trends for the national forests and for the entire U.S. elk range are for continued increases through the year 2040 (Flather and Hoekstra 1989).

Indeed, the future of elk populations in North America seems secure. Demand for hunting as well as the nonconsumptive values of elk will ensure the success of substantial populations. Elk populations will benefit from improved habitat conditions on arid portions of the range, improved livestock management, more effective integrated management of forested habitats, and continued implementation of fire management policies in the major wilderness areas and national parks (Peek 1995).

Reptiles and Amphibians

Yet when a child and barefoot, I more than once, at morn,
Have passed, I thought, a whiplash unbraided in the sun,
When, stooping to secure it, it wrinkled, and was gone.

—Emily Dickinson

There are more than 8,200 species of reptiles and more than 5,500 species of amphibians on Earth, including turtles, snakes, crocodiles, lizards (reptiles), frogs, toads, salamanders, and newts (amphibians). All reptiles have scales, but some are too small to be seen. Reptiles are ectothermic (i.e., they obtain heat from outside sources). Most lay eggs, but a few give birth to live young. For most amphibians, life begins in the water. They metamorphose, growing legs and changing in other ways to live on land. Like their "cold-blooded" reptile relatives, they depend on external energy sources (the sun) to maintain their body temperatures.

Did You Know?

Amphibians became the first vertebrates to live on land, and like their "cold-blooded" reptile relatives, depend on external energy sources to maintain their body temperatures. Reptiles are ectotherms. They obtain heat from outside sources, like the sun, and regulate their temperature through behaviors such as basking or seeking shade.

Amphibians and reptiles are crucial to the natural functioning of many ecological processes and key components for important ecosystems. In some areas certain species are economically consequential; others are aesthetically pleasing to many people and as a group they represent significant segments of the evolutionary history of North America. Knowledge gained from past study of amphibian development and metamorphosis has contributed immensely to our understanding of basic biological processes and has directly benefited humans (McDiarmid 1995).

Baseline information of the status and health of U.S. populations of amphibians and reptiles is remarkably sparse. No national program for monitoring populations of amphibians and reptiles is in operation. A recent publication (Heyer et al. 1994) recommended standard guidelines and techniques for monitoring amphibian populations and habitats; a similar volume on reptiles is planned.

Habitat degradation and loss seem to be the most important factors adversely affecting amphibian and reptile populations in North America. The drainage and loss of small aquatic habitat and their associated wetlands have had a major adverse effect on many amphibian species and some reptiles.

McDiarmid (1995) points out that many other factors in the decline of reptiles and amphibians have been implicated; most, perhaps all, are human-caused. For example, non-native species of gamefish introduced for sport have been implicated in the decline of frog populations in mountainous areas of some western states. Similarly, the introduction, accidental or intentional, of other non-native species (e.g., bullfrogs in western states, anoline lizards in south Florida, and snakes in Guam) has harmed native species in other parts of the country.

Turtles

> A turtle is at heart a misanthrope; its shell is in itself proof of its owner's distrust of this world. But we need not wonder at this misanthropy, if we think for a moment of the creatures that lived on this earth at the time when turtles first appeared. Almost any of us would have been glad of a shell in which to retire if we had been contemporaries of the smilodon [saber-toothed cat] and other monsters of earlier geologic times.
>
> —Anna Botsford Comstock

As Comstock (1986) points out, turtles have existed for a very long time; virtually unchanged for the last 200 million years (see Figure 49). Unfortunately, as Lovich (1995) points out, some of the same traits that allowed them to survive the ages often predispose them to endangerment. Delayed maturity

FIGURE 49.   **Turtles sunbathing in lily pad area of Green Lake, Seattle, Washington**
Photo by Frank R. Spellman

and low and variable annual reproductive success make turtles unusually susceptible to increased mortality through exploitation and habitat modifications (Brooks et al. 1991; Congdon et al. 1993).

Did You Know?

A turtle eats grasses, mushrooms, berries, insects, flowers, worms, water plants, crayfish, snails, fish, frogs, and dead animals.

In general, turtles are overlooked by wildlife managers in spite of their ecological significance and importance to humans. Turtles are, however, important as scavengers, herbivores, and carnivores, and often contribute significant biomass to ecosystems. In addition, they are an important link in ecosystems, providing dispersal mechanisms for plants, contributing to environmental diversity, and fostering symbiotic associations with a diverse array of organisms. Adults and eggs of many turtles have been used as a food resource by humans for centuries (Brooks et al. 1988; Lovich 1995). As use pressures and habitat destruction increase, management that considers the life-history traits of turtles will be needed.

Marine Turtles

Marine turtles have outlived almost all of the prehistoric animals with which they once shared the planet. Six species of marine turtles frequent the beaches and offshore waters of the southeastern Untied States (Escambia 2007):

- **Loggerhead**—is the most common turtle to nest in Florida. Over 50,000 loggerhead nests are recorded annually in Florida. This turtle is named for its disproportionately large head; it is a medium to large-sized sea turtle usually weighing 175 to 300 pounds as an adult. Loggerheads feed on crabs, mollusks, and jellyfish.
- **Green**—is the second most common turtle in Florida waters. Green sea turtles are the only herbivorous sea turtles. They feed on seagrasses in shallow areas through the Gulf of Mexico. An adult green turtle can reach more than three feet in length and weigh 300 to 400 pounds. The lower jaw is serrated to help cut the seagrasses it eats.
- **Kemp's Ridley**—is the rarest sea turtle in the world. They are small with adults reaching two to two and one-half feet in length and weighing 80 to 100 pounds. They primarily nest on one beach on the gulf coast of Mexico and are the smallest species of sea turtle. Scientists have been trying to transplant Kemp's Ridley eggs to Texas to establish a new nest-

ing colony. They are the only species of sea turtle known to lay their eggs during the day.

- **Leatherback**—is the largest sea turtle in the world and can be over 8 feet long and weigh 1400 pounds. It does not have a hard shell, but rather a leather-like carapace with bony ridges underneath the skin. The leatherback makes long migrations from its nesting beaches in the tropics as far north as Canada. Jellyfish are the favored prey to these turtles.
- **Hawksbill**—is a small to medium-sized turtle; they are usually found feeding primarily on sponges in the southern Gulf of Mexico and Caribbean. It gets its name from its distinctive hawk-like beak. The hawksbill sea turtle was hunted to near extinction for its beautiful shell which features overlapping scales.
- **Olive Ridley Sea Turtle**—is a bit larger than the Kemp's but is still a small turtle. Adult olive ridleys may reach three feet in length and will weight 100 to 110 pounds. The olive ridley is threatened except for the Mexican nesting population which is endangered.

All six are reported to nest, but only the loggerhead and green turtle do so in substantial numbers. Most nesting occurs from southern North Carolina to the middle west coast of Florida, but scattered nesting occurs from Virginia through southern Texas. The beaches of Florida, particularly in Brevard and Indian River counties, host what may be the world's largest population of loggerheads (Dodd 1995).

Marine turtles, especially juveniles and subadults, use lagoons, estuaries, and bays as feeding grounds. Areas of particular importance include Chesapeake Bay, Virginia (for loggerheads and Kemp's ridleys); Pamlico Sound, North Carolina (for loggerheads); and Mosquito Lagoon, Florida, and Laguna Madre, Texas (for greens). Offshore waters also support important feeding grounds such as Florida Bay and the Cedar Keys, Florida (for green turtles), and the mouth of the Mississippi River and the northeast Gulf of Mexico (for Kemp's ridleys).

Offshore reefs provide feeding and resting habitat (for loggerheads, greens, and hawksbills), and offshore currents, especially the Gulf Stream, are important migratory corridors (for all species, but especially leatherbacks). Olive ridley sea turtles occur in the eastern Pacific near Mexico.

Did You Know?

Raccoons destroy thousands of sea turtle eggs each year and are the single greatest cause of sea turtle mortality in Florida.

Most marine turtles spend only part of their lives in U.S. waters. For example, hatchling loggerheads ride oceanic currents and gyres (giant circular oceanic surface currents) for many years before returning to feed as subadults in southeastern lagoons. They travel as far as Europe and the Azores, and even enter the Mediterranean Sea, where they are susceptible to longline fishing mortality. Adult loggerheads may leave U.S. waters after nesting and spend years in feeding grounds in the Bahamas and Cuba before returning. Nearly the entire world population of Kemp's ridleys use a single Mexican beach for nesting, although juveniles and subadults, in particular, spend much time in U.S. offshore waters (Dodd 1995).

The biological characteristics that make sea turtles difficult to conserve and manage include a long life span, delayed sexual maturity, differential use of habitats both among species and life stages, adult migratory travel, high egg and juvenile mortality, concentrated nesting, and vast areal dispersal of young and subadults. Genetic analyses have confirmed that females of most species return to their natal beaches to nest (Bowen et al. 1992; Bowen et al. 1993). Nesting assemblages contain unique genetic markers showing a tendency toward isolation from other assemblages (Bowen et al. 1993); thus, Florida green turtles are genetically different from green turtles nesting in Costa Rica and Brazil (Bowen et al. 1992). Nesting on warm sandy beaches puts the turtles in direct conflict with human beach use and their use of rich off-shore waters subjects them to mortality from commercial fisheries (National Research Council 1990).

Marine turtles have suffered catastrophic declines since European discovery of the New World (National Research Council 1990). In a relatively short time, the huge nesting assemblages in the Cayman Islands, Jamaica, and Bermuda were decimated. In the United States, commercial turtle fisheries once operated in south Texas (Doughty 1984), the Cedar Keys, the Florida Keys, and Mosquito Lagoon; these fisheries collapsed from overexploitation of the mostly juvenile green turtle populations. Today, marine turtle populations are threatened worldwide and are under intense pressure in the Caribbean basin and Gulf of Mexico, including Cuba, Mexico, Hispaniola, the Bahamas, and Nicaragua. Marine turtles can be conserved only though international efforts and cooperation (Dodd 1995).

Amphibians

The toad hopped by us with jolting springs.

—Akers

Bury et al. (1995) point out that amphibians are ecologically important in most freshwater and terrestrial habitats in the United States; they can be numerous, function as both predators and prey, and constitute great biomass. Amphibians have certain physiological (e.g., permeable skin) and ecological (e.g., complex life cycle) traits that could justify their use as bioindicators of environmental health. For example, local declines in adult amphibians may indicate losses of nearby wetlands. The aquatic breeding habits of many terrestrial species result in direct exposure of egg, larval, and adult stages to toxic pesticides, herbicides, acidification, and other human-induced stresses in both aquatic and terrestrial habitats. Reported declines of amphibian populations globally have drawn considerable attention (Bury et al. 1980; Bishop and Petit 1992; Richards et al. 1993; Blaustein 1994; Pechmann and Wilbur 1994).

Approximately 230 species of amphibians, including about 140 salamanders and 90 anurans (frogs and toads) occur in the continental United States. Because of their functional importance in most ecosystems, declines of amphibians are of considerable conservation interest. If these declines are real, the number of listed or candidate species at federal, state, and local levels could increase significantly. Unfortunately, because much of the existing information on status and trends of amphibians is anecdotal, coordinated monitoring programs are greatly needed (Bury et al. 1995).

North American amphibian species exhibit two major distributional patterns, endemic and widespread. Endemic species tend to have small ranges or are restricted to specific habitats (e.g., species that occur only in one cave or in rock talus on a single mountainside). Declines are documented best for endemic species, partly because their smaller range makes monitoring easier. Populations of endemics are most susceptible to loss or depletions because of localized activities (Bury et al. 1980; Dodd 1991). Examples of endemic species affected by different local impacts include the Santa Cruz long-toed salamander in California, the Texas blind salamander in Texas, and the Red Hills salamander in Alabama; these three species are listed as federally threatened or endangered.

The number of endemic species that have suffered losses or are suspected of having severe threats to their continued existence has increased in the last 15 years. In part, the increase reflects descriptions of new species with restricted ranges, but the accelerating pace of habitat alteration is the primary threat.

The ranges of most endemics in the western states (26 species) are widely dispersed across the landscape. In contrast, endemics in the eastern and southeastern states (25 species) tend to be clustered in centers of endemism, such as in the Edwards Plateau (Texas), Interior (Ozark) Highlands (Arkansas, Oklahoma), Atlantic Coastal Plain (Texas to Virginia), and uplands or mountaintops in the Appalachians (West Virginia to Georgia).

Widespread species often are habitat generalists. Many were previously common, but have shown regional or range-wide declines (Hine et al. 1981; Corn and Fogelman 1984; Hayes and Jennings 1986). Reported declines of widespread species often lack explanation, perhaps because these observations have only recently received general attention or because temporal and spatial variations in population sizes of many amphibians are not well understood. Some reports are for amphibians in relatively pristine habitats where human impacts are not apparent (Bury et al. 1995).

No single factor has been identified as the cause of amphibian declines, and many unexpected declines likely result from multiple causes. Human-caused factors may intensify natural factors (Blaustein et al. 1994b) and produce declines from which local populations cannot recover and thus they go extinct. Known or suspected factors in those declines include destruction and loss of wetlands (Bury et al. 1980); habitat alteration, such as impacts from timber harvest and forest management (Corn and Bury 1989; Dodd 1991; Petranka et al. 1993); introduction of non-native predators, such as sport fish and bullfrogs, especially in western states (Hayes and Jennings 1986; Bradford 1989); increased variety and use of pesticides and herbicides (Hine et al. 1981); effects of acid precipitation, especially in eastern North America and Europe (Freda 1986; Beebee et al. 1990; Dunson et al. 1992); increased ultraviolet radiation reaching the ground (Blaustein et al. 1994a); and diseases resulting from decreased immune system function (Bradford 1991; Carey 1993; Pounds and Crump 1994).

Amphibian populations also may vary in size because of natural factors, particularly extremes in the weather (Bradford 1983; Corn and Fogelman 1984). The size of amphibian populations may vary, sometimes dramatically, from year to year, and what is perceived as a decline may be part of long-term fluctuations (Pechmann et al. 1991). The effect of global climate change on amphibians is speculative, but it has the potential for causing the loss of many species.

American Alligators in Florida

The American alligator is a large, semi-aquatic, armored reptile that is related to crocodiles. Their body alone ranges from 6–14 feet long. Almost black in color, it has prominent eyes and nostrils with coarse scales over the entire body. It has a large, long head with visible upper teeth using the edge of the jaws. Its front feet have 5 toes, while the rear feet have 4 toes that are webbed.

Woodard and Moore (1995) point out that, as members of the crocodile family, alligators are living fossils that can be traced back 230 million years. The American alligator is an integral component of wetland ecosystems in Florida. Alligators also provide aesthetic, educational, recreational, and economic benefits to humans. Because of the commercial value of alligator hides for making high-quality leather products, alligator hunting was a major economic and recreational pursuit of many Floridians from the mid-1800's to 1970. The

Florida alligator population varied considerably during the 1900's in response to fluctuating hunting pressure caused by unstable markets for luxury leather products. The declining abundance of alligators during the late 1950's and early 1960's led to the 1967 classification of the Florida alligator population as endangered throughout its range. Federal and international regulations imposed during the 1970's and 1980's have helped control trade of alligator hides, and illegal hunting of alligators was checked. The Florida alligator population responded immediately to protection and was reclassified as threatened in 1977 and as threatened because of its similarity in appearance to the American crocodile in 1985 (Neal 1985).

Native Ranid Frogs

Frogs Eat Butterflies. Snakes Eat Frogs. Hogs Eat Snakes. Men Eat Hogs.

—Wallace Stevens

Many recent declines and extinctions of native amphibians have occurred in certain parts of the world (Wake and Morowitz 1991). All species of native tree frogs have declined in the western United States over the past decade (Hayes and Jennings 1986). Most of these native amphibian declines can be directly attributed to habitat loss or modification, which is often exacerbated by natural events such as droughts or floods (Wake 1991). A growing body of research, however, indicates that certain native frogs are particularly susceptible to population declines and extinctions in habitats that are relatively unmodified by humans (e.g., wilderness areas and national parks in California; Bradford 1991; Fellers and Drost 1993; Kagarise Sherman and Morton 1993). To understand these declines, we must document the current distribution of these species over their entire historical range to learn where they have disappeared (Jennings 1995).

In 1988 the California Department of Fish and Game commissioned the California Academy of Sciences to conduct a 6-year study on the status of the state's amphibians and reptiles not currently protected by the Endangered Species Act. The study's purpose was to determine amphibians and reptiles most vulnerable to extinction and provide suggestions for future research, management, and protection by state, federal, and local agencies (Jennings and Hayes 1993).

Desert Tortoises

Desert tortoises, also known as desert turtles or gopher tortoises, are any of the land-dwelling turtles that are widespread throughout the southwestern

United States and Mexico. Berry and Medica (1995) point out that within the United States, desert tortoises live in the Mojave, Colorado, and Sonoran deserts of southeastern California, southern Nevada, southwestern Utah, and western Arizona. A substantial portion of the habitat is on lands administered by the U.S. Department of the Interior.

Did You Know?

Although 95% of the Desert tortoise's life is spent in underground burrows, it is able to live where ground temperature may exceed 140°F.

The U.S. government treats the desert tortoise as an indicator or umbrella species to measure the health and well-being of the ecosystems it inhabits. The tortoise functions well as an indicator because it is long-lived, takes 12–20 years to reach reproductive maturity, and is sensitive to changes in the environment. In 1990 the U.S. Fish and Wildlife Service listed the species as threatened in the northern and western parts of its geographic range because of widespread population declines and overall habitat loss, deterioration, and fragmentation.

Did You Know?

Ravens have caused more than 50% of juvenile desert tortoise deaths in some areas of the Mojave Desert.

Because some populations exhibit significant genetic, morphologic, and behavioral differences, the Desert Tortoise Recovery Team identified various population segments for critical habitat protection and long-term conservation within the Mojave and Colorado deserts (e.g., Lamb et al. 1989; USFWS 1994).

Fringe-toed Lizards

From tiny to gigantic, from drab to remarkably beautiful, from harmless to venomous, lizards are spectacular products of natural selection.

—Pianka and Vitt (2006)

The fringe-toed lizard (*Uma* spp.; order Squamata) is a medium sized, whitish lizard that inhabits many of the scattered windblown sand deposits of southeastern California and southwestern Mexico. These lizards have several specialized adaptations: elongated scales on their hind feet ("fringes") for added traction in loose sand, a shovel-shaped head and a lower jaw adapted to aid diving into and moving short distances beneath the sand, elongated scales covering their ears to keep sand out, and unique morphology (form or structure) of internal nostrils that allows them to breathe below the sand without inhaling sand particles (Barrows et al. 1995).

While these adaptations enable fringe-toed lizards to successfully occupy sand dune habitats, the same characteristics have restricted them to isolated sand "islands." Three fringe-toed lizard species live in the United States: The Mojave (*U. scoparia*), the Colorado Desert (*U. notata*), and the Coachella Valley (*U. inornata*). Of the three, the Coachella Valley fringe-toed lizard has the most restricted range and has been most affected by human activities. In 1980 this lizard was listed as a threatened species by the federal government.

In 1986 the Coachella Valley Preserve system was established to protect habitat for the Coachella Valley fringe-toed lizard. This action set several precedents: it was the first Habitat Conservation Plan established under the revised (1982) Endangered Species Act and the newly adopted section 10 of the act, it established perhaps the only protected area in the set aside for a lizard, and its design was based on a model of sand dune ecosystem processes, the sole habitat for this lizard. Three disjunct sites in California, each with a discrete source of windblown sand, were set aside to protect fringe-toed lizard populations: Thousand Palms, Willow Hole, and Whitewater River. Collectively, the preserves protect about 2% of the lizards' original range.

Barrows (unpublished data) points out that eight years after the establishment of the preserve system, few Coachella Valley fringe-toed lizards exist outside the boundaries of the three protected sites. Barrows recently identified scattered pockets of windblown sand occupied by fringe-toed lizards in the hill along the northern fringe of the valley, but only at low densities. Fringe-toed lizard populations within the protected sites have been monitored yearly since 1986. During this period, California experienced one of its most severe droughts, which ended in spring 1991. Numbers of fringe-toed lizards within the Thousand Palms and Willow Hole sites declined during the drought, but rebounded after 1991. By 1993, after three wet springs, lizard numbers had increased substantially.

Barrows et al. (1995) report that the lizards at the Whitewater River site were intensively monitored since 1985 by using mark-recapture methods to count the population on 2.25-ha (5.56-acre) plots. In 1986 this site had the highest population density of the three protected sites. As with the other two sites, the Whitewater River population declined throughout the drought, but only increased slightly after the drought broke in 1991. Compounding the drought effect, much of the fine sand preferred by fringe-toed lizards was blown off the site during the dry years. This condition was unique to the Whitewater River site; the other two protected sites have much deeper sand deposits and are less susceptible to wind erosion. New windblown sand was deposited on the Whitewater River site in 1993 after a period of high rainfall. The population appears to be increasing in response to these favorable conditions.

Tarahumara Frog

The Tarahumara frog is a medium-sized, drab green-brown frog with small brown to black spots on the body and dark crossbars on the legs. The hind feet are extensively webbed (USFWS 2007c).

Hale et al. (1995) point out that in the spring of 1983 the last known Tarahumara frog in the United States was found dead. Overall, the species seems to be doing well in Mexico, although the decline of more northern populations is of concern. The Tarahumara frog inhabits seasonal and permanent bedrock and bouldery streams in the foothills and main mountain mass of the Sierra Madre Occidental of northwestern Mexico. It ranges from northern Sinaloa, through western Chihuahua and eastern and northern Sonora, and until recently into extreme south-central Arizona.

USFWS (2007c) points out that the reasons why the Tarahumara frog disappeared from Arizona are not clear. However, the following hypotheses have been presented: 1) winter cold; 2) flooding or severe drought; 3) competition; 4) predation; 5) disease; and 6) heavy metal acid precipitation. Metals occur naturally in streamside deposits and may be mobilized by acid precipitation events.

Fishes

I have laid aside business, and gone afishing.

—Izaak Walton (1593–1683)

Izaak Walton "discovered that nature-study, fishing, and philosophy were akin and as inevitably related as the three angles of a triangle." One thing is certain; old Izaak Walton loved to fish. This point is made clear by Walton in his *The Compleat Angler* (1653):

It remains yet unresolved whether the happiness of a man in this World doth consist more in contemplation or action. Concerning which two opinions I shall forebear to add a third by declaring my own, and rest myself contented in telling you that both of these meet together, and do most properly belong to the most honest, ingenious, quiet and harmless art of angling. And first I tell you what some have observed, and I have found to be a real truth, that every sitting by the riverside is not only the quietest and the fittest place for contemplation, but will invite an angler to it.

Notwithstanding Walton's obvious love of fishing (he refers to it as angling, as many others do) and of the many others who reside throughout the globe and who share Walton's view on the subject, this text is not about fishing. It is, however, as I am sure you are well aware by now, all about the ecology of species and about ecology in general. Thus, in this section fishes are discussed. Specifically, from the material presented in this section the inescapable conclusion is that within historical time, native fish communities have undergone significant and adverse changes (Maughan 1995). These changes generally tend toward reduced distributions, lowered diversity, and increased numbers of species considered rare. These changes have been more inclusive and more dramatic in the arid western regions where there are primarily endemic (native) species, but similar, though more subtle changes, have occurred throughout the country. These trends are the same whether one focuses on faunas or on populations or genetic variation within a single species. Changes in fish communities may be indicative of the overall health of an aquatic system; some species have narrow habitat requirements.

It should not come as any great surprise that fish populations have changed over time; all things change with time. We have massively modified fish habitat through the very water demands that define our society (domestic, agricultural, and industrial water supplies; waste disposal; power generation; transportation; and flood protection). All of these activities have resulted in controlling or modifying the flow or degrading the quality of natural waters. In addition, almost all contaminants ultimately find their way into the aquatic system. Species of fishes that have evolved under the selection pressures imposed by natural cycles have often been unable to adapt to the changes imposed on them as a result of human activities.

Physical and chemical changes in their habitats are not the only stresses that fishes have encountered over time. Through fish management programs, the aquarium trade, and accidental releases, many aquatic species have been introduced to new areas far beyond their native ranges. Although these introductions were often done with the best of intentions, they have sometimes subjected native fish species to new competitors, predators, and disease agents that they were ill-equipped to withstand.

Did You Know?

Theoretically, the smaller the gene pool, the less likely a species may be able to adapt to changing environmental conditions.

It appears unlikely that the forces that have led to these changes in our fish fauna will lessen significantly in the immediate future. Therefore, if we are to preserve the diversity and adaptive potential of our fishes, we must understand much more of their ecology. Vague generalizations about habitat requirements or the results of biotic interactions are no longer enough. We must know quantitatively and exactly how fishes use habitat and how that use changes in the face of biotic pressures. Only when armed with such information are we likely to reduce the current trends among our native fishes (Maughan 1995).

Freshwater Fishes of the Contiguous United States

Up to 800 species of freshwater fishes are native to the United States (Lee et al. 1980; Moyle and Cech 1988; Warren and Burr 1994). These fishes range from old, primitive forms such as paddlefish, bowfin, gar, and sturgeon, to younger, more advanced fishes, such as minnows, darters, and sunfishes. They are not equally distributed across the nation, but tend to concentrate in larger, more diverse environments such as the Mississippi River drainage (375 species; Robison 1986; Warren and Burr 1994). Drainages that have not undergone recent geological change, such as the Tennessee and Cumberland rivers, are also rich in native freshwater fishes (250 species; Starnes and Etnier 1986). Fewer native fishes are found in isolated drainages such as the Colorado River (36 species; Carlson and Muth 1989). More arid states west of the 100th meridian average about 44 native fish species per state, while states east of that boundary average more than three times that amount (138 native species; Johnson 1995).

Extinction, dispersal, and evolution are naturally occurring processes that influence the kinds and numbers of fishes inhabiting our streams and lakes. More recent human-related impacts to aquatic ecosystems, such as damming of rivers, pumping of aquifers, addition of pollutants and introductions of nonnative species, also affect native fishes, but at a more rapid rate than natural processes. Some fishes are better able to withstand these rapid changes to their environments or are able to find temporary refuge in adjacent habitats; fishes that lack tolerance or are unable to retreat face extinction (Johnson 1995).

In 1979 the Endangered Species Committee of the American Fisheries Society (AFS) developed a list of 251 freshwater fishes of North America judged in danger of disappearing (Deacon et al. 1979), 198 of which are found in the United States. A decade later, AFS updated the list (Williams et al. 1989), noting 364 taxa of fishes in some degree of danger, 254 of which are native to the United States. Both AFS lists used the same endangered and threatened categories defined in the Endangered Species Act of 1973, and added a special concern category to include fishes that could become threatened or endangered with relatively minor disturbances to their habitat. These imperiled native fishes are the first to indicate changes in our surface waters; thus their status provides us with a method of judging the health of our streams and lakes.

Managed Populations: Loss of Genetic Integrity through Stocking

Philipp and Claussen (1995) point out that species are composed of genetically divergent units usually interconnected by some (albeit low) level of gene flow (Soule 1987). Because of this restriction in gene flow, natural selection can genetically tailor populations to their environments through the process of local adaptation (Wright 1931).

Because freshwater and anadromous (i.e., adults travel upriver from the sea to spawn) fishes are restricted by the boundaries of their aquatic habitats, genetic subdivisions may be more pronounced for these vertebrates than for others. Consequently, managers of programs for these species must realize that the stock (i.e., local discrete populations), and not the species as a whole, must be the units of primary management concern (Kutkuhn 1981).

Genetic variability in a species occurs both among individuals *within* population as well as among populations (Wright 1978). Variation *within* populations is lost through genetic drift (Allendorf et al. 1987), a process increased when population size becomes small. Variation *among* populations is lost when previously restricted gene flow between populations is increased for some reason (e.g., stocking, removal of natural barriers such as waterfalls); differentiation between populations is lost as a result of the homogenization of two previously distinct entities (Altukhov and Salmenkova1987; Campton 1987).

Beyond this loss of genetic variation, mixing two groups can result in outbreeding depression, which is the loss of fitness in offspring that results from the mating of two individuals that are too distantly related (Templeton 1987). This loss in fitness is caused by the disruption of the process that produced advantageous local adaptations through natural selection. Inbreeding depression, on the other hand, is the loss of fitness produced by the repeated crossing of related organisms. The area of optimal relatedness occurs between inbreeding depression and outbreeding depression.

Many sport fish populations are managed by using a combination of harvest regulation, habitat manipulation, and stocking. Jurisdiction for these activities falls to federal, state, tribal, and local governments, as well as private citizens. Many resource managers in the past were unaware of the long-term consequences that stocking efforts would have on the genetic integrity of local populations (Philipp et al. 1993).

Fish populations can be classified into three types: non-native introductions, in which a given species of fish is introduced into a body of water outside its native range (regardless of any political boundaries); stock transfers, in which fish from one stock are introduced into a water body in a different geographic region inhabited by a different stock of that same species, yet are still within their native range; and genetically compatible introductions, in which fish are removed from a given water body and they, or more often their offspring, are introduced back into that water body or another water body that is still within the boundaries of the genetic stock serving as the hatchery brood source (Philipp et al. 1993).

Although non-native introduction may often cause ecological problems for the environments in which they are introduced, they can also cause genetic problems if they hybridize with closely related native species. Examples of this are the hybridization of introduced small-mouth bass and spotted bass with native Guadalupe bass in Texas; and the hybridization of introduced rainbow trout with native Apache trout (Carmichael et al. 1993). The greatest degree of genetic damage, that is, the loss of genetic variation among populations, is caused by stock transfers, a common practice among fisheries management agencies and the private sector.

The genetic integrity of many other managed fish species is eroding as a result of management programs that inadvertently permit or deliberately promote stock transfers. This causes not only the loss of genetic variation among populations, but through outbreeding depression it is also probably negatively affecting the fitness of many native stocks involved. We need to address genetic integrity when restoring native populations (Philipp and Claussen 1995).

Colorado River Basin Fishes

Starnes (1995) explains that the Colorado River and its tributaries have undergone drastic alterations from their natural states over the past 125 years. These alterations include both physical change or elimination of aquatic habitats and the introduction of numerous non-native species, particularly fish. Ironically, several more species occur at most localities today than were historically present before these alterations. This situation complicates the use of biodiversity as a litmus test for monitoring trends of either the deterioration or the health of an aquatic ecosystem.

Over its entire basin, the Colorado River has been changed from its natural state perhaps as much as any river system in the world. The demands for water and power in the arid West have drastically altered the system by impoundments, irrigation diversions, diking, channelization, pollutants, and destruction of bank habitats by cattle grazing and other practices. Some reaches, ranging from desert spring runs to main rivers, have been completely dewatered or, seasonally, their flows consist almost entirely of irrigation return laden with silt and chemical pollutants. The Gila River of Arizona, one of the Colorado's largest tributaries, has not flowed over its lower 248 mi since the early 1900's. These alterations and their effects on the fish fauna have been discussed by several authors (Miller 1961; Minckley and Deacon 1968; Stalmaker and Holden 1973; Carlson and Muth 1989; Minckley and Deacon 1991). Only a few small tributaries, mostly at higher elevations, retain most of their natural characteristics.

Cutthroat Trout: Glacier National Park

The cutthroat trout is a species of freshwater fish in the salmon family. It is one of many fish species colloquially known as trout. Marnell (1995) points out that the indigenous fishery of Glacier National Park has been radically altered from its pristine condition during the past half-century through introductions of non-native fishes and the entry of non-native species from waters outside the park. These introductions have adversely affected the native westslope cutthroat trout (also known as Clark's trout, red-throated trout, and short-tailed trout) throughout much of its park range (see Figure 50).

The effects of non-native fishes on indigenous fisheries have been reviewed by Taylor et al. (1984), Marnell (1986), and Moyle et al. (1986). Effects of fish

FIGURE 50.   Elizabeth Lake, Glacier National Park, Montana
Photo by Frank R. Spellman

introductions in Glacier National Park include establishment of non-native trout populations in historically fishless waters, genetic contamination (i.e., hybridization) of some native westslope cutthroat trout stocks, and ecological interferences with various life-history stages of native trout.

Research conducted in the park during the 1980's addressed the genetic effects of fish introductions on native trout. Of 47 lakes known or suspected to contain cutthroat trout or trout hybrids, 32 lakes contained viable populations of cutthroat trout, rainbow trout, or hybrids. Trout introduced in the other waters were evidently unable to sustain themselves through natural reproduction. Research conducted includes:

- About 30 trout sampled from each lake underwent laboratory genetic analyses. Close agreement of the results from two analytical procedures yielded a high degree of confidence in the conclusions (Marnell et al. 1987).
- Fourteen pure strain populations of west slope cutthroat trout persist in 15 lakes (i.e., some interconnected lakes contain a single trout population) in the North and Middle Fork drainages of the Flathead River; the species was historically present in these waters.
- Pure strain native trout also inhabit four other Middle Fork lakes (i.e., Avalanche, Snyder, and Upper and Lower Howe lakes), but it is unclear whether they are indigenous or were transplanted from other park waters. Recent findings from sediment paleolimnology studies suggest that trout have been present in at least one other lake for more than 300 years (D. Verschuren, University of Minnesota, and author, unpublished data). Hence, trout populations in these four lakes are tentatively classified as indigenous.
- Introduced populations of Yellowstone cutthroat trout and trout hybrids including cutthroat-rainbow trout occur in 13 lakes distributed among the three continental drainages that form their headwaters in Glacier National Park. Native cutthroat trout were not found east of the Continental Divide in the Missouri River on South Saskatchewan River drainages within the park.

In addition to genetic concerns, ecological disturbances associated with the presence of introduced fishes have compromised the native westslope cutthroat fishery. Fish are no longer stocked in park waters; however several waters, including some that contain undisturbed native fisheries, remain vulnerable to invasion by non-native migratory species. Introduced kokanee salmon, a specialized planktivore, are believed to be competing with juvenile stages of native trout in some waters, especially during periods of winter ice cover when plankton may be limited. Predation by introduced lake trout has also been implicated in the decline of native cutthroat trout in several large glacial lakes in North and Middle Fork drainages (Marnell 1988). Native cutthroat trout have been compromised by fish introductions and invasions throughout about 84% of their historic range in Glacier National Park (Marnell 1988).

Although native cutthroat trout have been adversely affected throughout a large portion of their park range, the species has not been lost from any water where it was historically present. Glacier National Park remains one of the last strongholds of genetically pure strains of Lacustrine (i.e., lake-adapted) westslope cutthroat trout. This fact could have important implications for reestablishment of this unique subspecies throughout the central Rocky Mountains, where this trout has disappeared from most of its original range (Marnell 1995).

White Sturgeon

The white sturgeon, also known as the Pacific sturgeon, Oregon sturgeon, Columbia sturgeon, Sacramento sturgeon, and California white sturgeon, is the largest freshwater fish in North America and the third largest species of sturgeon. It has a slender, long body, head, and mouth. This fish has no scales; instead it has large bony scutes that serve as a form of armor.

Scott and Crossman (1973) explain that white sturgeon, again, the largest freshwater fish in North America, live along the west coast from the Aleutian Islands to central California. Genetically similar reproducing populations inhabit three major river basins: Sacramento-San Joaquin, Columbia, and Fraser. The greatest number of white sturgeon is in the Columbia River basin.

Did You Know?

Instead of scales the white sturgeon is covered with patches of miniscule dermal denticles and isolated rows of large bony plates (IBC 2007).

Historically, white sturgeon inhabited the Columbia River from the mouth upstream into Canada, the Snake River upstream to Shoshone Falls, and the Kootenai River upstream in Kootenai Falls (Scott and Crossman 1973). White sturgeon also used the extreme lower reaches of other tributaries, but not extensively. Current populations in the Columbia River basin can be divided into three groups: fish below the lowest dam, with access to the ocean (the lower Columbia River); fish isolated (functionally but not genetically) between dams; and fish in several large tributaries.

The Columbia River has supported important commercial, treaty, and recreational white sturgeon fisheries. A commercial fishery that began in the
1880's peaked in 1892 when 2.5 million kg (5.5 million lb) were harvested
(Craig and Hacker 1940). By 1899 the population had been severely depleted,
and annual harvest was very low until the early 1940's, but the population
recovered enough by the later 1940's that the commercial industry expanded.
A 6-ft maximum size restriction was enacted to prevent another population
collapse. Total harvest doubled in the 1970's and again in the 1980's because
of increased treaty and recreational fisheries. From 1983 to 1994, 15 substantial regulatory changes were implemented on the main stem Columbia River
downstream from McNary Dam as a result of increased fishing. Columbia
River white sturgeon are still economically important. Recreational, commercial, and treaty fisheries into the Columbia River downstream from McNary
Dam were valued at $10.1 million in 1992 (Tracy 1993).

Several factors make white sturgeon relatively vulnerable to overexploitation and changes in their environment. The fish may live more than 100 years
(Rieman and Beamesderfer 1990), and overexploitation is well documented
for long-lived, slow growing fish (Ricker 1963). Female white sturgeon are
slow to reach sexual maturity; in the Snake River they mature at age 15–32
(Cochnauer 1981). Mature females in the Columbia basin only spawn every
2–11 years (Stockley 1981; Cochnauer 1983; Welch and Beamesderfer 1993).
Sustainable harvest levels vary for impoundments in the Columbia River.
Several impoundments are managed as groups, making overexploitation
more likely in impoundments with low sustainable harvest levels.

White sturgeon populations in free-flowing and inundated reaches of the
Columbia River basin have been negatively affected by the abundant hydro-
power dams in most of the main stem Columbia and Snake rivers (Rieman
and Beamesderfer 1990). These dams have altered the magnitude and timing
of discharge, water depths, velocities, temperatures, turbidities, and substrates, and have restricted sturgeon movement within the basin. Sturgeons in
other river basins have declined in response to dam-induced habitat alterations (Artyukhin et al.1978).

Invertebrates

The "caraway worms" were the ones that revealed to us the mystery of the
pupa and butterfly. We saw one climb up the side of a house, and watched it
as with many slow, graceful movements of the head it wove for itself the
loop of the silk which we called the "swing" and which held it in place
after it changed to a chrysalis. We wondered why such a brilliant caterpillar should change to such a dull-colored object, almost the color of the

clapboard against which it hung. Then, one day, we found a damp, crumpled, black butterfly hanging to the empty chrysalis skin, its wings "all mussed" as we termed it; and we gazed at it pityingly; but even as we gazed, the crumpled wings expanded and then there came to our childish minds a dim realization of the miracle wrought within that little, dingy, empty shell.

—Anna Botsford Comstock (1911)

Invertebrates are among the most interesting and available of all living creatures for study. They are impressive in abundance and diversity, living on land and water and air. Many species are borne to distant places on air and water currents, and via modern transportation (Mason 1995).

Of the millions of species of animals worldwide, about 90% are invertebrates, that is, animals without backbones. The arthropods, or jointed-legged invertebrates such as beetles, account for 75% of his total. More than 90,000 described insect species inhabit North America; the Lepidoptera (butterflies and moths) alone account for about 11,500.

Mason (1995) points out that within an acre of land and water, hundreds of different invertebrates form an ecological web of builders, gatherers, collectors, predators, and grazers, all interacting with each other and each a necessary component of a healthy ecosystem. The large macroscopic invertebrates—like bees, beetles, butterflies, grasshoppers, snails, and earthworms—are well known, but other invertebrates are almost invisible because they are extremely tiny or camouflaged for protection. We have just begun to understand the ecology of some commercially important species, but we understand very little about the behavior, communication, and function of many other invertebrates within various ecosystems.

Each individual invertebrate is a highly complex, specialized animal. Some must change or metamorphose into several distinct life stages. For example, some insects transform from egg to larva, then to pupa, and finally emerge as a terrestrial winged adult. Some aquatic invertebrates do not have pupal stages, and the larvae (nymphs or naiads) grow progressively larger by molts. Earthworms bear cocoons that each contain about six miniature juveniles; they also reproduce by fragmentation (architomy).

Changes to the environment can disrupt basic interactions of invertebrate species, thereby affecting other organisms in the food chain. Disruptions of natural food cycles may cause drastic changes in the community structure and ecological web of life. This is especially true of the fauna that dwell in fragile ecosystems like caves and springs. Eventually even humans are affected by changes to food webs and destruction of beneficial habitat for wildlife.

Most invertebrates can survive extreme natural events like severe storms, blizzards, and flooding. When confronted by unnatural disturbances, however, such as excessive siltation from urban and highway developments, eutrophication (excessive nutrients) by runoff from agricultural lands, and contamination of aquatic habitats by toxic substances and acids, invertebrate populations can be severely damaged. Airborne toxicants like acid rain are harmful to the long-term well-being of insects. If disturbances are sufficient, natural fauna may be extirpated (removed or lost) and replaced by more tolerant kinds. This "unbalanced" situation usually results in a population explosion of a few species. Such a biological reaction makes these aquatic invertebrates excellent bioindicators of overall environmental conditions (Bartsch and Ingram 1959). The use of aquatic invertebrates for bioassay (testing the toxicity of substances to "standard" test organisms) has greatly helped to minimize adverse effects of contaminants on aquatic life.

Butterflies and moths are particularly susceptible to environmental disturbances, although their responses to mild disturbances and changes may be slow, lasting decades (Otte 1995; Swengel and Swengel 1995). McCabe (1995) concludes that some of the flux in biodiversity is likely due to the "edge effect" at the interface from one habitat to another, and not necessarily to anthropogenic (human-caused) disturbances.

Did You Know?

In regard to "edge effect," when habitat areas are fragmented the result is more edge area where patches interface with the surrounding environment. According to Cunningham & Cunningham (2002), these patch areas "with a relatively large ratio of edge to interior have some unique characteristics. They are often distinguished by increased predation when predators are able to hunt or forage along this edge more easily."

In the aquatic realm, organic chemicals and other toxic substances, acids and alkalis, and mine drainage can quickly decimate populations of mussels, mayflies, and stoneflies, whereas reduced water flow and introduction of pollutants like silt and excessive nutrients (Mason et al. 1995; Webb 1995) cause a slow, relentless destruction of the indigenous fauna.

In the past 50 years, nearly 72% of the United States 297 native mussel species have become endangered, threatened, or of special concern (Williams and Neves 1995). Their populations have been damaged because of siltation, point and nonpoint source pollution, and outright habitat destruction.

This zebra mussel and some other non-indigenous species represent "biological pollution" (Schloesser and Nalepa 1995), and should be considered much like toxic pollution for control and treatment. Non-native zebra mussels lack predators and have invaded nearly the full length of the Mississippi

River and its major tributaries, threatening the native mussel fauna of the eastern United States (Williams and Neves 1995).

It is important to point out that historical databases have traditionally focused on commercially important invertebrate species such as clams and oysters (Otte 1995). In contrast, little information exists on the status and trends of nonconsumptive, indigenous invertebrate life and existing data are often not in formats for use in modern decision-making tools (Messer et al. 1991).

An important, often-overlooked problem with providing scientifically credible data involves the taxonomy and systematics (identification and classification) of organisms. Today, our museum collections of invertebrates are often old and worn out, and there are few trained taxonomists to renew archival materials. In fact, many "type" specimens used for original species' descriptions in the early 1900's are unusable, making comparisons of recently collected specimens impossible.

Canada has been doing continuous biomonitoring for several decades, which has now resulted in status and trend analyses of subtle perturbations like acidification (Chmielewski and Hall 1993). It is clear that the success of future assessments in the United States will greatly depend on availability of and access to high-quality data; stop-gap measures are unlikely to prove successful because of inconsistencies caused by differing collection methods, taxonomy, and reporting units.

Diversity of Insects

> The mute insect, fix't upon the plant
> On whose soft leaves it hangs, and from
>   Whose cup
> Drains imperceptibly its nourishment,
> Endear'd my wanderings.

> —William Wordsworth

Insects are the most diverse group of organisms (Wheeler 1990); potentially they are highly indicative of environmental change through close adaptation to their environment; they represent the majority of links in the community food chain; and they have likely the largest biomass of the terrestrial animals (Holden 1989). Thus, knowledge about them is fundamental to studying the environment (Hodges 1995).

Grasshoppers

> *The grasshoppers weave their autumn song by the golden railing of the well.*

> —Chinese *Long Yearning*

A grasshopper is an amazing insect that can leap at least 20 times the length of its own body. If we could do that, we would be able to jump almost 120 feet! While it would be great to be able to jump so far, we might not be able to make as graceful a landing as the grasshopper (Grasshopper Facts 2007).

Did You Know?

Grasshoppers consume green forage roughly eight times as fast in proportion to their weight as beef animals on good range (Pohly 2007).

According to Otte (1995), grasshoppers are perhaps the most important grazing herbivores in the nation's grasslands, which from a human standpoint, are the most important food-producing areas. The damage that grasshoppers do to plants varies with the species. A few dozen species at most are highly injurious to crops, while those that feed on economically unimportant plants may have no measurable impact, and those that feed on detrimental plants are highly beneficial. Given such differences, it becomes important to distinguish properly between harmful and beneficial species. Grasshopper abundance in all kinds of grasslands means they are an important factor in the ecological equation. Their economic importance—positive and negative—means that they must be included in all studies of grassland and desert-grassland communities.

Lepidoptera: Butterflies and Moths

The man whispered "God speak to me" and the meadowlark sang. The Man did not hear. So the man yelled "God speak to me!" and the thunder rolled across the sky. The man did not listen. The man looked around and said, "God let me see you" and a star shown brightly, but the man did not notice. Then the man shouted "God show me a miracle" and a life was born, but the man did not know. So the man cried out in despair, "Touch me, God, and let me know you are here!" God reached down and touched the man, but the man brushed the butterfly away.

—Unknown

Winged Beauty

When attention is drawn to magnificence resting on a leaf,
We remain attentive prior to that pre-flight opening of wings;
The dense mosaic of tiny individually colored scales forms a sight beyond belief;
The ethereal beauty of form and design captivate us;
Upon gaining flight and exposing colors to light, a faint breeze sings.

—Frank R. Spellman

Whether it be the Fritillaries, Coppers, Monarchs, Painted Ladies, Sulphurs or Owl butterflies, Lepidoptera (Latin: scale wing; butterflies and moths) make up about 13% of the described and named 90,000 insect species of North America (11,500 named); they are among the better known large orders, although no complete inventory of the Lepidoptera species exists for any state, county, or locality in North America (Powell 1995).

Did You Know?

What is the difference between moths and butterflies? Moths fly at night, have feathered antennae, and rest with their wings open. In contrast, butterflies fly during the day, have knob-ended antennae, and rest with their wings closed.

The inventorying of butterflies, official (scientific) and unofficial (unscientific) has been going on for some time. For example, the Xerces Society started the Fourth of July Butterfly Count (FJC) in 1975, sponsoring it annually until 1993, when the North American Butterfly Association (NABA) assumed administration (Swengel 1995). The general methods of the butterfly count are patterned after the highly successful Christmas Bird Count (CBC), founded in 1900 and sponsored by the National Audubon Society (Swengel 1990).

Did You Know?

An interesting observation about wing patterns is referred to as Oudemans' principle. As you study the ventral wing pattern of a resting butterfly, you'll notice the pattern often smoothly translated from the hindwing to the visible tip portion of the forewing. In contrast, the covered portion of the forewing lacks the patternation and is often more brightly colored making for a disorientating flash of color as the butterfly launches into flight. Oudemans' principle can also be observed on the forewing patterns where design elements align between the fore and hind wings when the butterfly is displaying its dorsal surfaces (Bugbios 2007).

The results of the FJC, including butterfly data, count-site descriptions, and weather information on count day, are published annually. The count was designed as an informal program for butterfly enthusiasts and the general public. These counts can never substitute for more formal scientific censusing because data sets from the counts have flaws that impair scientific analysis. Nevertheless, the FJC program does provide data that, with considerable caution, can be useful for science and conservation (Swengel 1990). FJC data have been used to study the biology, status, and trends of both rare and widely distributed species (Swengel 1990; Nagel et al. 1991; Nagel 1992; Swengel unpublished data).

In regards to insect sampling (Opler 1995), butterflies and large moths are among the best-sampled insects and as such are excellent indicators of ecological conditions or environmental change. Because the caterpillars of most Lepidoptera are herbivorous, their species richness is most often a reflection of plant diversity (Brown and Opler 1990).

Butterflies and moths are particularly susceptible to environmental disturbances, although their responses to mild disturbances and changes may be slow, lasting decades (Swengel and Swengel 1995). This is no better demonstrated than by the butterfly community of the tall-grass prairie region of the United States.

The prairie biome, between the Missouri River and the Rocky Mountains, is a plant community dominated by grasses and non-grassy herbs (wildflowers or "forbs"), with some woody shrubs and occasional trees. Prairie is classified into three major types by rainfall and consequent grass composition. The easternmost and moistest division is the tall-grass prairie (Risser et al. 1981). Although tall-grass prairie once broadly covered the middle of the United States, this biome is now estimated to be at least 99% destroyed from presettlement by pioneers, who converted it for agricultural uses. Prairie loss continues through plowing, extreme overgrazing, and development, but at varying degrees. Prairie is also lost passively because the near-total disruption of previous ecological processes causes shifts in floristic composition and structure (Swengel and Swengel 1995).

As a result of this habitat destruction, butterflies and other plants and animals that are obligate to the prairie ecosystem are rare and primarily restricted to prairie preserves. The Dakota skipper and the regal fritillary are federal candidates for listing under the Endangered Species Act, and additional prairie

---

Did You Know?

Tallgrass prairie is a complex ecosystem, including flowers, trees, birds, mammals, insects, and microorganisms. But grass dominates. Like other grasses, tallgrasses do not form woody tissue nor increase in girth. Their stems are hollow except where the leaves join, leaves are narrow with parallel veins, and flowers are small and inconspicuous. Tallgrass prairie is so-named because the component grasses—big bluestem, little bluestem, indiangrass and switchgrass—can reach 8 or 9 feet (NPS 1995).

butterfly species are on state lists as officially threatened or endangered. Patches of original prairie vegetation remain in preserves, parks, unintensively used farmlands such as hayfields and pastures, and in unused land. These remnants of prairie, however, are isolated and often in some state of ecological degradation (Swengel and Swengel 1995).

The existence of prairie depends on the occurrence of certain climatic conditions and disturbance processes such as animal herbivory and fire. These natural processes, however, are severely disrupted today because of the destruction and fragmentation of the prairie biome. Without management intervention, the vegetational composition and structure of prairie sites are altered through invasion of woody species and smothering under dead plant matter. Prairie usually requires active management to maintain the ecosystem and its biodiversity, but it is difficult to know exactly which processes once naturally maintained the prairie ecosystem. Frequent fire, whether caused by lightning or set by native peoples, is usually considered the dominant prehistoric process that maintained prairie; thus management for tallgrass prairie in most states relies primarily or solely on frequent fire (e.g., Sauer 1950; Hulbert 1973; Vogl 1974). Other researchers, however, assert that prairie was the result of grazing by large herds of ungulates (England and DeVos 1969).

Despite this scientific conflict, it appears certain that successful management for maintaining the prairie landscape and its native species should be based on these natural processes, whatever they were. The vast diversity and specificity of insects to certain plants and habitat features make them finetuned ecological indicators. Thus, butterfly conservation is useful not only for maintaining these unique species, but also for helping us monitor and learn about the soundness of our general ecosystem management (Swengel and Swengel 1995).

Aquatic Insects and Biotic Indices

Swimming, clinging, or crawling on rocks
  Water bugs thrive in our stream
Telling us the importance of their water home.
Stoneflies and caddisflies are the ones we're glad to see.
  They mean we've got good water quality.
Mayflies and Dobsonflies are the one's we're glad to see.
  They mean we've got good water quality.
Midges and blackflies—we're not glad to see.
  They mean we have poor water quality. . . .

—Scott, 2001

Mason et al. (1995) point out that aquatic insects are among the most prolific animals on earth, but are highly specialized and represent less than 1% of the total animal diversity (Pennak 1978). Most people know the 12 orders and about 11,000 species of North American aquatic insects (Merritt and Cummins 1984) only by the large adults that fly around or near wetlands.

Aquatic insects are excellent overall indicators of both recent and long-term environmental conditions (Patrick and Palavage 1994). The immature stages of aquatic insects have a short life cycle, often several generations a year, and remain in the general area of propagation. Thus, when environmental changes occur, the species must endure the disturbance, adapt quickly, or die and be replaced by more tolerant species. These changes often result in an overabundance of a few tolerant species, and the communities become destabilized or "unbalanced" (Mason et al. 1995).

Spellman (1996) points out that environmental scientists find of interest and often use four different indicators of water quality—coliform bacteria count, concentration of dissolved oxygen (DO), biochemical oxygen demand (BOD), and the **biotic index**. The biota that exist at or near a stream, for example, are, as pointed out earlier, direct indicators (a biotic index) of the condition of the water. This biotic index is often more reliable than many of the laboratory chemical tests that environmental scientist/toxicologists use in attempting to determine the pollutant level in a stream. Indicator species help determine when pollutant levels are unsafe.

How does the biotic index actually work? How does it indicate pollution?

Certain common aquatic organisms, by indicating the extent of oxygenation of a stream, may be regarded as indicators of the intensity of pollution from organic waste. The responses of aquatic organisms in streams to large quantities of organic wastes are well documented. They occur in a predictable cyclical manner. For example, upstream from an industrial waste discharge point (an end-of-pipe, **point source** polluter), a stream can support a wide variety of algae, fish, and other organisms, but in the section of the stream where oxygen levels are low (below 5 ppm), only a few types of worms survive. As stream flow courses downstream, oxygen levels recover and those species that can tolerate low rates of oxygen (such as gar, catfish, and carp) begin to appear. Eventually, at some further point downstream, a clean water zone reestablishes itself and a more diverse and desirable community of organisms returns (Spellman 1996).

During this characteristic pattern of alternating levels of dissolved oxygen (in response to the dumping of large amounts of biodegradable organic material), a stream, as stated above, goes through a cycle. This cycle is called an oxygen sag curve. Its state can be determined using the biotic index as an indicator of oxygen content.

The biotic index is a systematic survey of invertebrate organisms. Since the diversity of species in a stream is often a good indicator of the presence of pollution, the biotic index can be used to correlate with water quality. A knowledgeable person (an environmental scientist or ecologist, for example) can easily determine the state of water quality of any stream simply through observation—observation of types of species present or missing used as an indicator of stream pollution. The biotic index, used in the determination of the types, species, and numbers of biological organisms present in a stream, is commonly used as an auxiliary to BOD determination in determining stream pollution. The disappearance of particular organisms tends to indicate the water quality of the stream.

The biotic index is based on two principles:

1. A large dumping of organic waste into a stream tends to restrict the variety of organisms at a certain point in the stream.
2. As the degree of pollution in a stream increases, key organisms tend to disappear in a predictable order.

Several different forms of the biotic index are commonly used. In Great Britain, for example, the Trent Biotic Index (TBI), the Chandler score, the Biological Monitoring Working Party (BMWP) score, and the Lincoln Quality Index (LQI) are widely used. Most forms use a biotic index that ranges from 0 to 10. The most polluted stream, which contains the smallest variety of organisms, is at the lowest end of the scale (0); the clean streams are at the highest end (10). A stream with a biotic index of greater than five will support game fish; a stream with a biotic index of less that four will not support game fish (Spellman 1996).

Because they are easy to sample, macroinvertebrates have predominated in biological monitoring. Macroinvertebrates are a diverse group. They demonstrate tolerances that vary between species. Discrete differences tend to show up, and often contain both tolerant and sensitive indicators. In addition, comparison with identification keys, which are portable and conveniently used in field settings, can easily identify invertebrates. Present knowledge of invertebrate tolerances and responses to stream pollution is well documented. In the United States, for example, the EPA has required states to incorporate narrative biological criteria into its water quality standards since 1993.

The biotic index provides a valuable measure of pollution, especially for species very sensitive to lack of oxygen. Consider the stonefly. Stonefly larvae live underwater and survive best in well-aerated, unpolluted waters with clean gravel bottoms. When the stream deteriorates from organic pollution, stonefly larvae cannot survive. The degradation of stonefly larvae has an exponential effect upon other insects and fish that feed off the larvae; when the stone fly larvae disappear, so in turn do many insects and fish (Spellman 1996). Table 17.1 shows a modified version of the BMWP biotic index. Since the BMWP biotic index indicates ideal stream conditions, this index takes into account the sensitivities of different macroinvertebrate species to stream contamination. Aquatic macroinvertebrate species are represented by diverse populations and are excellent indicators of pollution. These organisms are large enough to be seen by the unaided eye. Most aquatic macroinvertebrates live for at least a year. They are sensitive to stream water quality, on both a short-term and long-term basis. Mayflies, stoneflies, and caddis flies are aquatic macroinvertebrates considered clean-water organisms; they are generally the first to disappear from a stream if water quality declines and are therefore given a high score. Tubicid worms (tolerant to pollution) are given a low score.

In Table 17.1, a score from 1–10 is given for each family present. A site score is calculated by adding the individual family scores. The site score (total score) is then divided by the number of families recorded to derive the Average Score per Taxon (ASPT). High ASPT scores result from such taxa as stone flies, mayflies, and caddis flies present in the stream. A low ASPT score is obtained from heavily polluted streams dominated by tubicid worms and other pollution-tolerant organisms (Spellman 1996).

In using the biotic index, environmental scientists/ecologists make use of the fact that unpolluted streams normally support a wide variety of macroinvertebrates and other aquatic organisms with relatively few of one kind in making determinations about water quality in the field. While some aquatic species, such as mayflies and stoneflies, are more sensitive than others to certain pollutants, and succumb more readily to the effects of pollution, other species, such as mussels and clams, accumulate toxic materials in their tissues at sub-lethal levels. These species can be monitored (*must* be monitored to

TABLE 17.1
The BMWP Score System (modified for illustrative purposes)

Families	Common-Name Examples	Score
Hepatagenidae	Mayflies	
Leuctridae	Stoneflies	10
Aeshnidae	Dragonflies	8
Polycentropidae	Caddis flies	7
Hydrometridae	Water Strider	
Gyrinidae	Whirligig beetle	5
Chironomidae	Mosquitoes	2
Oligochaera	Worms	1

Note: Organisms with high scores, especially mayflies and stoneflies (the most sensitive), and others (dragonflies and caddis flies) are very sensitive to any pollution (deoxygenation) of their aquatic environment.

protect public health) to track pollution movement and buildup in aquatic systems.

Using a biotic index to determine the level of pollution in a water body demonstrates only one application. B. M. Levine et al. (1989) point out that similar determinations regarding soil quality can be made by observing and analyzing organisms (such as earthworms) in the soil. Studies conducted to assess the impact of sewage biosolids (sludge) treatments on old-field communities revealed that earthworms concentrate cadmium, copper, and zinc in their tissues at levels that exceed those found in the soil. Cadmium levels even exceed the concentrations found in the biosolids. Thus, earthworms may provide an "index" to monitor the effects of biosolids disposal on terrestrial communities.

Insect Macroinvertebrates

If one uses the biotic index as an indicator of environmental pollution in water bodies (lakes and streams), for example, then, obviously, one must have some understanding of insect macroinvertebrates. Insect macroinvertebrates are ubiquitous in streams and are often represented by many species. Although the numbers refer to aquatic species, a majority is to be found in streams (Spellman 1996).

The most important macroinvertebrate insect groups in streams are Ephemeroptera (mayflies), Plecoptera (stoneflies), Trichoptera (caddisflies), Diptera (true flies), Coleoptera (beetles), Hemiptera (bugs), Megaloptera (alderflies and dobsonflies), and Odonata (dragonflies and damselflies). The identification of these different orders is usually easy and there are many keys and specialized references (e.g., Merritt and Cummins, *An Introduction to the Aquatic Insects of North America*, 1996) available to help in the identification of species. In contrast, there are some genera and species that specialist taxonomists can often only diagnose, particularly the Diptera (Spellman 2003).

(1) Mayflies (Order: Ephemeroptera)

Streams and rivers are generally inhabited by many species of mayflies and, in fact, most species are restricted to streams. For the experienced freshwater ecologist who looks upon a mayfly nymph, recognition is obtained through trained observation: abdomen with leaf-like or feather-like gills, legs with a single tarsal claw, generally (but not always) with three cerci (three 'tails'; two cerci, and between them usually a terminal filament). The experienced ecologist knows that mayflies are hemimetabolous insects (i.e., where larvae or nymphs resemble wingless adults) that go through many postembryonic molts, often in the range between 20 and 30. For some species, body length increases about 15% for each instar.

Mayfly nymphs are mainly grazers or collector-gatherers feeding on algae and fine detritus, although a few genera are predatory. Some members filter particles from the water using hair-fringed legs or maxillary palps. Shredders are rare among mayflies. In general, mayfly nymphs tend to live mostly in unpolluted streams, where with densities of up to 10,000/sq. meters, they contribute substantially to secondary producers.

Adult mayflies resemble nymphs, but usually possess two pair of long, lacy wings folded upright; adults usually have only two cerci. The adult lifespan is short, ranging from a few hours to a few days, rarely up to two weeks, and the adults do not feed. Mayflies are unique among insects in having two winged stages, the subimago (winged and capable of flight but not sexually mature) and the imago (adult). The emergence of adults tends to be synchronous, thus ensuring the survival of enough adults to continue the species (Spellman 2003).

(2) Stoneflies (Order: Plecoptera)

Although many freshwater ecologists would maintain that the stonefly is a well-studied group of insects, this is not exactly the case. Despite their importance, less than 5–10% of stonefly species are well known with respect to life history, trophic interactions, growth, development, spatial distribution, and nymphal behavior.

Notwithstanding our lacking of extensive knowledge in regards to stoneflies, enough is known to provide an accurate characterization of these aquatic insects. We know, for example, that stonefly larvae are characteristic inhabitants of cool, clean streams (i.e., most nymphs occur under stones in well-aerated streams). While they are sensitive to organic pollution, or more precisely to low oxygen concentrations accompanying organic breakdown processes, stoneflies seem rather tolerant to acidic conditions. Lack of extensive gills at least partly explains their relative intolerance of low oxygen levels.

Stoneflies are drab-colored, small- to medium-sized ⅙ to 2¼ inches (4 to 60 mm), rather flattened insects. Stoneflies have long, slender, many-segmented antennae and two long narrow antenna-like structures (cerci) on the tip of the abdomen. The cerci may be long or short. At rest, the wings are held flat over the abdomen, giving a "square-shouldered" look compared to the roof-like position of most caddisflies and vertical position of the mayflies. Stoneflies have two pair of wings. The hindwings are slightly shorter than the forewings and much wider, having a large anal lobe that is folded fanwise when the wings are at rest. This fanlike folding of the wings gives the order its name: *"pleco"* (folded or plaited) and *"ptera"* (wings). The aquatic nymphs are generally very similar to mayfly nymphs except that they have only two cerci at the tip of the abdomen. The stoneflies have chewing mouthparts. They may be found anywhere in a non-polluted stream where food is available. Many adults, however, do not feed and have reduced or vestigial mouthparts.

Stoneflies have a specific niche in high-quality streams where they are very important as a fish food source at specific times of the year (winter to spring, especially) and of the day. They compliment other important food sources, such as caddisflies, mayflies, and midges (Spellman 2003).

(3) Caddisflies (Order: Trichoptera)

Trichoptera (Greek: *trichos*, a hair; *ptera*, wing) is one of the most diverse insect orders living in the stream environment, and caddisflies have nearly a worldwide distribution (the exception: Antarctica). Caddisflies may be categorized broadly into free-living (roving and net spinning) and case-building species.

Caddisflies are described as medium-sized insects with bristle-like and often long antennae. They have membranous hairy wings (explaining the Latin name "Trichos"), which are held tent-like over the body when at rest; most are weak fliers. They have greatly reduced mouthparts and five tarsi (end segments of legs). The larvae are mostly caterpillar-like and have a strongly sclerotized (hardened) head with very short antennae and biting mouthparts. They have well-developed legs with a single tarsi. The abdomen is usually 10-segmented, in case-bearing species the first segment bears three papillae, one dorsally and the other two laterally which help hold the insect centrally in its case, allowing a good flow of water past the cuticle and gills; the last or anal segment bears a pair of grappling hooks.

In addition to being aquatic insects, caddisflies are superb architects. Most caddisfly larvae live in self-designed, self-built houses, called *cases*. They spin out silk, and either live in silk nets or use the silk to stick together bits of whatever is lying on the stream bottom. These houses are so specialized that you can usually identify a caddisfly larva to genus if you can see its house (case). With nearly 1,400 species of caddisfly species in North America (north of Mexico), this is a good thing!

Caddisflies are closely related to butterflies and moths (Order: Lepidoptera). They live in most stream habitats and that is why they are so diverse (have so many species). Each species has special adaptations that allow it to live in the environment it is found in.

Mostly herbivorous, most caddisflies feed on decaying plant tissue and algae. Their favorite algae are diatoms, which they scrape off rocks. Some of them, though, are predacious.

Caddisfly larvae can take a year or two to change into adults. They then change into *pupae* (the inactive stage in the metamorphosis of many insects, following the larval stage and preceding the adult form) while still inside their cases for their metamorphosis. It is interesting to note that caddisflies, unlike stoneflies and mayflies, go through a "complete" metamorphosis.

Caddisflies remain as pupae for 2–3 weeks, and then emerge as adults. When they leave their pupae, splitting their case, they must swim to the surface of the water to escape it. The winged adults fly evening and night, and some are known to feed on plant nectar. Most of them will live less than a month: like many other winged stream insects, their adult lives are brief compared to the time they spend in the water as larvae.

Caddisflies are sometimes grouped into five main groups by the kinds of cases they make: free-living forms that do not make cases, saddle-case makers, purse-case makers, net-spinners and retreat-makers, and tube-case makers.

Caddisflies demonstrate their "architectural" talents in the cases they design and make. For example, a caddisfly might make a perfect, 4-sided box case of bits of leaves and bark, or tiny bits of twigs. It may make a clumsy dome of large pebbles. Others make rounded tubes out of twigs or very small pebbles. In our experience in gathering caddisflies, we have come to appreciate not only their architectural ability but also their flare in the selection of construction materials. For example, we have found many caddisfly cases constructed of silk, emitted through an opening at the tip of the labium, used together with bits of ordinary rock mixed with sparkling quartz and red garnet, green peridot, and bright fool's gold.

Besides the protection their cases provide them, the cases provide another advantage. The cases actually help caddisflies breathe. They move their bodies up and down, back and forth inside their cases, and this makes a current that brings them fresh oxygen. The less oxygen there is in the water, the faster they have to move. It has been seen that caddisflies inside their cases get more oxygen than those that are outside of their cases—and this is why stream ecologists think that caddisflies can often be found even in still waters, where dissolved oxygen is low, in contrast to stoneflies and mayflies (Spellman 2003).

(4) True Flies (Order: Diptera)

True or two- (*Di-*) winged (*ptera*) flies not only include the flies that we are most familiar with, like fruitflies and houseflies, they also include midges, mosquitoes, craneflies, and others. Houseflies and fruitflies live only on land, and we do not concern ourselves with them. Some, however, spend nearly their whole lives in water; they contribute to the ecology of streams.

True flies are in the order Diptera, and are one of the most diverse orders of the class Insecta, with about 120,000 species worldwide. Dipteran larvae occur almost everywhere except Antarctica and deserts where there is no running water. They may live in a variety of places within a stream: buried in sediments, attached to rocks, beneath stones, in saturated wood or moss, or in silken tubes, attached to the stream bottom. Some even live below the stream bottom.

True fly larvae may eat almost anything, depending on their species. Those with brushes on their heads use them to strain food out of the water that passes through. Others may eat algae, detritus, plants, and even other fly larvae.

The longest part of the true fly's life cycle, like that of mayflies, stoneflies, and caddisflies, is the larval stage. It may remain an underwater larva anywhere from a few hours to five years. The colder the environment, the longer it takes to mature. It pupates and emerges, then, and becomes a winged adult. The adult may live four months—or it may only live for a few days. While reproducing, it will often eat plant nectar for the energy it needs to make its eggs. Mating sometimes takes place in aerial swarms. The eggs are deposited back in the stream; some females will crawl along the stream bottom, losing their wings, to search for the perfect place to put their eggs. Once they lay them, they die.

Diptera are especially good "bioindicators" of aquatic environmental conditions because, in addition to the attributes of other aquatic insects, they occupy the full spectrum of habitats and conditions (Paine and Gaufin 1956; Roback 1957; Mason 1975; Hudson et al. 1990; Spellman 2003).

Moreover, Diptera serve an important role in cleaning water and breaking down decaying material, and they are a vital food source (i.e., they play pivotal roles in the processing of food energy) for many of the animals living in and around streams. However, the true flies most familiar to us are the midges, mosquitoes, and the craneflies, because they are pests. Some midge flies and mosquitoes bite; the cranefly, however, does not bite but looks like a giant mosquito.

Like mayflies, stoneflies and caddisflies, true flies are mostly in larval form. Like caddisflies, you can also find their pupae, because they are holometabolous insects (go through complete metamorphosis). Most of them are free-living; that is, they can travel around. Although none of the true fly larvae have the six, jointed legs we see on the other insects in the stream, they sometimes have strange little almost-legs—prolegs—to move around with.

Others may move somewhat like worms do, and some—the ones who live in waterfalls and rapids—have a row of six suction discs that they use to move much like a caterpillar does. Many use silk pads and hooks at the ends of their abdomens to hold them fast to smooth rock surfaces (Spellman 2003).

(5) Beetles (Order: Coleoptera) (Hutchinson 1981)

Of the more than one million described species of insect, at least one-third are beetles, making the Coleoptera not only the largest order of insects but also the most diverse order of living organisms. Even though they are the most speciose order of terrestrial insects, their diversity is not so apparent in running waters. Coleoptera belongs to the infraclass Neoptera, division Endpterygota. Members of this order have an anterior pair of wings (the *elytra*) that are hard and leathery and not used in flight; the membranous hindwings, which are used for flight, are concealed under the elytra when the organisms are at rest. Only 10% of the 350,000 described species of beetles are aquatic.

Beetles are holometabolous. Eggs of aquatic coleopterans hatch in one or two weeks, with diapause occurring rarely. Larvae undergo from 3 to 8 molts. The pupal phase of all coleopternas is technically terrestrial; making this life stage of beetles the only one that has not successfully invaded the aquatic habitat. A few species have diapausing prepupae, but most complete transformation to adults in two to three weeks. Terrestrial adults of aquatic beetles are typically short-lived and sometimes nonfeeding, like those of the other orders of aquatic insects. The larvae of Coleoptera are morphologically and behaviorally different from the adults, and their diversity is high.

Aquatic species occur in two major suborders, the Adephaga and the Polyphaga. Both larvae and adults of six beetle families are aquatic: Dytiscidae (predaceous diving beetles), Elmidae (riffle beetles), Gyrinidae (whirligig beetles), Halipidae (crawling water beetles), Hydrophilidae (water scavenger beetles), and Noteridae (burrowing water beetles). Five families, Chrysomelidae (leaf beetles), Limnichidae (marsh-loving beetles), Psephenidae (water pennies), Ptilodactylidae (toe-winged beetles), and Scirtidae (marsh beetles) have aquatic larvae and terrestrial adults, as do most of the other orders of aquatic insects; adult limnichids, however, readily submerge when disturbed. Three families have species that are terrestrial as larvae and aquatic as adults: Curculionidae (weevils), Dryopidae (long-toed water beetles), and Hydraenidae (moss beetles), a highly unusual combination among insects. (*Note:* Because they provide a greater understanding of a freshwater body's condition [i.e., they are useful indicators of water quality], we focus our discussion on the riffle beetle, water penny, and whirligig beetle.)

Riffle beetle larvae (most commonly found in running waters, hence the name Riffle Beetle) are up to ¾″ long. Their body is not only long but also hard, stiff, and segmented. They have six long segmented legs on the upper middle section of the body; the back end has two tiny hooks and short hairs. Larvae may take three years to mature before they leave the water to form a pupa; adults return to the stream.

Riffle beetle adults are considered better indicators of water quality than larvae because they have been subjected to water quality conditions over a longer period. They walk very slowly under the water (on stream bottom), and do not swim on the surface. They have small oval-shaped bodies and are typically about ¼" in length.

Both adults and larvae of most species feed on fine detritus with associated microorganisms that is scraped from the substrate, although others may be xylophagous, that is, wood eating (e.g., *Lara*, Elmidae). Predators do not seem to include riffle beetles in their diet, except perhaps for eggs, which are sometimes attacked by flatworms.

The adult *water penny* is inconspicuous and often found clinging tightly in a sucker-like fashion to the undersides of submerged rocks, where they feed on attached algae. The body is broad, slightly oval, and flat in shape, ranging from 4–6 mm (1/4") in length. The body is covered with segmented plates and looks like a tiny round leaf. It has six tiny jointed legs (underneath). The color ranges from light brown to almost black.

There are 14 water penny species in the U.S. They live predominately in clean, fast-moving streams. Aquatic larvae live one year or more (they are aquatic); adults (they are terrestrial) live on land for only a few days. They scrape algae and plants from surfaces.

Whirligig beetles are common inhabitants of streams and normally are found on the surface of quiet pools. The body has pincher-like mouthparts. There are six segmented legs on the middle of the body; the legs end in tiny claws. Many filaments extend from the sides of the abdomen. They have four hooks at the end of the body and no tail.

Did You Know?

When disturbed, whirligig beetles swim erratically or dive while emitting defensive secretions.

As larvae, they are benthic predators, whereas the adults live on the water surface, attacking dead and living organisms trapped in the surface film. They occur on the surface in aggregations of up to thousands of individuals. Unlike the mating swarms of mayflies, these aggregations serve primarily to confuse predators. Whirligig beetles have other interesting defensive adaptations. For example, the Johnston's organ at the base of the antennae enables them to echolocate using surface wave signals; their compound eyes are divided into two pairs, one above and one below the water surface, enabling them to detect both aerial and aquatic predators; and they produce noxious chemicals that are highly effective at deterring predatory fish (Spellman 2003).

(6) Water Strider ("Jesus bugs"; Order: Hemiptera)

The molecules of water are held together
By weak electrical forces
That extend in all directions.
At the water surface, the upward forces,
Having nowhere watery to go,
Lie along the surface
To form a thin skin, called a meniscus.
This is the skin that holds water drops,
Supports water striders,
Even skips stones.

—Stewart McKenzie

It is fascinating to sit on a log at the edge of a stream pool and watch the drama that unfolds among the small water animals. Among the star performers in small streams are the water bugs. These are aquatic members of that large group of insects called the "true bugs," most of which live on land. Moreover, unlike many other types of water insects, they do not have gills but get their oxygen directly from the air.

Most conspicuous and commonly known are the water striders or water skaters. These ride the top of the water, with only their feet making dimples in the surface film. Like all insects, the water striders have a three-part body (head, thorax, and abdomen), six jointed legs, and two antennae. They have long, dark, narrow bodies. The underside of the body is covered with water-repellent hair. Some water striders have wings, others do not. Most water striders are over 0.2 inch (5 mm) long.

Water striders eat small insects that fall on the water's surface and larvae. Water striders are very sensitive to motion and vibrations on the water's surface. It uses this ability in order to locate prey. It pushes its mouth into its prey, paralyzes it, and sucks the insect dry. Predators of the water strider, like birds, fish, water beetles, backswimmers, dragonflies, and spiders, take advantage of the fact that water striders cannot detect motion above or below the water's surface (Spellman 2003).

(7) Alderflies and Dobsonflies (Order: Megaloptera)

Larvae of all species of Megaloptera ("large wing") are aquatic and attain the largest size of all aquatic insects. Megaloptera is a medium-sized order with fewer than 5000 species worldwide. Most species are terrestrial; in North America 64 aquatic species occur.

In running waters, alderflies (Family: Sialidae) and dobsonflies (Family: Corydalidae; sometimes called hellgrammites or toe biters) are particularly important, as they are voracious predators, having large mandibles with sharp teeth.

Alderfly brownish-colored larvae possess a single tail filament with distinct hairs. The body is thick-skinned with six to eight filaments on each side of the abdomen; gills are located near the base of each filament. Mature body size: 0.5 to 1.25 inches. Larvae are aggressive predators, feeding on other adult aquatic macroinvertebrates (they swallow their prey without chewing); as secondary consumers, other larger predators eat them. Female alderflies deposit eggs on vegetation that overhangs water, larvae hatch and fall directly into water (i.e., into quiet but moving water). Adult alderflies are dark with long wings folded back over the body; they only live a few days.

Dobsonfly larvae are extremely ugly (thus, they are rather easy to identify) and can be rather large, anywhere from 25 to 90 mm (13″) in length. The body is stout, with eight pairs of appendages from their abdomen. Brush-like gills at the base of each appendage look like "hairy armpits." The elongated body has spiracles (spines) and has three pairs of walking legs near the upper body, and one pair of hooked legs at the rear. The head bears four segmented antennae, small compound eyes, and strong mouthparts (large chewing pinchers). Coloration varies from yellowish, brown, gray, and black, often mottled. Dobsonfly larvae, commonly known as hellgrammites, are customarily found along stream banks under and between stones. As indicated by the mouthparts, they are predators and feed on all kinds of aquatic organisms.

(8) Dragonflies and Damselflies (Order: Odonata)

In summer-noon flushes
When all the wood hushes,
Blue dragon-flies knitting
To and fro in the sun,
With sidelong jerk flitting,
Sink down on the rushes.
And motionless sitting,
Hear it bubble and run,
Hear its low inward singing
With level wings winging
On green tasselled rushes,
To dream in the sun.

—Lowell, *The Fountain of Youth*

The Odonata (dragonflies, suborder Anisoptera; and damselflies, suborder Zygoptera) is a small order of conspicuous, hemimetabolous insects (lack a pupal stage) of about 5,000 named species and 23 families worldwide. Odonata is a Greek word meaning toothed one. It refers to the serrated teeth located on the insect's chewing mouthparts (mandibles).

Characteristics of dragonfly and damselfly larvae include:

• Large eyes
• Three pairs of long segmented legs on upper middle section (thorax) of body
• Large scoop-like lower lip that covers bottom of mouth
• No gills on sides or underneath of abdomen

Did You Know?

Dragonflies and damselflies are unable to fold their four elongated wings back over the abdomen when at rest.

Dragonflies and damselflies are medium to large insects with two pairs of long equal-sized wings. The body is long and slender, with short antennae. Immature stages are aquatic, and development occurs in three stages (egg, nymph, adult).

Dragonflies are also known as darning needles. (*Note:* Myths about dragonflies warned children to keep quiet or else the dragonfly's "darning needles" would sew the child's mouth shut.) The nymphal stage of dragonflies consists of grotesque creatures, robust and stoutly elongated. They do not have long "tails." They are commonly gray, greenish, or brown to black in color. They are medium to large aquatic insects, size ranging from 15 to 45 mm; the legs are short and used for perching. They are often found on submerged vegetation and at the bottom of streams in the shallows. They are rarely found in polluted waters. Food consists of other aquatic insects, annelids, small crustacea, and mollusks. Transformation occurs when the nymph crawls out of the water, usually onto vegetation. There it splits its skin and emerges prepared for flight. The adult dragonfly is a strong flier, capable of great speed (>60 mph) and maneuverability (fly backward, stop on a dime, zip 20 feet straight up, and slip sideways in the blink of an eye!). When at rest the wings remain open and out to the sides of the body. A dragonfly's freely movable head has large, hemispherical eyes (nearly 30,000 facets each), which the insects use to locate prey with their excellent vision. Dragonflies eat small insects, mainly mosquitoes (large numbers of mosquitoes), while in flight. Depending on the species, dragonflies lay hundreds of eggs by dropping them into the water and leaving them to hatch or by inserting eggs singly into a slit in the stem of a submerged plant. The incomplete metamorphosis (egg, nymph, mature nymph, and adult) can take 2–3 years. Nymphs are often covered by algal growth (Spellman 2003).

Did You Know?

Adult dragonflies are sometimes called "mosquito hawks" because they eat such a large number of mosquitoes that they catch while they are flying.

Damselflies are smaller and more slender than dragonflies. They have three long, oar-shaped feathery tails, which are actually gills, and long slender legs. They are gray, greenish, or brown to black in color. Their habits are similar to those of dragonfly nymphs and emerge from the water as adults in the same manner. The adult damselflies are slow and seem uncertain in flight. Wings are commonly black or clear, and the body is often brilliantly colored. When at rest, they perch on vegetation with their wings closed upright. Damselflies mature in one to four years. Adults live for a few weeks or months. Unlike the dragonflies, adult damselflies rest with their wings held vertically over their backs. They mostly feed on live insect larvae (Spellman 2003).

Did You Know?

Relatives of the dragonflies and damselflies are some of the most ancient of the flying insects. Fossils have been found of giant dragonflies with wingspans up to 720 mm that lived long before the dinosaurs!

Other Invertebrates

In addition to the huge number of insects, invertebrates also include other backboneless animals. Among these are spiders and their relatives, centipedes and millipedes, crustaceans, mollusks or shelled animals, worms, and seashore creatures representing several other groups. Because many (including the author) feel that the freshwater mussel is one of America's hidden treasures, the freshwater mussel is the focus of the discussion that follows.

> They make no sound. They cannot see. Some may live for decades, but seldom move from a secure spot. Yet, freshwater mussels are causing a stir; becoming noticed and making us ponder their future as we make plans for our own.
>
> —USFWS 2007d

No other country in the world equals the U.S. in freshwater mussel diversity. Of the five families and roughly 1,000 species occurring globally, nearly 300 species and subspecies reside here (Turgeon et al. 1988). Unfortunately, these animals may be the most troubled natural resources in the country. It's estimated that 70% (Williams et al. 1993) of our freshwater mussels are extinct, endangered, or in need of special protection. Many of their problems stem from the changes that have occurred to their habitat during the past 200 years (USFWS 2007d).

Mussels were an important natural resource for Native Americans, who used them for food, tools, and jewelry. During the late 1800's and early 1900's, mussel shells supported an important commercial fishery; shells were used to manufacture pearl buttons until the advent of plastic buttons in the 1940's. Today the commercial harvest of freshwater mussel shells is exported to Asia for the production of spherical beads that are inserted into oysters, freshwater mussels, and other shellfish to produce pearls (Williams and Neves 1995).

There are no federal regulations on the harvest of mussels, except those species on the federal list of endangered or threatened species. Several states, however, regulate size, species, gear used, and season that mussels can be taken. Japanese demand for the high-quality U.S. mussel shells in recent years pushed the price to $6/lb in 1991. Shell exports peaked in 1991 at more than 9,000 tons, but demand declined in 1992 and 1993 and has leveled off to about 4,500 tons (Baker 1993).

An early indicator of adverse human effects on large open-water systems in North America was western Lake Erie, part of the Lake Huron–Lake Erie corridor of the Laurentian Great Lakes (Schloesser and Nalepa 1995). Local pollution of tributaries of western Lake Erie was recognized as early as 1890, when populations of whitefish and lake herring in the Detroit River declined (Beeton 1961). Waters of western Lake Erie stopped yielding whitefish and herring in the 1920s to 1930s, but not until the 1950s, after extensive biological investigations, were the open waters of western Lake Erie believed to have been polluted by human "local" activities (National Academy of Sciences 1978). Eutrophication (lake aging; the addition of nutrients) of Lake Erie created unsuitable conditions (primarily low dissolved oxygen concentrations) for fish and other animals in a major potion of Lake Erie—the world's twelfth largest lake. By the early 1960's, Lake Erie was declared "biologically dead" (Burns 1985).

Among the many ecosystem components affected by human-induced changes to western Lake Erie (Burns 1985) is the native mussel fauna. Reduced mussel populations that survived degraded conditions of the 1950's have been used in status and trends studies to evaluate traditional forms of pollution in western Lake Erie. Studies in the 1990's have focused on evaluating the effects of exotic species on mussel populations in the Lake Huron–Lake Erie corridor.

Exotic species have recently been characterized as "biological pollution,' a new concept in evaluating status and trends data. Again, for Nature's sake we repeat:

Plants

I wandered lonely as a cloud
That floats on high o'er vales and hills,
When all at once I saw a crowd,
A host, of golden daffodils;
Beside the lake, beneath the trees,
Fluttering and dancing in the breeze.

—William Wordsworth

The only way to begin a proper study of plants is to gaze upon, for example, a bouquet, a nosegay of white spider orchids surrounded by sprays of baby's breath with a merry mishmash of bee balm, cornflower, dill, flame lily, ginger, kansas feather, lilac, mimosa, safflower and yarrow—or a gaze at Wordsworth's daffodils—some pretty and some plain—all attention getters, for sure. Obviously Longfellow felt and understood the bliss and delight flowers bring us when he wrote the following:

In all places then and in all seasons,
Flowers expand their light and soul-like wings,
Teaching us by most persuasive reasons,
How akin they are to human things.

—Henry Wadsworth Longfellow

Though providing a thorough discussion and description of flowers is, at the moment of this writing, quite tempting, this section describes trends in just two of the major kingdoms of life on Earth: the Kingdom Plantae (which includes flowers, of course) and the molds, lichens, and mushrooms of the Kingdom Fungi (Guntenspergen 1995). In addition to the eye-catching beauty of most flowers, members of the plant and fungal kingdoms have both economic and ecological importance. Plants transform solar energy (photosynthesis) into usable economic products essential in our modern society, and provide the basis for most life on earth by generating the air we breathe, the food we eat, and the shelter and clothes that protect us from the elements. Fungi, because rarely seen, may be thought of as unimportant. But nothing could be further from the truth: Life on Earth could not exist without fungi. Fungi not only mediate critical biological and ecological processes including the breakdown of organic matter and recycling of nutrients, but they also play important roles in mutualistic associations with plants and animals. Members of the Kingdom Fungi also produce commercially valuable substances including antibiotics and ethanol, while other fungi are pathogenic and cause damage to crops and forest trees. Moreover, fungi are essential for proper plant root function. Because fungi and plants play such fundamental roles in our lives, it is important to have a comprehensive knowledge of the taxa comprising these groups. However, at a time when we are increasingly recognizing the importance of these groups, we are impoverishing our biological heritage. Rates of species loss are reaching alarming levels as ecosystems are degraded and habitat is lost. This erosion of biological diversity threatens the maintenance of long-term sustainable development and protection of the earth's biosphere (Gutenspergen 1995).

Fungi

As mentioned, fungi are divided into five classes:

- Myxomycetes, or slime fungi
- Phycomycetes, or aquatic fungi (algae)
- Ascomycetes, or sac fungi
- Basidiomycetes, or rusts, smuts, and mushrooms
- Fungi imperfecti, or miscellaneous fungi

Did You Know?

Although fungi are limited to only five classes, more than 80,000 known species exist.

Fungi differ from bacteria in several ways, including in their size, structural development, methods of reproduction, and cellular organization. They differ from bacteria in another significant way as well: their biochemical reactions (unlike the bacteria) are not important for classification, instead, their structure is used to identify them. Fungi can be examined directly, or suspended in liquid, stained, dried, and observed under microscopic examination where they can be identified by the appearance (color, texture, and diffusion of pigment) of their mycelia (Spellman 1996).

One of the tools available to environmental science students and specialists for use in the fungal identification process is the distinctive terminology used in mycology. Fungi go through several phases in their life cycle; their structural characteristics change with each new phase. Become familiar with the following defined terms.

Microfungi

Rossman (1995) and Spellman (1996) describe fungi as a group of organisms that constitute an extremely important and interesting group of eucaryotic, aerobic microbes ranging from the unicellular yeasts to the extensively mycelial molds. Not considered plants, they are a distinctive life form of great practical and ecological importance. Fungi are important because, like bacteria, they metabolize dissolved organic matter; they are the principal organisms responsible for the decomposition of carbon in the biosphere. Fungi, unlike bacteria, can grow in low moisture areas and in low pH solutions, which aid them in the breakdown of organic matter.

Did You Know?

Without fungi, the world would be completely covered with organic debris that would not rot, and nutrients would not be available for plant growth. All plants would die (Rossman 1995).

Microfungi comprise a large group of organisms that include such diverse forms as molds, slime molds, other molds, mushrooms, mildews, puffballs, yeasts as well as rusts and smuts, which cause plant diseases. They grow in all substrates, including plants, soil, water, insects, cows' rumen, hair, and skin. Because they lack chlorophyll (and thus are not considered plants), they must get nutrition from organic substances. They are either parasites, existing in or on animals or plants, or more commonly are saporytes, obtaining their food from dead organic matter. The fungi belong to the Kingdom Myceteae. The study of fungi is called mycology.

Did You Know?

Microfungi are said to be small because only part of the fungus is visible at one time, if at all. The viable parts produce thousands of tiny spores that are carried by the air, spreading the fungus (Rossman 1995).

As mentioned previously, McKinney (1962), in *Microbiology for Sanitary Engineers*, complains that the study of mycology has been directed solely toward classification of fungi and not toward the actual biochemistry involved with fungi. McKinney goes on to point out that for those involved in the sanitary field it is important to recognize the "sanitary importance of fungi . . . and other steps will follow" (p. 40). For students of ecology understanding the role of fungi is important. Ecologists, for example, need knowledge and understanding of the organism's ability to function and exist under extreme conditions, which make them important elements in biological wastestream treatment processes and in the degradation that takes place during waste-composting processes.

Among the multitudinous molds are humble servants such as *Penicillium notatum*, the source of penicillin, and *Tolyposporium niveum*, a product of cyclosporin, the immune-system suppressant used for organ transplant operations. In sustainable agriculture the fungal performers are agents of biological control and crop nutrition, helping the environment through the reduced use of chemical pesticides and fertilizers. Fungi can stop a hoard of locusts by attacking the chitinous insect exoskeleton or control nematodes that destroy the roots of crop plants (CAB 1993). Although strains of fungi can degrade plastics and break down hazardous wastes such as dioxin (Jong and Edwards 1991), only a fraction of these fungi have been screened as beneficial organisms.

Microfungi can also be harmful, causing the irritating human affliction known as athlete's foot as well as disastrous diseases of crops and trees. The potato famine in Ireland during the mid- to late 1800's was caused by a fungus called Phytophthora infestans that rotted the potato crops for several years (Large 1962). Because of this disease, nearly a million Irish starved to death; others immigrated to the United States. Once the nature of the disease was determined, a solution based on fungus control was found. Knowing what fungi exist, where they occur, and what they do is essential.

Macrofungi

According to Mueller (1995), Macrofungi are a diverse, commonly encountered, and ecologically important group of organisms. Macrofungi may be unicellular or filamentous. They are large, 5–10 microns wide, and can be identified by a microscope. The distinguishing characteristics of the group, as a whole, include: (1) they are non-photosynthetic, (2) lack tissue differentiation, (3) have cell walls of polysaccharides (chitin), and (4) propagate by spores (sexual or asexual).

Did You Know?

Macrofungi are vitally significant in forests; many species help break down dead organic material, such as dead tree trunks and leaves, into simple compounds usable by rowing plants. Thus, they act as nature's recyclers, without which forests could not function (Mueller 1995).

Fungi can be grown and studied by cultural methods. However, when culturing fungi, use culture media that limits the growth of other microbial types—controlling bacterial growth is of particular importance. This can be accomplished by using special agar (culture media) that depress pH of the culture medium (usually Sabouraud glucose or maltose agar) to prevent the growth of bacteria. Antibiotics can also be added to the agar that will prevent bacterial growth.

As part of their reproductive cycle, fungi produce very small spores that are easily suspended in air and widely dispersed by the wind. Insects and other animals also spread fungal spores. The color, shape, and size of spores are useful in the identification of fungal species.

Reproduction in fungi can be either sexual or asexual. The union of compatible nuclei accomplishes sexual reproduction. Most fungi form specialized asexual and/or sexual spore-bearing structures (fruiting bodies). Some fungal species are self-fertilizing and other species require outcrossing between different but compatible vegetative thalluses (mycelia).

Most fungi are asexual. Asexual spores are often brightly pigmented and give their colony a characteristic color (green, red, brown, black, blue—the blue spores of ***Penicillium roquefort*** are found in blue or Roquefort cheese).

Asexual reproduction is accomplished in several ways:

1. Vegetative cells may **bud** to produce new organisms. This is very common in the yeasts.
2. A parent cell can divide into two daughter cells.
3. The most common method of asexual reproduction is the production of **spores**. Several types of asexual spores are common:
    a. A hypha may separate to form cells (arthrospores) that behave as spores.
    b. If a thick wall before separation encloses the cells, they are called chlamydospores.
    c. If budding produces the spores, they are called **blastospores**.
    d. If the spores develop within sporangia (sac), they are called **sporangiospores**.
    e. If the spores are produced at the sides or tips of the hypha, they are called **conidiospores**.

Fungi are found wherever organic material is available. They prefer moist habitats and grow best in the dark. Most fungi are saprophytes, acquiring their nutrients from dead organic matter, gained when the fungi secrete hydrolytic enzymes, which digest external substrates. They are able to use dead organic matter as a source of carbon and energy. Most fungi use glucose and maltose (carbohydrates) and nitrogenous compounds to synthesize their own proteins and other needed materials. Knowing from what materials fungi synthesize their own protein and other needed materials in comparison to what bacteria is able to synthesize is important to those who work in the environmental disciplines for understanding the growth requirements of the different microorganisms (Spellman 1996).

Lichens

Bennett points out that lichens are a unique life form because they are actually two separate organisms, a fungus and an alga, living together in a symbiosis. Lichens seem to reproduce sexually, but what appears to be a fruiting structure is actually that of the fungal component. Consequently, lichens are classified by botanists as fungi, but are given their own lichen names.

The best estimate of the number of U.S. lichen species is between 3,500 and 4,000, grouped in about 400 genera. The current checklist for the United States and Canada is probably in excess of 3,600 (Egan 1987).

Lichens are small plant-like organisms that grow just about everywhere: soils, tree trunks and branches, rocks and artificial stones, roofs, fences, walls, and even underwater. They are famous for surviving climatic extremes and are even the dominant vegetation in those habitats. Some lichens, however, are only found in very specialized habitats. The diversity of lichens in an area, therefore, is highly dependent on habitat diversity. Many special habitats across the United States are declining or disappearing because of human activities, and some lichen species are consequently in decline (Bennett 1995).

Bryophytes

The beauty there is in mosses must be considered from the holiest, quietist nook.

—Henry David Thoreau (1842)

Merrill (1995) points out that non-vascular bryophytes (mosses, liverworts, and hornworts) are small green plants that reproduce by means of spores (or vegetatively) instead of seeds. They are not considered to have given rise to the vascular plants but they probably were the earliest land plants (Qui and Palmer 1999). They evolved, like the rest of the land plants, from green algal ancestors. They are a group of simple land plants, usually only a few centimeters high that are well-adapted to moist habitats. Although often small and inconspicuous, bryophytes are remarkably resilient and successful. They are sensitive indicators of air and water pollution, and play important roles in the cycling of water and nutrients and in relationships with many other plants and animals. With its varied landscape habitats, it is not surprising that about one third of the world's Bryophyte species are found in Tropical America, with high levels of endemism (Gradstein et al. 2001). Information about bryophytes and their ecology is essential to develop comprehensive conservation and management policies and to restore degraded ecosystems (Merrill 1995).

There are three groups of bryophytes: mosses (~10,000 species), liverworts (hepatics), and hornworts. Bryophytes rank second (after the flowering plants) among major groups of green land plants, with an estimated 15,000–18,000 species worldwide.

Mosses are the best known and most abundant and conspicuous in moist habitats, but are also found in grasslands and deserts, where they endure prolonged dry periods. Hepatics also include some arid-adapted species, but most are plants of humid environments. In mosses and leafy hepatics, the conspicuous plant body is leafy; in some liverworts and all hornworts, the plant is a flattened, ribbon-like "thallus" that lies flat on the ground. Bryophytes have no roots but are anchored by slender threads called rhizoids, which also play a role in the absorption of water and mineral nutrients.

Bryophytes have successfully exploited many environments, perhaps partly because they are rarely in direct competition with higher plants (Anderson 1980). For such small organisms, the climate near the ground (microclimate) is often very different from conditions recorded by standard meteorological methods, and shifts in temperature and humidity are often extreme. A remarkable adaptation of bryophytes is their ability to remain alive for long periods without water, even under high temperatures, and then resume photosynthesis within seconds after being moistened by rain or dew (Merrill 1995).

Bryophytes perform many environmental roles: (1) In forest ecosystems they act like a sponge retaining and slowly releasing water; (2) many are pioneer plants, growing on bare rock and contributing to soil development; (3) in bogs and mountain forests they form a thick carpet, reducing erosion; (4) they provide habitat for other plants and small animals as well as microorganisms like nitrogen-fixing blue-green bacteria; (5) lacking a cuticle and transport tissue they readily absorb whatever is around them and can serve as bioindicators of pollution and environmental degradation; and (5) they are also closely associated with organisms as diverse as protozoa, rotifers, nematodes, earthworms, mollusks, insects, and spiders (Gerson 1982), as well as plants and fungi. Direct interactions of bryophytes include providing food, shelter, and nesting materials for small mammals and invertebrates; indirectly, they serve as a matrix for a variety of interactions between organisms (Spellman 1996).

Native Vascular Plants

Morse et al. (1995) point out that most of the familiar flora of the American landscape, such as trees, shrubs, herbs, vines, grasses, and ferns, are known as vascular plants. These plants have systems for transporting water and photosynthetic products and are differentiated into stems, leaves, and roots (nonvascular plants include algae, fungi, and mosses and lichens). Except in arctic and alpine areas, vascular plants dominate nearly all of North America's natural plant communities. About 17,000 species of vascular plants are native to one or more of the 50 U.S. states, along with several thousand additional native subspecies, varieties, and named natural hybrids (Kartesz 1994).

Human activities have expanded the geographical distribution of many plant species, particularly farm crops, timber trees, garden plants, and weeds. When a non-native plant species is found growing outside cultivation, it is considered an exotic species in that area. About 5,000 exotic species are known outside cultivation in the United States. While many exotic plant species are desirable in some contexts (such as horticulture), hundreds of invasive nonnatives have become major management problems when established in places valued as natural areas (McKnight 1991; U.S. Congress 1993). A few particularly troublesome non-natives are regulated under specific federal or state laws as noxious weeds.

References and Recommended Reading

A-B.com. 2007. *American Badgers*. Accessed 03/16/07 @ http://www.american.badgers .com.

Ainley, D.G., Boekelheide, R.J., eds. 1990. *Seabirds of the Farallon Islands: Ecology, dynamics, and structure of an upwelling-system community.* Stanford, CA: Stanford University.

Ainley, D.G., and G.W. Hunt Jr. 1991. The status and conservation of seabirds in California. Pages 103–114 in J.P. Croxall, ed. Seabird status and conservation; a supplement. *International Council of Bird Preservation Tech. Bull.* 11.

Ainley, D.G., and T.J. Lewis. 1974. The history of Farallon Island marine bird population, 1854–1972. *Condor* 76:432–446.

Aldrich, J.W. 1993. Classification and distribution. *In* T.S. Baskett, M.W. Sayre, R.E. Tomlinson, and R.E. Mirarchi, eds. *Ecology and management of the mourning dove.* Harrisburg, PA: Stackpole Books.

Allendorf. F.W., N. Ryman, and F.M. Utter. 1987. Genetics and fishery management past, present, and future. Pages 1–19 in N. Ryman and F.M. Utter, eds. *Population Genetics and Fishery Management.* Washington Sea Grant Program, Seattle.

Altukhov, Y.P., and E.A. Salmenkova. 1987. Stock transfer relative to natural organization, management and conservation of fish populations. Pages 333–344 *in* N. Ryman and F.M. Utter, eds. *Population Genetics and Fishery Management.* Washington Sea Grant Program, Seattle.

Anderson, L.E. 1980. Cytology and reproductive biology of mosses. Pages 37–76 *in* R.J. Taylor and A.E. Leviton, eds. The mosses of North America. Pacific Division of the American Association for the Advancement of Science, San Francisco.

Anderson, M.G. 1989. *Species closure: A case study of wintering waterfowl on San Francisco Bay, 1988–1990.* M.S. thesis, Acadia, CA: Humboldt State University.

Anderson, M.G., Rhymer, J.M., and Rohwer, F.C. 1992. Philopatry, dispersal, and the genetic structure of waterfowl populations. Pages 365–395 *in* B.D. Batt, A.D. Kadlee, and G.L. Krapu, eds. *Ecology and Management of Breeding Waterfowl.* Minneapolis: University of Minnesota Press.

Artyukhin, Y.K., A.D. Sukhoparova, and L.G. Fimukhira. 1978. The gonads of sturgeon in the littoral zone below the dam of the Volograd water engineering system. *Journal of Ichthyology* 18: 912–923.

Austin, J.E., and M.R. Miller. 1995. Northern Pintail. *In* The Birds of North America, No. 163, A. Poole and F. Gill, eds. The Academy of Natural Sciences of Philadelphia, and The American Ornithologist's Union, Washington, DC.

Avianweb. 2007. Dabbling Ducks. Accessed 03/07/07 @ http://www.avainweb.com/dabblingducks.html.

Baker, P.M. 1993. Resource Management: A shell exporter's perspective. Pages 69–71 *in* K.S. Cummings , A.C. Buchanan, and L.M. Koch, eds. Conservation and Management of Freshwater Mussels. Proceedings of a symposium. Illinois Natural History survey, Champaign.

Barrows, C., A. Muth, M. Fisher, and J. Lovich. 1995. Coachella Valley Fringe-toed Lizards. *In* Our Living Resources. Washington, DC: USGS.

Bartonek, J.C., and Nettleship, D.N., eds. 1979. Conservation of marine birds of northern North America. U.S. Fish and Wildlife Service, *Wildlife Res. Rep.* 11.

Bartsch, A.F., and W.M. Ingram. 1959. Stream life and the pollution environment. *Public Works* 90:104–110.

Beebee, T.J.C, F.J. Flower, A.C. Stevenson, S.T. Patrick, P.G. Appleby, C. Fletcher, C. Marsh, J. Natkanski, B. Rippey, and R.W. Battarbee. 1990. Decline of the natterjack toad *Bufo calamita* in Britain; palaeoecological, documentary, and experimental evidence for breeding site acidification. *Biological Conservation* 53:1–20.

Beeton, A.M. 1961. Environmental changes in Lake Erie. *Transactions of the American Fisheries Society* 90:153–159.

Bellrose, F.C. 1980. *Ducks, geese and swans of North America,* 3rd ed. Harrisburg, PA: Stackpole Books.

Benedict, A. 2007. *Of Fog Larks and Sea Quail: Sea Birds at Risk in a Changing World.* Accessed 03/10/07 @ http://seawolf-adventures.com/seabirds.html.

Bennett, J.P. 1995. Lichens. *In* Our Living Resources. Washington DC: USGS.

Berry, K.H., and P. Medica. 1995. Desert Tortoises in the Mojave and Colorado Deserts. *In* Our Living Resources. Washington, DC: USGS.

BFFRP. 2005. *Ferret Facts.* Black-footed ferret recovery program. Accessed 03/16/07 @ http://www.blackfootedferret.org/facts.html.

Biggins, D., and J. Godbey. 1995. Black-footed Ferrets. *In* Our Living Resources. Washington, DC: USGS.

Biggins, D., B.J. Miller, L. Hanebury, R. Oakleaf, A. Farmer, R. Crete, and A. Dood. 1993. A technique for evaluating black-footed ferret habitat. Pages 73–88 *in* J.L. Oldemeyer, D.E. Biggins, B.J. Miller, and R. Crete, eds. *Management of Prairie Dog Complexes for the Reintroduction of the Black-Footed Ferret*. U.S Fish and Wildlife Service Biological re. 13.

Biggins, D., M.H. Schroeder, S.C. Forrest, and L. Richardson. 1985. Movements and habitat relationships of radio-tagged black-footed ferrets. Pages 11.1–11.17 *in* S.H. Anderson and D.B. Inkley, eds. Proceedings of the Black-footed Ferret Workshop. Cheyenne: Wyoming Game and Fish Department.

Bishop, C.A., and K.E. Petit, eds. 1992. Declines in Canadian amphibian populations: designing a national monitoring strategy. *Canadian Wildlife Service Occasional Paper* 76:1–120.

Blaustein, A.R. 1994. Chicken Little or Nero's fiddle? A perspective on declining amphibian populations. *Herpetogolgica* 50:85–97.

Blaustein, A.R., P.D. Hoffman, D.G. Hokit, J.M. Kiesecker, S.C. Walls, and J.B. Hays. 1994a. UV repair and resistance to solar UV-B in amphibian eggs: a link to population declines? *Proceedings of the National Academy of Sciences* 91:1791–1795.

Blaustein, A.R., D.B. Wake, and W.P. Sousa. 1994b. Amphibian declines: judging stability, persistence, and susceptibility of populations to local and global extinction. *Conservation Biology* 8:60–71.

Blohm, R.J. 1989. Introduction to harvest understanding surveys and season setting. Pages 118–129 *in* K.H. Beattie, ed. Washington, DC: Sixth International Waterfowl Symposium.

Boarman, W.I., and K.H. Berry. 1995. Common Ravens in the Southwestern United States, 1968–92. *In* Our Living Resources. Washington, DC: USGS.

Bowen, B.W., A.B. Meylan, J.P. Rose, C.J. Limpus, G.H. Balazs, and J.C. Avise. 1992. Global population structure and natural history of the green turtle in terms of matriarchal phylogeny. *Evolution* 46:865–881.

Bowen, F., J.C. Avise, J.I. Richardson, A.B. Meylan, D. Margaritoulis, and S.R. Hopkins-Humpy. 1993. Population structure of loggerhead turtles in the northwestern Atlantic Ocean and Mediterranean Seas. *Conservation Biology* 7:834–844.

Bradford, D.F. 1983. Winterkill, oxygen relations, and energy metabolism of a submerged dormant amphibian, *Rana muscosa*. *Ecology* 64:1171–1183.

Bradford, D.F. 1989. Allotopic distribution of native frogs and introduced fishes in high Sierra Nevada lakes of California: implication of the negative effect of fish introductions. *Copeia* 1989:775–778.

Bradford, D.F. 1991. Mass mortality and extinction in a high-elevation population of *Rana mucosa*. *Journal of Herpetology* 5:174–177.

Brady, J.T., R.K. LaVal, T.H. Kuntz, M.D. Tuttle, D.E. Wilson, and R.L. Clawson, 1983. *Recovery plan for the Indiana bat*. Washington, DC: U.S Fish and Wildlife Service.

Brooks, R.J., G.P. Brown, and D.A. Galbraith. 1991. Effects of a sudden increase in natural mortality of adults on a population of the common snapping turtle. *Canadian Journal of Zoology* 69:1314–1320.

Brooks, R.J., D.A. Galbraith, E.G. Nancekivell, and C.A Bishop. 1988. Developing guidelines for managing snapping turtles. Pages 174–179 *in* R.C. Szaro, K.E. Severson, and

D.R. Patton, tech. cords. Management of amphibians, reptiles, and small mammals in North America. U.S. Forest Service Gen. Tech. Rep. RM-166.

Brown, J.W., and P.A. Opler. 1990. Patterns of butterfly species density in peninsular Florida. *Journal of Biogeography* 17:615–622.

Bugbios. 2007. *Butterfly wing patterns.* Accessed 04/19/07 @ http://www.insects.org/ class.

Burger, G.V. 1978. Agriculture and wildlife. Pages 89–107 *in* H.P. Brokaw, ed. Wildlife and America. Washington, DC: Council on Environmental Quality.

Burns, N.M. 1985. *Erie: The Lake That Survived.* Totowa, NJ: Rowman and Allenhead.

Bury, R.B., O. S. Corn, C.K. Dodd Jr., R.W. McDiarmid, and N.J. Scott. 1995. Amphibians. *In* Our Living Resources. Washington, DC: USGS.

Bury, R.B., C.K. Dodd Jr., and G.M. Fellers. 1980. Conservation of the Amphibia of the United States: a review. *U.S. Fish and Wildlife Service Res. Publ.* 134:1–34.

Byrd, G.V., R.H. Day, and E.P. Knudson. 1983. Patterns of colony attendance and censusing of auklets at Buldir Island, Alaska. *Condor* 85:274–280.

CAB. 1993. Locust project enters phase two: Commonwealth Agricultural Bureau (CAB) International News. June. P. 4.

Caithamer, D.E., and G.W. Smith. 1995. North American Ducks. *In* Our Living Resources. Washington, DC: USGS.

Callahan, E.V. 1993. *Indiana bat summer habitat requirements.* M.S. thesis, Columbia: University of Missouri.

Campton, D.E. 1987. Natural hybridization and introgression in fishes: methods of detection and interpretation. Pages 161–192 *in* N. Ryman and F.M. Utter, eds. *Population Genetics and Fishery Management.* Washington Sea Grant Program, Seattle.

Carey, C. 1993. Hypothesis concerning the causes of the disappearance of boreal toads from the mountains of Colorado. *Conservation Biology* 7:355–362.

Carlson, C.A., and B.T. Muth. 1989. The Colorado River: Lifeline of the American Southwest. Pages 220–239. In D.P. Dodge, ed. Proceedings of the International Large Rivers Symposium. *Canadian Journal of Fisheries and Aquatic Sciences.* Special Publ. 106.

Carmichael, G.J., J.N. Hanson, M.E. Schmidt, and D.C. Morizot. 1993. Introgression among Apache, cutthroat, and rainbow trout in Arizona. *Transactions of the American Fisheries Society* 122:121–130.

Carpenter, J.W., and C.N. Hillman. 1978. *Husbandry, reproduction, and veterinary care of captive ferrets.* Proceedings of the American Association of Zoo Veterinarians Workshop, Knoxville, TN.

Carter, H.R., D.S. Gilmer, J.E. Takekawa, R.W. Lowe, and U.W. Wilson. 1995. Breeding seabirds in California, Oregon, and Washington. Pp. 43–49 in E.T. LaRoe, G.S. Farris, C.E. Puckett, P.D. Doran, and M.J. Mac, eds., *Our Living Resources: A Report to the Nation on the Distribution, Abundance, and Health of U.S. Plants, Animals, and Ecosystems.* National Biological Service, Washington, D.C. [VMML e 1268].

Carter, H.R., G.J. McChesney, D.L. Jaques, C.S. Strong, M.W. Parker, J.E. Takekawa, D.L. Jory, and D.L .Whitworth. 1992. *Breeding Populations of Seabirds in California, 1989–1991,* vol. I. Dixon, CA: U.S. Fish and Wildlife Service.

Carter, H.R., and M.I. Morrison, eds. 1992. *Status and conservation of the marbled idweste in North America.* Proceedings of a 1987 Pacific Seabird Group Symposium. Camarillo, CA: Proceedings of the Western Foundation of Vertebrate Zoology.

Chmielewski, C.M., and R.J. Hall. 1993. Changes in the emergence of blackflies over 50 years from Algonquin Park streams: Is acidification the cause? *Canadian Journal of Fisheries and Aquatic Sciences* 50:1517–1529.

Christensen, N.I. 1981. *Fire Regimes in Southeastern ecosystems.* U.S. Forest Service Gen. Tech. Rep. WO-26.

Clarke, D.C 1988. Prairie dog control—a regulatory viewpoint. Pages 119–120 *in* D.W. Uresk and G. Schenbeck, eds. Eight Great Plains Wildlife Damage Control Workshop Proceedings. U.S. Forest Service Gen. Tech. Rep. RM-154.

Cochnauer, T.G. 1981. Survey status of white sturgeon populations in the Snake River, Bliss Dam to C.J. Strike Dam. Idaho Department of Fish and Game, River and Stream Investigations, Job Performance Rep., Project F-73-R-2, Job 1-b, Boise.

Cochnauer, T.G. 1983. *Abundance, distribution, growth, and management of white sturgeon in the middle Snake River, Idaho.* Ph.D. dissertation, University of Idaho, Moscow. 52 pp.

Comstock, A.B. 1986. *Handbook of Nature Study.* Ithaca: Cornell University Press.

Congdon, J.D., A.E. Dunham, and R.C. Van Loben Sels. 1993. Delayed sexual maturity and demographics of Blanding's turtles: implications for conservation and management of long-lived organisms. *Conservation Biology* 7:826–833.

Corn, P.S., and R.B. Bury. 1989. Logging in western Oregon: Response of headwater habitats and stream amphibians. *Forest Ecology and Management* 29:39–57.

Corn, P.S., and J.C. Fogelman. 1984. Extinction of montane populations of the northern leopard (*Rana pipiens*) in Colorado. *Journal of Herpetology* 18:147–152.

Cornell U. 1999. *Mourning Dove.* Accessed 03/10/07 @ http://www.birds.cornell.edu.BOW.

Craig, J.A., and R.L. Hacker. 1940. The history and development of the fisheries of the Columbia River. *U.S. Bureau of Fisheries Bull.* 49(32):132–216.

Cunningham, W.P., and Cunningham, M.A. 2002. *Principles of Environmental Science: Inquiry and Applications.* New York: McGraw-Hill.

Deacon, J.E., G. Kobetich, J.D. Williams, S. Contreras, et al. 1979. Fishes of North America endangered, threatened, or of special concern: 1979. *Fisheries* 4(2):30–44.

Delacour, J.T. 1954. *The waterfowl of the world, Vol. 1.* London: Country Life, Ltd.

Dickson, J.G. 1995. Return of Wild Turkeys. *In* Our Living Resources. Washington, DC: USGS.

Dodd, C.K., Jr. 1991. The status of the Red Hills salamander *Phaeognathus hubrichti,* Alabama, USA, 1976–1988. *Biological Conservation* 55:57–75.

Dodd, C.K. 1995. Marine turtles in the Southeast. *In* Our Living Resources. Washington, DC: USGS.

Dolton, D.D. 1995. Morning doves. *In* Our Living Resources. Washington, DC: USGS.

Doughty, R.W. 1984. Sea turtles in Texas: A forgotten commerce. *Southwestern Historical Quarterly* 88:43–70.

Drobney, R.D., and R.L. Clawson. 1995. Indiana bats. *In* Our Living Resources. Washington, DC: USGS.

Droege, S., and J.R. Sauer Jr. 1990. North American Breeding Bird Survey annual summary 1989. *U.S. Fish and Wildlife Service Biological Rep.* 90(8). 16 pp.

Ducks Unlimited. 2007. *Prairie Pothole Region.* Accessed 03/07/07 @ http://www.ducks .Org/conservation/initiative 45.aspx.

Dunson, W.A., R.L. Wyman, and E.S. Corbett. 1992. A symposium on amphibian declines and habitat acidification. *Journal of Herpetology* 16:349–352.

Egan, R.S. 1987. A fifth checklist of the lichen-forming, lichenocolous and allied fungi of the continental United States and Canada. *Bryologist* 90:77–173.

England, R.E., and A. DeVos. 1969. Influence of animals on pristine conditions on the Canadian grasslands. *Journal of Range Management* 22:87–94.

Erwin, R.M., 1995. Colonial Waterbirds. *In* Our Living Resources. Washington, DC: USGS.

Erwin, R.M., Frederick, P.C., and Trapp, J.L. 1993. Monitoring of colonial waterbirds in the United States: Needs and priorities. Pages 18–22 in M. Moser, B.C Prentice, and J. van Vessem, eds. Waterfowl and wetland conservation in the 1990's—a global perspective. Proceeding of the International Waterfowl and Wetlands Research Bureau Symposium, St. Petersburg Beach, FL. Slimbridge, UK: International Waterfowl and Wetlands research bureau Special Publ.

Escambia. 2007. *Turtle Types.* Accessed 03/18/07 @ http://escambia.ifas.ufl.edu/ marine/types_of_sea_turtles.htm.

Estes, J.A., R.J. Jameson, J.L. Bodkin, and D.R. Carlson. 1995. California Sea Otters. *In* Our Living Resources. Washington, DC: USGS.

FaunaWest Wildlife Consultants.1989. *Relative abundance and distribution of the common raven in the deserts of southern California and Nevada during spring and summer of 1989.* Riverside, CA: Bureau of Land Management.

Federal Register. 35:16047. *Federal Endangered Species.* USFWS.

Federal Register. 1993. Proposal to list the southwestern willow flycatcher as an endangered species, and to designate critical habitat. U.S. Fish and Wildlife Service. *Federal Register* 58:39495–39522.

Fellers, G.M., and C.A. Drost. 1993. Disappearance of the Cascades frog *Rana Cascadae* at the southern end of its range. California, USA. *Biological Conservation* 65(2):177–181.

Finch, D.M., and P.W Stangel. 1993. *Status and management of Neotropical migratory birds*; 1992 September 21–25; Estes Park, CO, Gen. Tech. Rep. RM-229, U.S. Forest Service, Rocky Mountains Forest and Range Experiment Station, Fort Collins, CO.

Flather, C.H., and T.W. Hoekstra. 1989. *An analysis of the wildlife and fish situation in the United States: 1989–2040.* U.S. Department of Agriculture Forest Service Gen. Tech. Rep. RM-178.

Forrest, S.C., D.E. Biggins, L. Richardson, T.W. Clark, T.M. Campbell II, K.A. Fagerstone, and E.T. Thorne. 1988. Population attributes for the black-footed ferret at Meeteetse, Wyoming, 1981–1985. *Journal of Mammalogy* 69(2):261–273.

Franson, J.C., Sileo, L., Thomas, N.J. 1995. Causes of eagle deaths. *In* Our Living Resources. Washington, DC: USGS.

Freda, J. 1986. The influence of acidic pond water on amphibians: A review. *Water, Air, and Soil Pollution* 30:439–450.

Fuller, M.R., Henny, C.J. and Wood P.B., 1995. Raptors. *In* Our Living Resources. Washington, DC: USGS.

Gardner, J.E., J.D. Garner, and J.E. Hofmann. 1991. Summer roost selection and roosting behavior of *Myotis idweste* in Illinois. Final report. Illinois Natural History Survey. Champaign, Illinois: Department of Conservation.

Gee, G.G., and Hereford, S.G. 1995. Mississippi Sandhill Cranes. *In* Our Living Resources. Washington, DC: USGS.

Gerson, U. 1982. Bryophytes and invertebrates. Pages 291–322 *in* A.J.E. Smith, ed. *Bryophyte Ecology*. New York: Chapman and Hall.

Gill, R.E., Jr., Handel, C.M., and Page, G.W. 1995. Western North American Shorebirds. *In* Our Living Resources. Washington, DC: USGS.

Gould, P.J., Forsell, D.J., and Lensink, C.J. 1982. Pelagic distribution and abundance of seabirds in the Gulf of Alaska and eastern Bering Sea. U.S. Fish and Wildlife Service FWS/OBS-8248.

Gradstein, S.R., Churchill, S.P., and Salazar Allen, N. 2001. Guide to the Bryophytes of tropical America. *Mem. N.Y. Bot. Gard.* 86:1–577.

Grasshopper Facts. 2007. Accessed 04/09/07 @ http://www.thaibugs.com/Articles/grasshoppers.

Guntenspergen, G.R. 1995. Plants. *In* Our Living Resources. Washington, DC: USGS.

Haig, S. 1992. The piping plover. Pages 1–18 *in* A. Poole, P. Stettenheim, and F. Gill, ed. *Birds of North America*. Washington, DC: American Ornithologists Union.

Haig, S. and J.H. Plissner. 1995. Piping plovers. In Our Living Resources. Washington, DC: USGS.

Hale, S.F., C.R. Schwalbe, J.L. Jarchow, C.J. May, C.H. Lowe, and T. B. Johnson. 1995. Disappearance of Tarahumara frog. *In* Our Living Resources. Washington, DC: USGS.

Halls, L.K., ed. 1984. *White-tailed deer: Ecology and management*. Harrisburg, PA: Stackpole Books.

Harrington, B.A., 1995. Shorebirds: East of the 105th Meridian. *In* Our Living Resources. Washington, DC: USGS.

Hatch, S.A. 1993a. Population trends of Alaskan seabirds. *Pacific Seabird Group Bull.* 20:3–12.

Hatch, S.A. 1993b. Ecology and population status of northern fulmars *Fulmarus glacialis* of the North Pacific. Pages 82–92 *in* K. Vermeer, Briggs, K.T., Morgan, K.H., and Siegel-Causey, D., eds. *The status, ecology and conservation of marine birds of the North Pacific*. Ottawa: Canadian Wildlife Service.

Hatch, S.A., Byrd, G.V., Irons, D.B., and Hunt, G.L, Jr. 1993a. Status and ecology of kittiwakes in the North Pacific. Pages 140–153 *in* K. Vermeer, Briggs, K.T., Morgan, K.H., and Siegel-Causey, D., eds. *The status, ecology and conservation of marine birds of the North Pacific*. Ottawa: Canadian Wildlife Service.

Hatch, S.A. and Hatch, M.A., 1978. Colony attendance and population monitoring of black-legged kittiwakes on the Semidi Islands, Alaska. *Condor* 90:613–620.

Hatch, S.A., and Hatch, M.A. 1989. Attendance patterns of murres at breeding sites: implications for monitoring. *Journal of Wildlife Management* 53:483–493.

Hatch, S.A., and Platt, J.F. 1995. Seabirds in Alaska. *In* Our Living Resources. Washington, DC: USGS.

Hatch, S.A., Roberts B.D., and Fadely, B.S. 1993b. Adult survival of black-legged kittiwakes in a Pacific colony. *Ibis* 135:247–254.

Hayes, M.P., and M.R. Jennings. 1986. Decline of ranid frog species in western North America: are bullfrogs (*Rana catesbeiana*) responsible? *Journal of Herpetology* 20:490–509.

Heinrich, B. 1989. *Ravens in winter.* New York: Summit Books.

Hestback, J.B. 1995a. Canada Geese in the Atlantic Flyway. *In* Our Living Resources. Washington, DC: USGS.

Hestback, J.B. 1995b. Decline of Northern Pintails. *In* Our Living Resources. Washington, DC: USGS.

Heyer, W.R., M.A. Donnelly, R.W. McDiarmid, L. Hayek, and M.S. Foster, eds. 1994. *Measuring and monitoring biological diversity: standard methods for amphibians.* Washington, DC: Smithsonian Institution Press.

Hillman, C.N., and R.L. Linder. 1973. The black-footed ferret. Pages 10–20 *in* R.L. Linder and C.N. Hillman, eds. Proceedings of the Black-footed Ferret and Prairie Dog Workshop. South Dakota State University Publications, Brookings.

Hine, R.L., B.L. Les, and B.F. Hellmich. 1981. Leopard frog populations and mortality in Wisconsin, 1974–76. *Wisconsin Department of Natural Resources Tech. Bull.* 122:1–39.

Hochbaum, G.S., and E.F. Bossenmaier. 1972. Response of pintail to improved breeding habitat in southern Manitoba. *Canadian Field-Naturalist* 86:79–81.

Hodges, R.W. 1995. Diversity and Abundance of Insects. *In* Our Living Resources. Washington, DC: USGS.

Hohman, W.L., Haramis, G.M., Jorde, D.G., Korschgen, C.E., Takehawa, J. 1995. *In* Our Living Resources. Washington, DC: USGS.

Holden, C. 1989. Entomologists wane as insects wax. *Science* 246:754–756.

Hudson, P.L., D.R. Lenat, B.A. Caldwell, and D. Smith. 1990. Chironomidae of the southeastern United States: A checklist of species and notes on biology, distribution, and habitat. *U.S. Fish and Wildlife Service Fish and Wildlife Res.* 7. 46 pp.

Huff, W.R. 1993. Biological indices define water quality standards. *Water Environment and Technology* 5, 21–22.

Hulbert, L.C. 1973. Management of Konza Prairie prior to approximate pre-whiteman influences. Pages 14–19 *in* L.C. Hulbert, ed. Third Midwest prairie conference proceedings. Manhattan: Kansas State University.

Humphrey, S.R., A.R. Richter, and J.B. Cope. 1977. Summer habitat and ecology of the endangered Indiana bat. *Myotis idweste. Journal of Mammalogy* 58:334–346.

Hunt, G.L., Pitman, R.L., Naughton, M., Winnett, K., Newman, A., Kelly, P.R., and Briggs, K.T. 1979. Reproductive ecology and foraging habits of breeding seabirds. Pages 1–399 *in* Summary of marine mammal and seabird surveys of the southern California Bight area 1975–1978, vol. 3—Investigators' reports. Part 3. Seabirds—Book 2. University of California-Santa Cruz. For U.S. Bureau of Land Management, Los Angeles, CA. Contract AA550-CT7-36.

Hupp, J.W., J.A. Schmutz, and C.D. Ely. 2007. Moult migration of emperor geese. *Journal of Avian Biology.*

Hutchinson, G.E. 1981. Thoughts on aquatic insects. *Bioscience* 31, 495–500.

IBC. 2007. *White Sturgeon.* Accessed 03/31/07 @ http://www.beadventure.com/adventure/angling/game_fish/sturgeon.phtml.

Jefferies, M., and Mills, D. 1990. *Freshwater Ecology: Principles and Applications.* London: Belhaven Press.

Jennings, M.R. 1995. Native Ranid Frogs in California. *In* Our Living Resources. Washington, DC: USGS.

Jennings, M.R., and M.P. Hayes. 1993. Amphibian and reptile species of special concern in California. Final Report submitted to the California Department of Fish and Game. Inland Fisheries Division, Rancho Cordova, under Contract (8023), 336 pp.

Johnson, D.H., and Grier, J.W. 1988. Determinants of breeding distributions of ducks. *Wildlife Monograph* 100:1–37.

Johnson, J.E. 1987. *Protected Fishes of the United States and Canada.* American Fisheries Society, Bethesda, MD.

Johnson, J.E. 1995. Imperiled Freshwater Fishes. *In* Our Living Resources. Washington, DC: USGS.

Jong, S.C., and J.J. Edwards. 1991. *American type culture collection catalogue of filamentous fungi,* 18th ed. Rockville, MD.

Kagarise Sherman, C., and M.L. Morton. 1993. Population declines of Yosemite toads in the eastern Sierra Nevada of California. *Journal of Herpetology* 27(2):186–198.

Kaminski, R.M., and Weller, M.W. 1992. Breeding habits of nearctic waterfowl. Pages 568–589 *in* B.J. Batt, A.D. Afton, M.G. Anderson, C.D. Ankney, D.H. Johnson, J.A. Kadlec, and G.L. Krapu, eds. *Ecology and Management of Breeding Waterfowl.* Minneapolis: University of Minnesota Press.

Kamrin, M.A. 1989. *Toxicology.* Chelsea, MI: Lewis Publishers.

Kappes, J.J., 1993. *Interspecific interactions associated with red-cockaded woodpecker cavities at a north Florida site.* M.S. thesis, Gainesville, FL:

Kartesz, J.T. 1994. *A synonymized checklist of the vascular flora of the United States, Canada, and Greenland,* 2nd ed. Portland, OR: Timber Press.

Kenyon, K.W. 1969. The sea otter in the eastern Pacific Ocean. *North American Fauna* 68:1.3.

Kiel, W.H., Jr. 1959. Mourning dove management units—a progress report. *U.S. Fish and Wildlife Service Special Sci. Rep.—Wildlife* 42.

Kinsinger, A. 1995. Marine mammals. *In* Our Living Resources. Washington, DC: USGS.

Knight, R., and M. Call. 1980. *The common raven.* Bureau of Land Management Tech. Note 344.

Knight, R., and J. Kawashima. 1993. Responses of raven and red-tailed hawk populations to linear right-of-ways. *Journal of Wildlife Management* 57:266–271.

Kurta, A., D. King, J.A. Teramino, J.M. Stribley, and K.J. Williams. 1993. Summer roosts of the endangered Indiana bat (*Myotis idweste*) on the northern edge of its range. *American Midland Naturalist* 129:132–138.

Kutkuhn, J.H. 1981. Stock definition as a necessary basis for cooperative management of Great Lakes fish resources. *Canadian Journal of Fisheries and Aquatic Sciences* 3:1476–1478.

Lamb, T., J. Avise, and J.W. Gibbons. 1989. Phylogeo-graphic patterns in mitochondrial DNA of the desert tortoise and evolutionary relationships among the North American gopher tortoises. *Evolution* 43(1): 76–87.

Large, E.C. 1962. *Advance of the Fungi*. New York: Dover.

Larsen, K.H., and J.H. Dietrich. 1970. Reduction of a raven population on lambing grounds with DRC-1339. *Journal of Wildlife Management* 34:200–204.

Lee, D.S., C.B. Gilbert, C.H. Hocutt, R.E. Jenkins, D.E. McAlister, and J.R. Stauffer Jr. 1980. *Atlas of North American Freshwater Fishes*. North Carolina State Museum of Natural History.

Lennartz, M.R., Hooper, R.G., and Harlow, R.F. 1987. Sociality and cooperative breeding of red-cockaded woodpeckers. *Behavioral Ecology and Sociobiology* 20:77–88.

Levine, M. B., Hall, A. T., Barret, G. W., and Taylor, D. H. 1989. Heavy-Metal Concentration During Ten Years of Sludge Treatment to an Old-Field Community. *Journal of Environmental Quality* 18, No. 4: 411–418.

Linder, R.L., R.B. Dahlgren, and C.N. Hillman. 1972. Black-footed ferret-prairie dog interrelationships. Pages 22–37 *in* Proceedings of the Symposium on Rare and Endangered Wildlife of the Southeastern U.S. Santa Fe, NM: New Mexico Department of Game and Fish.

Lindzey, F.G. 1982. The North American badger. Pages 653–663 *in* J.A. Chapman and G.A. Feldhammer, eds. *Wild Mammals of North America*. Baltimore, MD: Johns Hopkins University Press.

Lovich, J.E. 1984. Biodiversity and zoogeography of non-marine turtles in Southeast Asia. Pages 381–391 *in* S.K. Majumdar, F.J. Brenner, J.E. Lovich, J.F. Schalles, and E.W. Miller, eds. *Biological Diversity: Problems and Challenges*. Pennsylvania Academy of Science.

Lovich, J.E. 1995. Turtles. *In* Our Living Resources. Washington, DC: USGS.

Macdonald, D.W., ed. 2006. *The Princeton Encyclopedia of Mammals*. Princeton, NJ: Princeton University Press.

Marnell, L.F. 1986. Impacts of hatchery stocks on wild fish populations. Pages 339–347 *in* R.H. Stroud, ed. *Fish Culture in Fisheries Management*. Bethesda, MD: American Fisheries Society.

Marnell, L.F. 1988. Status of the westslope cutthroat trout in Glacier National Park, Montana. *American Fisheries Society Symposium* 4:61–70. Bethesda, MD.

Marnell, L.F. 1995. Cutthroat Trout in Glacier National Park, Montana. *In* Our Living Resources. Washington, DC: USGS.

Marnell, L.F., R.J. Behnke, and F.W. Allendorf. 1987. Genetic Identification of cutthroat trout in Glacier National Park, Montana. *Canadian Journal of Fisheries and Aquatic Sciences* 44:1830–1839.

Marsh, R.E. 1984. Ground squirrels, prairie dogs and marmots as pests on rangeland. Pages 195–208 *in* Proceedings of the Conference for Organization and Practice of Vertebrate Pest Control. ICI Plant Protection Division, Fernherst, England.

Mason, C.F. 1990. Biological aspects of freshwater pollution. Pollution: Causes, Effects, and Control. R. M. Harrision, ed. Cambridge: The Royal Society of Chemistry.

Mason, W.T., Jr. 1975. Chironomidae (Diptera) as biological indicators of water quality. Pages 40–51 *in* C.C. King and L.E. Elfner, eds. Organisms and Biological communities as indicators of environmental quality. Circular 8. Ohio Biological Survey, Columbus.

Mason, W.T., Jr. 1995. Invertebrates. *In* Our Living Resources. Washington, DC: USGS.

Mason, W.T., Jr., C.R. Fremling, and A.V. Nebeker. 1995. Aquatic insects as indicators of environmental quality. *In* Our Living Resources. Washington, DC: USGS.

Mattson, D.J., R.G. Wright, K.C. Kendall, and C.J. Martinka. 1995. Grizzly Bears. *In* Our Living Resources. Washington, DC: USGS.

Maughan, O.E. 1995. Fishes. *In* Our Living Resources. Washington, DC: USGS.

McCabe, T.L. 1995. The changing insect fauna of Albany's Pine Barrens. *In* Our Living Resources. Washington, DC: USGS.

McDiarmid, R.W. 1995. Reptiles and amphibians. *In* Our Living Resources. Washington, DC: USGS.

McKenzie, S. 2007. *Poems on/about water.* Accessed 04/19/07 @ http://www.poemsabout.com.

McKinney, R.E. 1962. *Microbiology for Sanitary Engineers.* New York: McGraw-Hill.

McKnight, B.N., ed. 1991. *Biological pollution: the control and impact of invasive exotic species.* Indianapolis: Indiana Academy of Science.

Mech, L.D., D.H. Pletscher and C.J. Martinka. 1995. Gray Wolves. *In* Our Living Resources. Washington, DC: USGS.

Merrill, G.L.S. 1995. Bryophytes. *In* Our Living Resources. Washington, DC: USGS.

Merritt, R.W., and K.W. Cummins. 1984. Introduction. Pages 1–3 *in* R.W. Merritt and R.W. Cummins, eds. *An Introduction to the Aquatic Insects of North America,* 2nd ed. Dubuque, IA: Kendall/Hunt Publishing Company.

Messer, J.J., R.A. Linthurst, and W.S. Overton. 1991. An EPA program for monitoring ecological status and trends. *Environmental Monitoring and Assessment* 17:67–78.

Meyer, E. 1989. *Chemistry of Hazardous Materials,* 2nd ed. Englewood Cliffs, NJ: Prentice-Hall.

Miller, A.I., T.D. Counihan, M.J. Parsley, L.G. Beckman. 1995. Columbia River Basin white sturgeon. *In* Our Living Resources. Washington, DC: USGS.

Miller, R.R. 1961. Man and the changing fish fauna of the American Southwest. *Papers of the Michigan Academy of Science, Arts, and Letters* 46:365–404.

Miller, R.R. 1972. Threatened freshwater fishes of the United States. *Transactions of the American Fisheries Society* 101(2):239–252.

Minckley, W.I., and J.E. Deacon. 1968. Southwestern fishes and the enigma of endangered species. *Science* 159:1424–1432.

Minckley, W.I., and J.E. Deacon, eds. 1991. *Battle against Extinction: Native Fish Management in the American West.* Tucson: University of Arizona Press.

Missouri Department of Conservation 1991. *Endangered bats and their management in Missouri.* Jefferson City: Missouri Department of Conservation.

Morse, L.E., J.T Kartesz, and L.S. Kutner. 1995. Native vascular plants. *In* Our Living Resources. Washington, DC: USGS.

Moyle, P.B., H. Li, and B.A. Baron. 1986. The Frankenstein effect: impact of introduced fishes in North America. Pages 415–426 *in* R.H. Stroud, ed. Fish culture in fisheries management. American Fisheries Society. Bethesda, MD.

Moyle, P.B., and J.J. Cech, Jr. 1988. *Fishes: an introduction to ichthyology.* Englewood Cliffs, NJ: Prentice Hall.

Mueller, G.M. 1995. Macrofungi. *In* Our Living Resources. Washington, DC: USGS.

Nagel, H. 1992. The link between Platte River flows and the regal fritillary butterfly. *The Braided River* 4:10–11.

Nagel, H.G., T. Nightengale, and N. Dankert. 1991. Regal fritillary butterfly population estimation and natural history on Rowe Sanctuary, Nebraska. *Prairie Naturalist* 23:145–152.

National Academy of Sciences. 1978. *Eutrophication: causes, consequences, correctives.* Proceedings of a symposium. Washington, DC.

National Research Council. 1990. *Decline of the Sea Turtles: Causes and Prevention.* Washington, DC: National Academy Press.

Nature. 2007. *Ravens.* Accessed 4/15/07 @ http://www.pbs.org/wnet/nature/Ravens.

Neal, W. 1985. Endangered and threatened wildlife and plants: Reclassification of the American alligator in Florida to threatened due to similarity of appearance. *Federal Register* 50(119): 25,672–25,678.

Nettleship, D.N., Sanger, G.A., and Springer, P.F., eds. 1984. *Marine birds: their feeding ecology and commercial fisheries relationships.* Proceedings of the Pacific Seabird Group Symposium. Ottawa, Ontario: Canadian Wildlife Service Spec. Publ.

NOAA 2007. *Pinnipeds, Whales, Dolphins, Porpoises and Sirenia Order.* Accessed 03/13/07 @ http://mnmml.afsc.noaa.gov/education/sierenia.htm.

Novak, R.M. 1994. Another look at wolf taxonomy. In L.D. Carbyn, S.H. Fritts, and D.R. Seip, eds. *Ecology and conservation of wolves in a changing world.* Edmonton, Alberta: Canadian Circumpolar Institute.

NPS. 1995. *The Tall Grass Prairie.* National Park Service. Washington, DC: U.S. Department of Interior.

NWR. 2007. *Mississippi Sandhill Cranes.* Accessed 03/11/07 @ http://www.fws.gov/Mississippiandhillcrane/mscranes.

Obbard, M.E., J.G. Jones, R. Newman, A. Booth, A.J. Satterthwaite, and G. Linscombe. 1987. Furbearer Harvests in North America. Pages 1007–1038 *in* M. Novak, J.A. Baker, M.E. Obbard, and B. Malloch, eds. Wild furbearer management and conservation in North America. The Ontario Trappers Association and the Ministry of Natural Resources, Toronto, Ontario.

O'Brien, S., J.S. Martenson, M.A. Eichelberger, E.T. Thorne, and F. Wright. 1989. Biochemical genetic variation and molecular systematic of the back-footed ferret. Pages 21–33 in *Conservation Biology and the Black-Footed Ferret.* New Haven, CT: Yale University Press.

Opler, P.A. 1995. Species richness and trends of western butterflies and moths. *In* Our Living Resources. Washington, DC: USGS.

O'Toole, C (Ed.). 1986. *The Encyclopedia of Insects.* New York: Facts on File, Inc.

Otte, D. 1995. Grasshoppers. *In* Our Living Resources. Washington, DC: USGS.

Paine, G.W., and A.R. Gaufin. 1956. Aquatic Diptera as indicators of pollution in a midwestern stream. *The Ohio Journal of Science* 56:291–304.

Patrick, R., and D.M. Palavage. 1994. The value of species as indicators of water quality. *Proceedings of the Academy of Natural Sciences Philadelphia* 145:55–92.

Pattee, O.H., and Mesta, R. 1995. California condors. *In* Our Living Resources. Washington, DC: USGS.

Pattee, O.H., and Wilbur, S.R. 1989. Turkey vulture and California condor. Pages 61–65 *in* Proceedings of the Western Raptor Management Symposium and Workshop. Washington, DC: National Wildlife Federation.

Pechmann, J.H.K., D.E. Scott, R.D. Semlitsch, J.P. Caldwell, L.J. Vitt, and J.W. Gibbons. 1991. Declining amphibian populations; the problem of separating human impacts from natural fluctuations. *Science* 253:892–895.

Pechmann, J.H.K., and H.M. Wilbur. 1994. Putting declining amphibian populations in perspective: natural fluctuations and human impacts. *Herpetologica* 50:65–84.

Peek, J.M. 1995. North American Elk. *In* Our Living Resources. Washington, DC: USGS.

Pennak, R.W. 1978. *Fresh-water invertebrates of the United States,* 2nd ed. New York: John Wiley & Sons.

Peterjohn, B.G., and Sauer, J.R., 1993. North American Breeding Bird Survey Annual Summary 1990–1991. *Bird Populations* 1:52–67.

Petranka, J.W., M.E. Eldridge, and K.E. Haley. 1993. Effects of timber harvesting on southern Appalachian salamanders. *Conservation Biology* 7:363–370.

Philipp, D.P., and J.E. Clausen. 1995. Loss of genetic diversity among managed populations. *In* Our Living Resources. Washington, DC: USGS.

Philipp, D.P., J.M. Epifanio, and J.J. Jennings.1993. Conservation genetics and current stocking practices: are they compatible? *Fisheries* 18:14–16.

Pianka, E., and L. Vitt. 2006. *Lizards: Windows to the Evolution of Diversity.* Los Angeles: University of California Press.

Pohly, J. *Grasshopper facts.* Colorado State University. Accessed 04/09/07 @ http://www.colostate.edu/Depts/CoopeExt/4DMG/Gardent/Amazing/grasfact.htm.

Porter, W.F. 1991. *White-tailed deer in eastern ecosystems: implications for management and research in national parks.* National Resources Report NPS/NRSUNY/NRR-91/05. Denver, CO: National Park Service.

Pounds, J.A., and M.L. Crump. 1994. Amphibian declines and climate disturbance: the case of the golden toad and harlequin frog. *Conservation Biology* 8:72–85.

Powell, J.A. 1995. Lepidoptera inventories in the Continental United States. *In* Our Living Resources. Washington, DC: USGS.

Pulliam, H.R. 1988. Sources, sinks, and population regulations. *American Naturalist* 132:652–661.

Qui, Y.-L., and J.D. Palmer. 1999. Phylogeny of early land plants: Insights from genes and genomes. *Trends in Plant Science* 4:26–30.

Richards, S.J., K.R. McDonald, and R.A. Alford. 1993. Declines in populations of Australia's endemic tropical rainforest frogs. *Pacific Conservation Biology* 1:66–77.

Ricker, W.E. 1963. Big effects from small causes: Two examples from fish population dynamics. *Journal of the Fisheries Research Board of Canada* 20:257–264.

Rieman, B.E., and Beamesderfer, R.C. 1990. White sturgeon in the lower Columbia River: is the stock overexploited? *North American Journal of Fisheries Management* 10:388–396.

Risser, P.G., E.C. Birney, H.D. Blocker, S.W. May, W.J. Parton, and J.A. Wiens. 1981. *The true prairie ecosystem.* Stroudsburg, PA: Hutchinson Ross Publishing.

Roback, S.S. 1957. The immature tendipedids of the Philadelphia area. *Academy of Natural Sciences of Philadelphia Monograph* 9.

Robbins, C.S., Bruun, B., and Zim, H.S. 1966. *Birds of North America.* New York: Golden Press.

Robison, H.W. 1986. Zoogeographic implications of the Mississippi River basin. Pages 267–285 *in* C.H. Hocutt and E.O. Wiley, eds. The zoogeography of North American freshwater fishes. New York: John Wiley and Sons.

Rocky Mountain Elk Foundation. 1989. Wapiti across the West. *Bugle* 6:138–140.

Rogers, L.L. 2002. *Black Bear Facts.* Accessed 03/15/07 @ http://www.bear.org/Black/Black_Bear_Facts.html.

Root, T.L., and McDaniel, L., 1995. Winter population trends of selected songbirds. *In* Our Living Resources. Washington, DC: USGS.

Rossman, A.Y. 1995. Microfungi: Molds, mildews, rusts, and smuts. *In* Our Living Resources. Washington, DC: USGS.

Rusch, D.H., R.E. Malecki, and R. Trost. 1995. Canada Geese of North America. *In* Our Living Resources. Washington, DC: USGS.

Sargeant, A.B., and Raveling, D.G. 1992. Mortality during the breeding season. Pages 396–422 *in* B.J. Batt, A.D. Afton, M.G. Anderson, C.D. Ankney, D.H. Johnson, J.A. Kadlec, and G.I. Krapu, eds. *Ecology and management of breeding waterfowl.* Minneapolis: University of Minnesota Press.

Sauer, C. 1950. Grassland climax, fire and management. *Journal of Range Management* 3:16–20.

Schloesser, D.W., and R.F. Nalepa. 1995. Freshwater Mussels in the Lake Huron-Lake Erie Corridor. *In* Our Living Resources. Washington, DC: USGS.

Scott, B. 2001. *Macroinvertebrates.* Accessed 04/19/07 @ http://www.wavcc.org/wvc/cadre/Water Quality/Indes.html.

Scott, W.B., and E.J. Crossman. 1973. Freshwater fishes of Canada. *Fisheries Research Board of Canada Bull.* 184, Ottawa.

Seabirds in California, Oregon, and Washington. *In* Our Living Resources. Washington, DC: USGS.

Sealy, S.G., ed. 1990. Auks at sea. Proceedings of an International symposium of the Pacific Seabird Group. *Studies in Avian Biology* 14.

Serie, J. 1993. *Waterfowl harvest and population survey data.* U.S. Fish and Wildlife Service. Laurel, MD: Office of Migratory Bird Management.

Sharp, J. 1988. Politics, prairie dogs, and the sportsman. Pages 117–118 *in* D.W. Uresk and G. Schenbeck, eds. Eight Great Plains Wildlife Damage Control Workshop Proceedings. U.S. Forest Service Gen. Tech. Rep. RM-154.

Smith, R.I. 1968. The social aspects of reproductive behavior in the pintail. *Auk* 85:381–396.

Smith, R.I., 1970. Response of pintail breeding populations to drought. *Journal of Wildlife Management* 34:943–946.

Sogge, M.K. 1995. Southwestern Willow Flycatchers in the Grand Canyon. *In* Our Living Resources. Washington, DC: USGS.

Soule, M.E., ed. 1987. *Conservation Biology: The Science of Scarcity and Diversity.* Sunderland, MA: Sinauer Associates, Inc.

Sowls, A.L., DeGange, A.R., Nelson, J.W., and Lester, G.S. 1980. *Catalog of California seabird colonies.* U.S. Fish and Wildlife Service. FWS/OBS 37/80.

Sowls, A.L., Hatch, S.A., and Lensink, C.J. 1978. *Catalog of Alaskan seabird colonies.* U.S. Fish and Wildlife Service FWS/OBS-7878.

Speich, S.M, and Wahl, T.R. 1989. Catalog of Washington seabird colonies. U.S. Fish and Wildlife Service. *Biological Rep* 88(6).

Spellman, F.R. 1996. *Stream Ecology and Self-Purification: An Introduction for Wastewater and Water Specialists.* Lancaster, PA: Technomic Publishing Company.

Spellman, F.R. 2003. *Handbook of Water/Wastewater Treatment Plant Operations.* Boca Raton, FL: CRC Press.

SSMHS. 2007. *Wooly Bear Caterpillars.* Sault Ste. Marie Horticultural Society. Accessed 04/17/07 @ http://www.bacyardwildlifehabitat.info/index.htm.

Stalmaker, C.B., and P.B. Holden. 1973. Changes in native fish distribution in the Green River system, Utah-Colorado. *Utah Academy of Science, Arts, Letters Proceedings* 50:25–32.

Starnes, W.C. 1995. Colorado River Basin fishes. *In* Our Living Resources. Washington, DC: USGS.

Starnes, W.C., and D.A. Etnier. 1986. Drainage evolution and fish biogeography of the Tennessee and Cumberland rivers drainage realm. Pages 325–361 *in* C.H. Hocutt and E.O. Wiley, eds. *The zoogeography of North American Freshwater Fishes.* New York: John Wiley and Sons.

Steeg, B.V., and R.E. Warner. 1995. American badgers in Illinois. *In* Our Living Resources. Washington, DC: USGS.

Stevens, Wallace. 2004. Frogs eat butterflies. Snakes eat frogs. Hogs eat snakes. Men eat hogs. *Poetry X,* Edited by Jough Dempsey. Accessed 24 March 2007 from http://poetry.poertyx.com/Poems/5301/.

Stewart, R. E., and H. A. Kantrud. 1971. Classification of natural ponds and lakes in the glaciated prairie region. Resource Publication 92, Bureau of Sport Fisheries and Wildlife, U.S. Fish and Wildlife Service Washington, DC.

Stockley, C. 1981. *Columbia River Sturgeon.* Washington Department of Fisheries Prog. Rep. 150, Olympia.

Storm, G.L., and Palmer, G.L. (1995). White-tailed deer. *In* the Northeast. In Our Living Resources. Washington, DC: USGS.

Swengel, A.B. 1990. Monitoring butterfly populations using the 4th of July butterfly count. *American Midland Naturalist* 124:395–406.

Swengel, A. 1993. Permutations of painted ladies. *American Butterflies* 1(2):34.

Swengel, A.B. 1995. Fourth of July butterfly count. *In* Our Living Resources. Washington, DC: USGS.

Swengel, A. B., and S.R. Swengel. 1995. The Tall-grass Prairie butterfly community. *In* Our Living Resources. Washington, DC: USGS.

Taylor, J.N., W.R. Courtenay Jr., and J.A. McCann. 1984. Known impacts of exotic fishes in the continental United States. Pages 322–373 *in* W.R. Courtenay Jr. and J.R. Stauffer, eds. *Distribution, Biology and Management of Exotic Fishes.* Baltimore, MD: Johns Hopkins University Press.

Teisl, M.F., and Southwick, R. 1995. *The Economic Contributions of Bird and Waterfowl Recreation in the United States during 1991.* Arlington, VA: Southwick Associates.

Templeton, A.R. 1987. Coadaptation and outbreeding depression. Pages 105–116 *in* M.E. Soule, ed. *Conservation Biology: The Science of Scarcity and Diversity.* Sunderland, MA: Sinauer Associates.

Tomlinson, R.E., D.D. Dolton, H.M. Reeves, J.D. Nichols, and L.A. McKibben. 1988. Migration, harvest, and population characteristics of mourning doves banded in the Western management Unit, 164–77. *U.S. Fish and Wildlife Service Tech. Rep.* 13.

Tomlinson, R.E., and J.H. Dunks. 1993. Population characteristics and trends in the Central Management Unit. Pages 305–340 *in* T.S. Baskett, M.W. Sayre, R.E. Tomlinson, and R.E. Mirarchi, eds. *Ecology and Management of the Mourning Dove.* Harrisburg, PA: Stackpole Books.

TPW. 2007. *Stocking Public Waters.* Texas Parks and Wildlife. Accessed 03/29/07 @ http://www.tpwd.state.tx.us/fishboat/fish/management/stocking/.

TPWD. 2007. *Red-cockaded Woodpecker.* Texas Parks and Wildlife Department. Accessed 03/11/07 @ http://www.tpwd.state.tx.us/huntwild/wild/birding/redcockaded_woodpecker/.

Tracy, C.A. 1993. *Status of white sturgeon resources in the mainstem. Columbia River.* Final Report. Dingell/Johnson-Wallop/Breaux Project F-77-R, Washington Department of Fisheries, battleground.

Tuggle, B.N. 1995. Mammals. *In* Our Living Resources. Washington, DC: USGA.

Turgeon, D.D., A.E. Bogan, E.V. Coan, W.K. Emerson, W.G. Lyons, W.L. Pratt, C.F.E. Roper, A. Scheltema, F.G. Thompson, and J.D. Williams. 1988. Common and scientific names of aquatic invertebrates from the United States and Canada: Mollusks. *American Fisheries Society Special Publ.* 16.

U.S. Congress, Office of Technology Assessment. 1993. *Harmful non-indigenous species in the United States.* U.S. Government Printing Office OTA-F-565, 391 pp.

USEPA. 1986. *Superfund Public Health Evaluation Manual.* Washington, DC: Office of Emergency and Remedial Response.

USFWS. 1982. *Mexican wolf recovery plan.* Albuquerque, NM.

USFWS. 1987. *Northern Rocky Mountain wolf recovery plan.* Denver, CO.

USFWS 1992. *Alaska seabird management plan.* Anchorage, AK: Division of Migratory Birds.

USFWS. 1992. *Recovery plan for the eastern timber wolf.* Twin Cities, MN.

USFWS. 1994. *Desert tortoise (Mojave population) recovery plan.* Portland, OR. 77 pp. + appendix.

USFWS. 1998. *Gray wolf.* Accessed 09/14/10 @ http://ww.fws.gov/species/species_accounts/ bio_gwol.html.

USFWS. 2002. *Colonial-nesting waterbirds.* Accessed 4/15/07 @ http://www.fws.gov/birds/waterbirds.

USFWS. 2002. *Songbirds: A colorful chorus.* Accessed 11/14/10 @ http://www.fws.gov.

USFWS. 2004. *Southwestern Willow Flycatcher.* Accessed 03/11/07 @ http://www.fws.gov.

USFWS. 2007a. *Seabirds.* Accessed 03/09/07 @ http://alaska.fws.gov/mbsp/mbm/seabirds/seabirds.htm.

USFWS. 2007b. *All About Piping Plovers.* Accessed 03/11/07 @ http://www.fws.gov/plover/facts.html.

USFWS. 2007c. *Tarahumara Frog.* Accessed 03/26/07 @ www.fws.gov/southwest/es.arizona/T_Frog_SpeciesAccount.htm.

USFWS. 2007d. *Discover Freshwater Mussels: America's Hidden Treasure.* Accessed 04/11/07 @ http://www.fws.gov.news/mussels.html.

USFWS. 2008. *Red-cockaded Woodpecker.* Accessed 11/16/10 @ http://www.fws.gov/northeast/.

USFWS and Canadian Wildlife Service. 1986. *North American Waterfowl Management Plan.* Washington, DC: U.S. Fish and Wildlife Service.

USFWS and Canadian Wildlife Service. 1994. *North American waterfowl management plan 1994 update, expanding the commitment.* Washington, DC: U.S. Fish and Wildlife Service.

USGS. 1995. *Our Living Resources: Birds.* Accessed 03/06/07 @ http://biology.usgs.gov/t+s/index.

USGS. 2007. *Pintails! What Are They?* Accessed 4/15/07 @ http://www.fws.gov.

Vaughan, M.R., and M.R. Pelton. 1995. Black Bears in North America. *In* Our Living Resources. Washington, DC: USGS.

Vermeer, K., Briggs, K.T., Morgan, K.H., and Siegel-Causey, D., eds. 1993. *The status, ecology, and conservation of marine birds of the North Pacific.* Proceedings of a Pacific Seabird group Symposium. Ottawa, Ontario: .Canadian Wildlife Service Spec. Publ.

Vogl, R.J. 1974. Effect of fire on grasslands. Pages 139–194 *in* T.T. Kozlowski and C.E. Ahlgren, eds. *Fire and Ecosystems.* New York: Academic Press.

Wake, D.B. 1991. Declining amphibian populations. *Science* 253(5022):860.

Wake, D.B., and H.H. Morowitz. 1991. Declining amphibian populations—a global phenomenon? Findings and recommendations. *Alytes* 9(1):33–42.

Warren, M., and H.M. Burr. 1994. Status of freshwater fishes of the United States: Overview of an imperiled fauna. *Fisheries* 19(1):6–18.

Webb, D.W. 1995. Biodiversity degradation in Illinois stoneflies. *In* Our Living Resources. Washington, DC: USGS.

Welch, D.W., and R.C. Beamesderfer. 1993. Maturation of female white sturgeon in lower Columbia River impoundments. Pages 89–108 *in* R.C. Beamesderfer and A.A. Nigro, eds. Status and habitat requirements of the white sturgeon populations in the Columbia River downstream form McNary Dam, Vol 2. Final Report to Bonneville Power Administration, Portland, OR.

Wheeler, Q.D. 1990. Insect diversity and cladistic constraints. *Annals of the Entomological Society of America* 83(6):1031–1047.

Wilbur, S.R. 1978. *The California condor, 1966–76: A look at its past and future.* U.S. Fish and Wildlife Service North American Fauna 72.

Williams, E.S., E.T. Thorne, M.J.G. Appel, and D.W. Belitsky. 1988. Canine distemper in black-footed ferrets from Wyoming. *Journal of Wildlife Diseases* 24:385–398.

Williams, J.D., and R.J. Neves. 1995. Freshwater mussels: A neglected and declining aquatic resource. *In* Our Living Resources. Washington, DC: USGS.

Williams, J.D., M.L. Warren Jr., K.S. Cummings, J.L. Harris, and R.J. Neves. 1993. Conversation status of freshwater mussels of the United States and Canada. *Fisheries* 18(9):6–22.

Williams, J.E., J.E. Johnson, D.A. Hendrickson, S. Contreras-Balderas, J.D. Williams, M. Navarro-Mendoza, D.E. McAllister, and J.E. Deacon. 1989. Fishes of North America endangered, threatened, or of special concern. 1989. *Fisheries* 14(6):2–20.

Wilson, D.E., M.A. Bogan, R.L. Brownell Jr., A.M. Burdin, and M.K. Maminov. 1991. Geographic variation in sea otters. *Enhydra Iuris. Journal of Mammalogy* 2(1):22–26.

Woodward, A.R., and Moore, C.T. 1995. American Alligators in Florida. *In* Our Living Resources. Washington, DC: USGS.

Wooten, A. 1984. *Insects of the World*. New York: Facts on File, Inc.

Wright, S. 1931. Evolution in Mendelian populations. *Genetics* 16:97–159.

Wright, S. 1978. *Evolution and the genetics of populations*. Vol. 4, Variability within and among natural populations. Chicago: University of Chicago Press.

VI

WILDERNESS

How vain it is to sit down to write when you have not stood up to live.

—Henry David Thoreau (1817–1862)

18

Land Organism

The outstanding scientific discovery of the twentieth century is not the television, or radio, but rather the complexity of the land organism. Only those who know the most about it can appreciate how little is known about it.

—Aldo Leopold, *A Sand County Almanac* (1949)

After lunch one fall day, Aldo Leopold and his crew of surveyors opened fire on an old mother wolf and her six adolescent pups at the foot of a mountain. "In those days we had never heard of passing up a chance to kill a wolf," Leopold later said. "I thought that because few wolves meant more deer, that no more wolves would mean a hunters' paradise." But after seeing the "fierce green fire" in the wolf's eyes die out, he wrote, "I sensed that neither the wolf nor the mountain agrees with such a view."

—James William Gibson, on Aldo Leopold's 1909 wolf hunt (2009)

Maintaining contact with wild nature is the essence of wilderness.

—Frank R. Spellman, *Ecology for Non-Ecologists* (2008)

What Is Wilderness?

T HIS IS ONE of those questions philosophers, scientists, conservationists, and writers have struggled with for years. Personally, having spent countless hours, days, weeks, and even months at a time in what others have little difficulty in describing as the "wilderness," we can attest to the difficulty others have in defining the term—the concept, that is. This is the case, of course, because what others describe as wilderness we describe as paradise.

As a general matter, wilderness (derived from Old English, meaning not controllable by humans) is defined as a natural environment on Earth that has not been significantly modified by human activity. That is, an area which has not been subjected to the heavy footprint of humankind. How heavy is humankind's footprint? Well, as an example we can point out that in 1992, E.O. Wilson estimated that there were a billion billion insects alive around the planet, comprising a biomass of around a trillion kilograms. That's a lot—but not much more than the weight of all humankind; keep in mind, however, that we've continued to breed heavily since then. The point is it is staggering to try to comprehend that a single species weighs as much as all the members of another species. Indeed, the footprint of humankind is heavy; it is large, damaging, and sometimes irreversible.

Probably the best definition of wilderness is provided by the WILD Foundation: "The most intact, undisturbed wild natural areas left on our planet—those last truly wild places that humans do not control and have not developed with roads, pipelines or other industrial infrastructures."

If we accept Leopold's description of the wilderness as the "land organism," or more properly stated that wilderness is one part of the land organism, then the master mosaic of its entire component parts jell in harmony with each other. Leopold's concept of wilderness does not overlook the significant role it plays in the overall health of interrelated ecosystems. Rare and endangered animal and plant species require habitats that are relatively undisturbed so gene pools can be sustained, adaptations made, and populations maintained (Wildlink 2010). This can be accomplished whenever a location is biologically intact and/or whenever a place is legally protected (like a national park, monument, or forest area) so that it remains free of industrial infrastructure, and wild, and open to low impact recreation.

Well, when you consider our description above of what we and others define a wilderness area as being, notice we did not state what a wilderness area is not. One might perceive that a prime wilderness area such as Yellowstone National Park, Glacier National Park, Zion National Park, Sequoia, Yosemite, Death Valley, The Blue Ridge Parkway, and Appalachian Trail are biologically pristine locations. However, this is not necessarily the case. Any place on earth that in some way, shape, or form has been impacted by humans is not pristine. On the other hand, areas such as those listed above are mainly biologically intact, with minor human impact or minor historical human activity not disqualifying them from being considered wilderness. Even the presence of human inhabitants within its boundaries does not necessarily disqualify it from being wilderness: many wild areas are inhabited by indigenous populations—these play a key role in keeping an area free of development; thus, keeping the wilderness intact and wild.

In the sections that follow, we deliberately select and describe wilderness topics and locations that encompass the scientific value, watershed benefits, life support systems, historic, cultural, spiritual, aesthetic, recreational, refuge, and educational values that truly characterize wilderness.

U.S. Forest Resources

I wonder if they like it—being trees?
I suppose they do.
It must feel so good to have the ground so flat,
And feel yourself stand straight up like that.
So still in the middle, and then branch at ease.
Big boughs that arch, small ones that bend and blow,
And all those fringy leaves that flutter so.
You'd think they'd break off at the lower end.
When the wind fills them, and their great heads bend.
But when you think of all the roots they drop,
As much at bottom as there is on top,
A double tree, widespread in earth and air,
Like a reflection in the water there.

—Charlotte Perkins Stetson

According to USDA (2004), it is estimated that in 1630 the area of **forest land** in the United States was just over 1 billion acres or about 46% of the total land area. By 1907, the area of forest land had declined to an estimated 759 million acres or 34% of the total land area. Forest area has been relatively stable since 1907. In 2002, forest land comprised 749 million acres, or 33% of the total land area of the United States. Since 1630, there has been a net loss of 297 million acres of forest land, predominantly due to agricultural conversions. Nearly two-thirds of the net conversion to other uses occurred in the last half of the nineteenth century when an average of 13 square miles of forest was cleared every day for 50 years.

Darr (1995) points out that the secretary of agriculture is directed by law to make and keep current a comprehensive inventory and analysis of the present and prospective conditions of and requirements for the renewable resources of U.S. forests and rangelands. This inventory includes all forests and rangelands, regardless of ownership. The work is carried out by people in the

Forest Inventory and Analysis program of the U.S. Department of Agriculture Forest Service (USFS).

Did You Know?

Inventories provide key forest resource information for planners and policy makers. Increasingly, people turn to these inventories for information on biological diversity, forest health, and developmental decisions.

According to USFS (1992), information is collected from over 130,000 permanent sample plots selected to assure statistical reliability. Vegetation on the plots is measured on average about every 10 years. Characteristics of the vegetation and land are measured, including ownership, productivity for timber production, the kinds and sizes of trees, how fast trees are growing, whether any trees have died from natural causes, and whether they have been cut.

It is important to note that stable forest area does not mean that there has been no change in the character of the forest. There have been shifts from agriculture to forests and vice versa. Some forest land has been converted to more intensive uses, such as urban. Even where land has remained in forest use, there have been changes as forests respond to human manipulation, aging, and other natural processes (USDA 2004).

As mentioned, over the years, the U.S. forest cover has changed because of the way people use and manage forest land. Today, about 33% of the U.S. land area or 298 million ha (737 million acres), is forest land, about two-thirds of the forested area in 1600. Since 1600, some 124 million ha (307 million acres) of forest land have been converted to other uses, mainly agricultural. More than 75% of this conversion occurred in the nineteenth century, but by 1920, clearing forests for agriculture had largely halted (Darr 1995).

Some 34% of all forest land is federally owned and managed by the U.S. Forest Service, the Bureau of Land Management, and other federal agencies. The rest is owned by nonfederal public agencies, forest industry, farmers, and other private individuals. About 19 million ha (47 million acres; 6% of all U.S. forest land) are reserved from commercial timber harvest in wilderness, parks, and other land classifications.

Forest land is widely but unevenly distributed. North Dakota has the smallest percentage of forest cover (1%) and Maine has the greatest (8%). Forest areas vary greatly from sparse scrub forests of the arid interior West to the highly productive forests of the Pacific coast and the South, and from pure **hardwood** forests to multispecies mixtures and coniferous forests. In total, 57% of the forest land is east of the Great Plains states. In the East, the oak-hickory forest type group is the most common. Figure 51 shows Mabry Mill in a portion of the Blue Ridge Parkway, Virginia. The forest near and around Mabry Mill is home to a mix of plant life, from algae to oak trees. The northern Blue Ridge Mountains have about 1600 different species of higher plants. Fewer than one hundred of these are trees and shrubs that make up the dominant vegetation that is visible year round. The Blue Ridge region provides the requirements of deciduous trees such as oaks, hickories, and maples. Much of Shenandoah today is an oak-hickory forest. The forest would be incomplete, however, without rose azalea, jack-in-the-pulpit, interrupted fern, lady slipper orchid, and British soldiers lichen (Shenandoah 2002). In the West, the category referred to as "other **softwoods**" is most common (Darr 1995).

Timberland forests are logged for timber, plywood, and paper products. This timberland is generally the most productive and capable of producing at least 1.4 m³ of industrial wood per hectare a year (20 ft³/acre) and is not reserved from timber harvest (Powell et al. 1993). Two-thirds of the nation's forested ecosystems (198 million ha or 490 million acres) are classified as timberland. Because of historical interest in timber production, more information

FIGURE 51. Mabry Mill, Blue Ridge Parkway, Virginia
Photo by Frank R. Spellman

is available for the characteristics of timber inventories on timberland than for other forest land.

USDA (2004) points out that U.S. forests provide wildlife habitats and thereby support biodiversity; take carbon out of the air and thus serve as carbon sinks; and provide the outdoor environments desired by many people for recreation. Moreover, gathering nontimber forest products is a significant use of the nation's forests that affects forest ecosystems. These products include medicinals, food and forage species, floral and horticultural species, resins and oils, art and craft species, game animals, and fur bearers. Harvest of these products from forest ecosystems is a significant and very important activity for many Americans, for recreational, commercial, subsistence, and cultural uses.

> Maybe a vision of the original longleaf pine flatwoods has been endowed to me through genes, because I seem to remember their endlessness. I seem to recollect when these coastal plains were one big, brown-and-tan, daybreak-to-dark longleaf forest. It was a monotony one learned to love . . . with the passing years. A forest never reveals its secrets but reveals them slowly over time, and a longleaf forest is full of secrets.

> —Janisse Ray, *Ecology of a Cracker Childhood* (1999)

Forested Wetlands

If the two words, forested wetlands, are separated they are somewhat distinct; if they separate, we tend to define each in stand alone fashion. On the surface, forested is rather easy to comprehend and to define and to understand and to accept without explanation: a place with a lot of trees. Wetland, however, is a matter (a place or region) of a much different sort. The term wetland conjures up many internal manifestations of descriptive thoughts (mostly not pleasant): Our human micro-chip warehouse displays for us, in regards to wetlands, a scene of mosquito- and yellow fly-infested, dreary, dismal, and soggy piece of ground too deep for a human to wade in, too shallow for a boat to draw, too tangled for passage (see Figure 52). As Ray (2005) puts it, "a wetland is simply a natural feature full of natural features." The problem is most of us have difficulty recognizing, understanding, and/or defining natural features.

European settlers encountered these "natural features" in many parts of the southern United States where the landscape is largely composed of forested wetlands. These wetlands were a major feature of river floodplains and isolated depressions or basins or pocosins from Virginia to Florida, west to eastern Texas and Oklahoma, and along the Mississippi River to southern Illinois. Based on the accounts of pre-twentieth-century naturalists such as Audubon,

FIGURE 52.   Wetland Mill Pond, Blue Ridge Parkway, Virginia
Photo by Frank R. Spellman

Banister, John and William Bartram, Brickell, and Darby, the flora and fauna of many wetlands were unusually rich even by pre-colonial standards (Wright and Wright 1932). These early travelers described vast unbroken forests of oaks, ashes, maples, and other tree species, many with an almost impassable under-story of saplings, shrubs, vines, switch cane, and palmetto. Low swampy areas with deep, long-term flooding were dominated by bald cypress and tupelo and typically had sparse understories (Keeland et al. 1995).

Most southern forested wetlands fall in the broad category of bottomland hardwoods, characterized and maintained by a natural hydrological regime of alternating annual wet and dry periods and soils that are saturated or inundated during a portion of the growing season. Variations in elevation, hydroperiod, and soils result in a mosaic of plant communities across a floodplain. Wharton et al. (1982) classified bottomland hardwoods into 75 community types, including forested wetland types such as Atlantic white cedar bogs, red maple and cypress-tupelo swamps, pocosins, hydric hammocks, and Carolina bays.

Realistic estimates of the original extent of forested wetlands are not available because accurate records of wetlands were not maintained until the early twentieth century, and many accounts of wetland size were little more than speculation (Dahl 1990). Klopatek et al. (1979) estimated the pre-colonial forested wetland area of the United States to be about 27.2 million ha (67.2 million acres), but Abernathy and Turner (1987) suggested that this figure was low because it ignored small isolated wetlands.

Did You Know?

Albemarle/Pamlico Coastal region in North Carolina is home to an *Atlantic White Cedar Bog region.* This is a unique habitat that has naturally acidic waters and is cooler than surrounding hardwood swamps or pinelands. Cedar bogs support large breeding bird populations (USFWS 1998). *Pocosin,* or swamp on a hill, is a seemingly flat area that rises slightly in the center, forming a raised bog (U.S. Army Corps of Engineers 1998). *Hydric hammocks* are forested wetlands (swamps) that are dominated by a mixture of primarily hardwood tree species with sabal palms. Hydric hammocks are typically found in areas where limestone is close to the soil surface. Soils are variable; often a clay layer or limestone layer helps keep the soil saturated for long periods (Spellman 1998).

Fire Regimes within Fire-adapted Ecosystems

Fires ignited by people or through natural causes have interacted over evolutionary time with ecosystems, exerting a significant influence on numerous ecosystem functions (Pyne 1982). Fire recycles nutrients, reduces biomass, influences insect and disease populations, and is the principal change agent affecting vegetative structure, composition, and biological diversity. As humans alter fire frequency and intensity, many plant and animal communities experience a loss of species diversity, site degradation, and increases in size and severity of wildfires (Ferry et al. 1995). This section examines the role fire plays in the ecological process around which most North American ecosystems evolved.

In order to understand the influence fire can have on an ecosystem, it is important to know the basics of fire regime. Fire regimes are considered as the total pattern of fires over time that is characteristic of a region or ecosystem (Kilgore and Heinselman 1990). A fire regime describes the pattern that fire follows in a particular ecosystem. It consists of the following components (Bond and Keeley 2005):

1. **Fuel Consumption and Spread Patterns**—Fire can burn at three levels. Ground fires burn through soil that is rich inorganic matter. Surface fires burn through dead plant materials that are on the ground. Crown fires burn in the tops of shrubs and trees. Ecosystem may experience mostly one level of fire or a mix of the three.
2. **Intensity**—Defined as the energy release per unit length of fireline. Can be estimated as the product of linear spread rate, low heat of combustion, and combusted fuel mass per unit area.
3. **Severity**—Term used by ecologists to refer to impact that a fire has on an ecosystem (estimate of plant mortality).
4. **Frequency**—This is a measure of how common fires are in a given ecosystem (interval between fires at a given site, or the amount of time it takes to burn an equivalent of a specified area).
5. **Seasonality**—This refers to the time of year during which fires are most common.

In the following, five plant communities are discussed: the sagebrush steppe, juniper woodlands, ponderosa pine forest, lodgepole pine forest, and the southern pineland. Status and trends of altered fire regimes in fire-adapted ecosystems highlight the role that fire plays in wildland stewardship. Fire regimes are considered as the total pattern of fires over time that is characteristic of a region or ecosystem (Kilgore and Heinselman 1990).

1. **Sagebrush-grass Plant Communities**—Ferry et al. (1995) point out greater frequency of fire has seriously affected the sagebrush steppe during the last 50 years. One such community, the semi-arid intermountain sagebrush steppe, encompasses about 45 million ha (112 million acres). After repeated fires, nonnative European annual grasses such as cheatgrass and medusa-head now dominate the sagebrush steppe (West and Hassan 1985). It is unclear whether cheatgrass invasion, heavy grazing pressure, or shorter fire return intervals initiated the replacement of perennial grasses and shrubs by the non-native annual grasses. It is clear, however, that wildfires aid in replacing native grasses with cheatgrass, as well as causing the loss of the native shrub component (Whisenant 1990). Inventories show that cheatgrass is dominant on about 6.8 million ha (17 million acres) of the sagebrush steppe and that it could expand into an additional 25 million ha (62 million acres) in the sagebrush steppe and the great basin sagebrush type (Pellant and Hall 1994).
2. **Western Juniper Woodlands**—Juniper woodlands occupy 17 million ha (42 million acres) in the intermountain region (West 1988). Juniper species common to this region are western juniper, Utah juniper, single-

seeded juniper, and Rocky Mountain juniper. Presettlement juniper woodlands were usually savanna-like or confined to rocky outcrops not typically susceptible to fire (Nichol 1937).

Juniper woodlands began increasing in both density and distribution in the late 1800's (R.F. Miller unpublished data) because of climate, grazing, and lack of fire (Miller and Waigand 1994). Warm and wet climate conditions then were ideal for juniper and grass seed production. Fire frequency had decreased because the grazing of domestic livestock had greatly reduced the grasses and shrubs that provided fuel, and relocation of Native Americans eliminated an important source of ignition. Continued grazing and 50 years of attempted fire exclusion have allowed juniper expansion to go unchecked.

3. **Ponderosa Pine Forest**—Decreases in fire frequency are also seriously affecting ponderosa pine forests, a common component on about 16 million ha (40 million acres) in the western United States. Historically, the ponderosa pine ecosystem had frequent, low-intensity, surface fires that perpetuated park-like stands with grassy undergrowth (Barrett 1980). Ponderosa pine is a tree that is well adapted to fire. As ponderosa pines get older, their bark gets thicker. Thick bark protects the cambium layer from being burned. They generally have fewer branches low on their trunks so fire can't burn into the tops of the trees. The trees are spaced widely so one tree can't catch another on fire, and have a deep taproot (BNF 2004). For six decades, humans attempted to exclude fire on these sites (OTA 1993). Fifty years ago, Weaver (1943) stated that complete prevention of forest fires in the ponderosa pine region had undesirable ecological effects and that already-deplorable conditions were becoming increasingly serious. Today, many ponderosa pine forests are overstocked, plagued by epidemics of insects and disease, and subject to severe stand-destroying fires (Mutch et al. 1993).

4. **Lodgepole Pine Forest**—Like ponderosa pine forests, lodgepole pine forests are experiencing a change in structure, distribution, and functioning of natural processes because of fire exclusion and increases in disease. Wildfire may be the most important factor responsible for establishment of existing stands (Wellner 1970). Historical stand-age distributions in lodgepole pine forests indicated an abundance of younger age classes resulting from periodic fires. Fire exclusion, by precluding the initiation of new stands, is responsible for a marked change in distribution of age classes in these forests.

Dwarf mistletoe, the primary disease of lodgepole pine, also has a profound effect on forest structure and function, although it occurs slowly. Data show that chronic increases of dwarf mistletoe are partly due to the exclusion of fire (Zimmerman and Laven 1984) because fire is the natural control of dwarf mistletoe and has played a major role in the distribution and abundance of current populations and infection intensities (Alexander and Hawksworth 1975). As the frequency and extent of fire have decreased in lodgepole pine stands over the last 200 years, dwarf mistletoe infection intensity and distribution are clearly increasing (Zimmerman and Laven 1984).

5. **Southern Pinelands**—In contrast to the juniper, ponderosa pine, and lodgepole pine communities, fire frequencies have not drastically decreased in the 78 million ha (193 million acres) of southern pinelands. These pinelands are composed of diverse plant communities associated with long-leaf, slash, loblolly, and shortleaf pines. Fire has continued on an altered basis as an ecological process in much of the southern pinelands; historically, fire burned 10%–30% of the forest annually (Wright and Bailey 1982); the southern culture never effectively excluded fire for its pinelands (Pyne 1982), although human-ignited fires have partially replaced natural fires. Consequently, the number of fires has been reduced and the season of burns has changed from predominately growing-season to dormant-season (fall or winter) fires (Robbins and Myers 1992). Altering the burning season and frequency has significantly affected southern pineland community structure, composition, and biological diversity.

The role of fire becomes more complex as it interacts with land management. Maintaining interactions between disturbance processes and ecosystem functions is emphasized in ecosystem management. It is vital for managers to recognize how society influences fire as an ecological process. In addition, managers must uniformly use information on fire history and fire effects to sustain the health of ecosystems that are both fire-adapted and fire-dependent. Managers must balance the suppression program with a program of prescribed fire applied on a landscape scale if we are to meet our stewardship responsibilities (Ferry et al. 1995).

Vegetation Change in National Parks

Natural ecosystems are always changing, but recent changes in the United States have been startlingly rapid, driven by 200 years of disturbances accompanying settlement by an industrialized society. Logging, grazing, land clearing, increased or decreased frequency of fire, hunting of predators, and other changes have affected even the most remote corners of the continent. Recent trends can be better understood by comparison with more natural past trends of change, which can be reconstructed from fossil records. Conditions before widespread impacts in a region are termed "presettlement"; conditions after the impacts are "postsettlement" (Cole 1995a).

Fossil plant materials from the last few thousand years are used to study past changes in many natural areas. Pollen buried in wetlands, for example, can reveal past changes in vegetation (Faegri and Iverson 1989), and larger fossil plant parts can be studied in deserts where the fossilized plant collections of packrats, called packrat middens, have been preserved (Betancourt et al. 1990).

This section summarizes the rates of vegetation change in four national park areas over the last 5,000 years as reconstructed from fossil pollen and packrat middens. These four national park areas from different ecological regions demonstrate the flexibility of these paleoecological techniques and display similar results (Cole 1995a).

1. **Northern Indiana Prairie**—A 4,500-year history of vegetation change was collected from Howes Prairie Marsh, a small marsh surrounded by prairie and oak savanna in the Indiana Dunes Nations Lakeshore near the southern tip of Lake Michigan. Only 40 km (25 mi) from Chicago, this area has been affected by numerous impacts from settlements but still supports comparably pristine tall-grass prairie vegetation as well as the endangered Karner blue butterfly (an endangered species native to the Great Lakes region of the U.S.). Although this site has experienced more disturbances than any of the others described here, it is a most valuable site because of its many species (Wilhelm 1990) and its tall-grass prairie vegetation that has been nearly eliminated elsewhere.

2. **Northern Michigan Forest**—A similar analysis was carried out on pollen from a small bog (12-Mile Bog) surrounded by pine forest along the southern shore of Lake Superior. This site, within Pictured Rocks National Lakeshore, was more severely affected by logging and slash burning in the 1890's than by the periodic wildfires that characterized this forest earlier, but it has been protected for the last 80 years. The magnitude of change caused by the crude logging and slash burning of the logging era was far greater than any recorded during the 2,500 years since Lake Superior receded to create the forest of white and red pine (Cole 1995a).

3. **California Coastal Sage Scrub**—Fossil pollen was analyzed from an estuary on Santa Rosa Island off the coast of southern California (Cole and Liu 1994). The semi-arid landscape around the estuary is covered with coastal sage scrub, chaparral, and grassland. This site, within Channel Islands National Park, is one of the least affected areas in this region of

rapidly expanding urbanization, although the island's native plants and animals were not well adapted to withstand the grazing of the large animals introduced with the ranching era of the 1800's. This island, which had no native large herbivores, became populated with thousands of sheep, cattle, horses, goats, pigs, deer, and elk. The National Park Service is removing many of the large herbivores, although most of the island remains an active cattle ranch (Cole 1995a).

4. **Southern Utah Desert**—Because fossil pollen is usually preserved in accumulating sediments of wetlands, different paleoecological techniques are necessary in arid areas. In western North America, fossil deposits left by packrats have proven a useful source of paleoecological data (Betancourt et al. 1990). Past desert vegetation can be reconstructed by analyzing bits of leaves, twigs, and seeds collected by these small rodents and incorporated into debris piles in rocks, shelters, or caves. These debris piles can be collected, analyzed, and radiocarbon dated (Cole 1995a).

The vegetation history of a remote portion of Capitol Reef National Park (Harnett Draw) was reconstructed through the analysis of eight packrat middens ranging in age from 0 to 5,450 years (Cole 1995b). The vegetation remained fairly stable throughout this period until the last few hundred years. The most recent deposits contain many plants associated with overgrazed areas such as whitebark rabbitbrush, snakeweed, and greasewood, which were not recorded at the site before settlement.

Conversely, other plants that are extremely palatable to grazing animals were present throughout the last 5,450 years, only to disappear since settlement. Plant species preferred by sheep and cattle, such as winterfat and rice grass, disappeared entirely, while many other palatable plant species declined in abundance after 5,000 years of comparative stability.

The past rates of vegetation change for this site were calculated in a manner similar to the fossil pollen records. Although the rate of change calculation is less precise than the fossil pollen records because there were fewer samples, the results show a similar pattern. The rate of vegetation change is highest between the two most recent records (Cole 1995a).

Although this area is still grazed by cattle today through grazing leases to private ranchers from the National Park Service, much of the severe damage was probably done by intensive sheep grazing during the late 1800's when the entire region was negatively affected by open-land sheep ranching. We cannot yet demonstrate whether the grazing effects are continuing or if the site is improving, although reinvasion of palatable species is unlikely in the face of even light grazing. Severe overgrazing is required to eliminate abundant palatable species, but once they are eliminated, even light grazing can prevent their restoration (Cole 1995a).

To successfully meet human needs, competing demands for the use of the land's resources must be resolved. Experience has shown that wise land management decisions are more likely to be made if land managers understand a site and are able to place the status quo into a historical perspective. Because the ultimate goal for the management of many areas is to mitigate settlement impacts and return the land to its presettlement status, detailed knowledge of the effects of settlement is imperative.

Whitebark Ecosystem

USGS (2003) reports whitebark pine trees produce large seeds high in fats that are an important source of food for many animal species. Red squirrels and Clark's nutcrackers usually harvest the lion's share of whitebark pine seeds. Grizzly bears and black bears raid squirrel caches that contain whitebark cones to get pine seeds, one of their favorite foods (Kendall and Arno 1990). Other mammals, large and small, and many species of birds also feed on whitebark pine seeds, pine nuts, as they commonly are called. Because whitebark pine are long-lived and can grow large trunks, they provide valuable cavities for nesting squirrels, northern flickers (a type of woodpecker; it is the only ground feeding woodpecker), and mountain bluebirds.

In addition to its importance as a wildlife resource, whitebark pine trees play other pivotal roles in the ecosystems where they occur. Because whitebark pine can grow in dry, windy, and cold sites where no other trees can establish, it pioneers many harsh **subalpine** and alpine sites and commonly dominates treeline communities in the northern Rocky Mountains. Its presence can modify the microclimate in alpine ecotone communities enough to allow other trees to establish, such as subalpine fir. Whitebark pine communities also influence the **hydrology** of the drainages where they occur. Because they grow on windy ridges and in other mountainous areas and have broad crowns, whitebark pine trees tend to act as snow fences and are responsible for significant amounts of high-elevation snow accumulation. This results in delayed snow melt and extends ephemeral stream flow periods (USGS 2003).

Whitebark pine is well-suited to harsh conditions and populates high-elevation forests in the northern Rocky Mountains, North Cascade, and Sierra Nevada ranges. These pine trees are adapted to cold, dry sites and pioneer burns and other disturbed areas. At timberline, they grow under conditions tolerated by no other tree species, thus playing an important role in snow accumulation and persistence. Because few roads occur in whitebark pine eco-systems and because the tree's wood is of little commercial interest, informa-tion on the drastic decline of this picturesque tree has only recently emerged (Kendall 1995).

Whitebark pine is threatened by an introduced disease and fire suppression. In its northern range, many whitebark pine stands have declined by more than 90%. The most serious threat to the tree is from white pine blister rust, a non-native fungus that has defied control. Less than one whitebark pine tree in 10,000 is rust-resistant. Mortality has been rapid in areas like western Montana, where 42% of whitebark pine trees have died from the disease in the last 20 years; 89% of the remaining trees are infected with rust (Keane and Arno 1993). Although drier conditions have slowed the spread of blister rust in the white-bark pine's southern range, infection rates there are increasing and large die-offs are eventually expected to occur.

Before fire suppression, whitebark pine stands burned every 50–300 years. Under current management, they will burn at 3,000-year intervals. Without fire, serial whitebark pine trees are replaced by shade-tolerant conifers and become more vulnerable to insects and disease (Kendall 1995).

The alarming loss of whitebark pine has broad repercussions: mast for wild-life is diminished and the number of animals the habitat can support is re-duced. Such results hinder grizzly bear recovery and may be catastrophic to Yellowstone grizzlies for whom pine seeds are a critical food. Predicted changes in whitebark pine communities include the absence of reforestation of harsh sites after disturbance and lower treelines. In addition, stream flow and timing will be altered as snowpack changes with vegetation.

Whitebark pine will be absent as a functional community component until rust-resistant strains evolve. Natural selection could be speeded with a breed-ing program like that developed for western white pine, which also suffers from rust. In some areas where whitebark pine is regenerating, its competitors should be eliminated. To perpetuate whitebark pine at a landscape scale, fires must be allowed to burn in whitebark pine ecosystems.

Wisconsin Oak Savanna

The term *savanna* has never been well defined; maybe loosely at best. Cole (1960) summed up the situation this way: "Perhaps of all types of vegetation the savanna is the most difficult to define, the least understood, and the one whose distribution and origin is the most subject to controversy." Henderson and Epstein (1995) point out that even though not well-defined, oak savanna is well-recognized. It is a class of North American plant communities that was part of a large transitional complex of communities between the vast treeless prairies of the West and the deciduous forests of the East. This system or mosaic was driven by frequent fires and possibly influenced by large herbivores (ungulates) such as bison and elk. Oaks were the dominant trees, hence the term *oak savanna*. A wide rage of community types was found within this transitional complex; collectively, they represented a continuum from prairie to forest, with no clear dividing lines between savanna and other community types (Curtis 1959).

Savannas all have a partial canopy of open grown trees and a varied ground layer of prairie and forest herbs, grasses, and shrubs, as well as plants restricted to the light shading and mix of shade and sun so characteristic of savanna. Oaks were clearly the dominant trees, and, hence, the common use of the term oak savanna. Definitions of savanna tree cover range from 5% to 80% canopy; however, the lower canopy covers of 5%–50% or 5%–30% are more widely used criteria. Savanna types range from those associated with dry, gravelly, or sandy soils; those on rich, deep soils; and those on poorly drained, moist soils (Henderson and Epstein 1995).

Oak savannas have probably been in North America for some 20–25 million years (Barry and Spicer 1987), expanding and contracting with climatic changes and gaining and losing species (on a geologic time scale) through evolution and extinction. For the past several thousand years, such savannas have existed as a relatively stable band of varying width and continuity, from northern Minnesota to central Texas.

At the time of European settlement (ca. 1830), oak savanna covered many millions of hectares. It varied somewhat in species composition from north to south and east to west, but structure and functions were probably similar throughout. In the upper Midwest (Minnesota, Wisconsin, Michigan, Iowa, Illinois, Indiana, and Missouri) there were an estimated 12 million ha (29.6 million acres; Curtis 1959). As the Midwest's rich soil was used for agriculture and fire was suppressed, this ecosystem all but disappeared from the landscape throughout its range. Today, oak savanna is a globally endangered ecosystem (Henderson and Epstein 1995).

References and Recommended Reading

Abernathy, Y., and R.E. Turner. 1987. U.S. forested wetlands: 1940–1980. *BioScience* 37:721–727.

Alexander, M.E., and F.G. Hawksworth. 1975. *Wildland fires and dwarf mistletoe: a literature review of ecology and prescribed burning.* U.S. Forest Service Gen. Tech. Rep. RM-14, Rocky Mountain Forest Range Experiment Station, Fort Collins, CO. 12 pp.

Arno, S.F., and R.J. Hoff. 1989. *Silvics of whitebark pine*. U.S. Forest Service Gen. Tech. Rep. INT-235. 11 pp.

Barrett, S.W. 1980. Indians and fire. *Western Wildlands* 6(3):17–21.

Barry, A.T., and R.A. Spicer. 1987. *The evolution and palacobiology of land plants*. Croom Helm, London. 309 pp.

Betancourt, J.L., T.R. Van Devender, and P.S. Martin, eds. 1990. *Packrat middens: the last 40,000 years of biotic change*. The University of Arizona Press, Tucson, 467 pp.

BNF. 2004. *Trees: Ponderosa Pine. Boise Nation Forest*. Accessed 05/04/07 @ http://www.Fs.fed.us/r4/boise/local-resources/trees/p-pine.shtml.

Bond, W.J., and J.E. Keeley. 2005. Fire as a global 'herbivore': the ecology and evolution of flammable ecosystems. *Trends in Ecology and Evolution* 20:387–394.

Boyd, R.J. 1995. Terrestrial Ecosystems. *In* Our Living Resources. Washington, DC: USGS.

Burgess, R.I., and D.M Sharpe, eds. 1981. *Forest Island dynamics in man-dominated landscapes*. Springer-Verlag, New York. 310 pp.

Campbell, N.A. 1996. *Biology*, 4th ed. The Benjamin/Cummings Publishing Company, Inc., Menlo Park, CA.

Cole, K. 1995a. Vegetation change in National Parks. *In* Our Living Resources. Washington, DC: USGS.

Cole, K. 1995b. *A survey of the fossil packrat middens and reconstruction of the pregrazing vegetation of Capitol Reef National Park*. National Park Service Res. Rep. In press.

Cole, K.L., D.R. Engstrom, R.P. Futyma, and R. Stottlemyer. 1990. Past atmospheric deposition of metals in northern Indiana measured in a peat core from Cowles Bog. *Environmental Science and Technology* 24:543–549.

Cole, K.L., and G. Liu. 1994. Holocene paleoecology of an estuary on Santa Rosa Island, California. *Quaternary Res.* 41:326–335.

Cole, M.M. 1960. Cerrado, caating and pentanal: the distribution and origin of the savanna vegetation of Brazil. *Geogaph. J.* 126:168–79.

Curtis, J.T. 1959. *The vegetation of Wisconsin*. Univ. Wis. Press, Madison. 657 pp.

Dahl, T.E. 1990. Wetlands losses in the United States 1780's to 1980's. U.S. Fish and Wildlife Service, Washington, DC, 21 pp.

Darr, D.R. 1995. U.S. Forest Resources. *In* Our Living Resources. Washington, DC: USGS.

Elfring, C. 1989. Yellowstone: fire storm over fire management. *BioScience* 39(10):667–672.

Faegri, R., and I. Iversen. 1989. *Textbook of pollen analysis*. Wiley and Sons, New York. 328 pp.

Ferry, G.W., R.G. Clark, R. E. Montgomery, R.W. Mutch, W.P. Leenhouts, G. T. Zimmerman. 1995. Altered fire regimes within fire-adapted ecosystems. *In* Our Living Resources. Washington, DC: USGS.

Gibson, James William. 2009. Lessons from Aldo Leopold's historic wolf hunt. Los Angeles Times, December 13. http://articles.latimes.com/2009/dec/13/opinion/la-oe-gibson13-2009dec13

Grossman, D.H., and K. L. Goodin. 1995. Rare Terrestrial ecological communities of the United States. *In* Our Living Resources. Washington, DC: USGS.

Harris, L.D. 1984. *The fragmented forest*. The University of Chicago Press, Chicago. 211 pp.

Henderson, R.A., and E.J. Epstein. 1995. Oak Savannas in Wisconsin. *In* Our Living Resources. Washington, DC: USGS.

Jacobson, G.L., Jr., and E.C. Grimm. 1986. A numerical analysis of Holocene forest and prairie vegetation in central Minnesota. *Ecology* 67:958–966.

Keane, R.R., and S.F. Arno. 1993. Rapid decline of whitebark pine in western Montana: evidence from 20-year remeasurements. *Western Journal of Applied Forestry* 8(2):44–47.

Keeland, B.D., J.A. Allen, and V.V. Burkett. 1995. Southern Forested Wetlands. *In* Our Living Resources. Washington, DC: USGS.

Kendall, K.C. 1995. Whitebark Pine: Ecosystem in Peril. *In* Our Living Resources. Washington, DC: USGS.

Kendall, K.C., and S.F. Arno. 1990. Whitebark pine—an important but endangered wildlife resource. Pp. 264–273 in U.S. Forest Service Gen. Tech. Rep. INT-270.

Kerry, G. W., R.G. Clark, R.E. Montgomery, R.W. Mutch, W.P. Leenhouts, and G.T. Zimmerman. 1995. Altered Fire Regimes Within fire-adapted Ecosystems. *In* Our Living Resources. Washington, DC: USGS.

Kilgore, B.M., and M.L. Heinselman. 1990. Pp. 297–335 in J.C. Hendee, G.M. Stankey, and R.C. Lucas, eds. Fire in wilderness ecosystems. Wilderness management, 2nd ed. North American Press, Golden, CO.

Klopatek, J.M., R.J. Olson, C.J. Emerson, and J.L. Jones. 1979. Land use conflicts with natural vegetation in the United States. *Environmental Conservation* 6:192–200.

Miller, R.F., and P.E. Waigand. 1994. Holocene changes in semi-arid pinyon-juniper woodlands: response to climate, fire and human activities in the great Basin. *Biological Science* 44(7). In press.

Mutch, R.W., S.F. Arno, J.K. Brown, C.E. Carlson, R.D. Ottmar, and J.L. Peterson. 1993. *Forest health in the Blue Mountains: a management strategy for the fire-adapted ecosystems*. U.S. Forest Service Gen. Tech. Rep. PNW-310. 14 pp.

Nash, B.L., K. Tonnessen, and D.J.M. Flores.1995. Air Quality in the National Park System. *In* Our Living Resources. Washington, DC: USGS.

Nichol, A.A. 1937. *The natural vegetation of Arizona*, University of Arizona Tech. Bull. 68.

Olgilvie, R.T. 1990. Distribution and ecology of whitebark pine in western Canada. Pp. 54–60 in U.S. Forest Service Gen. Tech. Rep. INT-270.

OTA. 1993. *Preparing for an uncertain climate*. 2 vols. U.S. Congress, Office of Technology Assessment. Washington, DC.

Pellant, M., and C. Hall. 1994. Distribution of two exotic grasses on public lands in the Great Basin: status in 1992. *In* Proceedings of a Symposium on Ecology, Management and Restoration of Intermountain Annual rangelands. Intermountain Forest and Range Experiment Station, Ogden, UT.

Peterson, D.L. 1995. Air Pollution Effects on Forest Ecosystems in North America. *In* Our Living Resources. Washington, DC: USGS.

Powell, D.S., J.L. Faulkner, D.R. Darr, Z. Zhu, and D.W. MacCleery. 1993. *Forest resources of the United States, 1992.* Gen. Tech. Rep. RM-234. U.S. Forest Service, Rocky Mountain Forest and Range Experiment Station, Fort Collins, CO. 132 pp.

Pyne, S.J. 1982. *Fire in America: A cultural history of wildland and rural fire.* Princeton University Press, Princeton, N.J. 653 pp.

Ray, J. 1999. *Ecology of a Cracker Childhood.* Minneapolis, MN: Milkweed Productions.

Ray, J. 2005. *Pinhook: Finding Wholeness in a Fragmented Land.* White River Junction, VT: Chelsea Green Publishing.

Robbins. L.E., and R.L. Myers. 1992. *Seasonal effects of prescribed burning in Florida: a review.* Tall Timbers Research, Inc. Miscellaneous Publ. 8, 96 pp.

Shenandoah. 2002. Shenandoah National Park. Accessed 05/02/07 @ http://www .shenandoah.national-park.com/nat.hm.

Shugart, H.H., and D.C. West. 1977. Development of an Appalachian deciduous forest succession model and its application to assessment of the impact of the chestnut blight. *Journal of Environmental Management* 5:161–179.

Smith, T.M., and R.L. Smith. 2006. *Elements of Ecology,* 6th Edition. The Benjamin/ Cummings Publishing Company, Inc. Menlo Park, CA.

Spellman, F.R. 1998. *The Science of Environmental Pollution.* Lancaster, PA: Technomic Publishing Co.

Tansley, A.G. 1935. *Changing the face of the earth,* 2nd ed. Oxford: Blackwell.

US Army Corp of Engineers. 1998. *Management of Peatland Shrub—and Forest-dominated Communities for threatened and endangered species.* Washington, DC.

USDA. 2004. *U.S. Forest Resource Facts and Historical Trends.* Accessed 04/27/07 @ http://fia.fs.fed.us.

USFS. 1992. *Forest Services resources inventories: an overview. Forest inventory, economics, and recreation research.* Washington, DC. 39 pp.

USFWS. 1998. *Restoring an Atlantic White Cedar Bog.* Accessed 05/03/07 @ http:// www.fws. Gov/nc-es/coastal/awhitecedar.htm.

USGS. 2003. *Whitebark Pine Communities.* Accessed 05/06/07 @ http://www.nmsc .usgs.gov.

Waldrop, T.A., D.H. Van Lear, R.T. Lloyd, and W.R. Harms. 1987. *Long-term studies of prescribed burning of loblolly pine forests of the southeastern Coastal Plain.* U.S. Forest Service Southeastern Forest Experiment Station Gen. Tech. Rep. SE-45. Asheville, NC. 23 pp.

Weaver, H. 1943. Fire as an ecological and silvicultural factor in the ponderosa-pine region of the Pacific Slope. *Journal of Forestry* 41:7–14.

Wellner, C.A. 1970. Fire history in the northern Rocky Mountains. Pp. 41–64 in The role of fire in the Intermountain West. Proceedings of the Intermountain Fire Research Council Symposium, Missoula, MT.

West, N.E. 1988. Intermountain deserts, shrub steppes, and woodlands. Pp. 209–230 in M.B. Barbour and W.D. Billings, eds. North American terrestrial vegetation. Cambridge University Press, New York.

West, N.E., and M.A. Hassan. 1985. Recovery of sagebrush-grass vegetation following wildlife. *Journal of Range Management* 38:131–134.

Wharton, C.H., W.M. Kitchens, and T.W. Sipe. 1982. *The ecology of bottomland hardwood swamps of the Southeast: a community profile.* U.S. Fish and Wildlife Service. FWS/OBS-81/37. 133 pp.

Whisenant, S.G. 1990. Changing fire frequencies on Idaho's Snake River Plains: ecological and management implications. Pp. 4–10 in E.D. McArthur, E.M. Romney, S.D. Smith, and P.T. Tueller, eds. Proceedings of a Symposium on Cheatgrass Invasion, Shrub Die-off, and Other Aspects of Shrub Biology and Management. U.S. Forest Service Gen. Tech. Rep. INT-276. Intermountain Forest and Range Experiment Station, Ogden, UT.

Wild. 2010. *What is a Wilderness Area?* Accessed 11/22/10 @ http://www.widl.org/main?about/what-is-a-wilderness-area/.

Wildlink 2010. *Wilderness.* Accessed 11/22/10 @ http://wildlink.wilderness.net/wildnerss.html.

Wilhelm, G.S. 1990. *Special vegetation of the Indiana Dunes National Lakeshore.* Indiana Dunes National Lakeshore Research Program, Rep. 90-02, Porter, IN.

Wright, H.A., and S.W. Bailey. 1982. *Fire ecology: United States and southern Canada.* John Wiley and Sons, New York. 501 pp.

Wright, A.H., and A.A. Wright. 1932. The habitats and composition of the vegetation of Okefenokee Swamp, Georgia. *Ecological Monographs* 2:109–232.

Zimmerman, G.T., and R.D. Laven. 1984. Ecological interrelationship of dwarf mistletoe and fire in lodgepole pine forests. Pp. 123–131 in F.G. Hawksworth and R.F. Scharpf, technical coordinators. Biology of dwarf mistletoes: Proceedings of the Symposium. U.S. Forest Service Gen. Tech. Rep. RM-111. Rocky Mountain Forest Range Experiment Station, Fort Collins, CO. 131 pp.

Glossary

Aerobic—occurring or living only in the presence of oxygen.

Algae—large and diverse assemblages of eukaryotic organisms that lack roots, stems, and leaves but have chlorophyll and other pigments for carrying out oxygen-producing photosynthesis.

Algology or **Phycology**—the study of algae.

Amino acids—building blocks of proteins.

Anabolism—the utilization of energy and materials to build and maintain complex structures from simple components.

Anaerobic—active, living, occurring, or existing in the absence of free oxygen.

Annular space—the space between the casing and the wall of the hole.

Antheridium—special male reproductive structures where sperm are produced.

Apical meristem—consists of meristematic cells located at tip (apex) of a root or shoot.

Aplanospore—nonmotile spores produced by sporangia.

Appendage—any extension or outgrowth from the body.

Aquifer—a porous, water-bearing geologic formation.

Asexual reproduction—requires one parent cell.

Atmosphere—is the gaseous mantle enveloping the hydrosphere and lithosphere which is 78% nitrogen by volume.

Autotroph (primary producer)—any green plant that fixes energy of sunlight to manufacture food from inorganic substances.

Autotrophic—using light energy (photosynthesis) or chemical energy (chemosynthesis).

Bacteria—among the most common microorganisms in water. Bacteria are primitive, single-celled microorganisms (largely responsible for decay and decomposition of organic matter) with a variety of shapes and nutritional needs.

Benthic—algae attached and living on the bottom of a body of water.

Binary fission—nuclear division followed by division of the cytoplasm.

Biochemical Oxygen Demand (BOD)—a widely used parameter of organic pollution applied to both wastewater and surface water, involving the measurement of the dissolved oxygen used by microorganisms in the biochemical oxidation of organic matter.

Biogeochemical cycles—are cyclic mechanisms in all ecosystems by which biotic and abiotic materials are constantly exchanged.

Biome—regional land-based ecosystem type such as a tropical rainforest, taiga, temperate deciduous forest, tundra grassland, or desert. Biomes are characterized by consistent plant forms and are found over a large climatic area.

Biotic factor (community)—the living part of the environment composed of organisms that share the same area, are mutually sustaining, interdependent, and constantly fixing, utilizing, and dissipating energy.

Biotic Index—the diversity of species in an ecosystem is often a good indicator of the presence of pollution. The greater the diversity, the lower the degree of pollution. The biotic index is a systematic survey of invertebrate aquatic organisms which is used to correlate with river quality.

Blastospore or **bud**—spores formed by budding.

Bole—trunk of a tree above the root collar and extending along the main axis.

Budding—process by which yeasts reproduce.

Caisson—large pipe placed in a vertical position.

Cambium—the lateral meristem in plants.

Carbohydrates—main source of energy for living things such as sugar and starch.

Carbon—is an essential part of all organic compounds; photosynthesis is a source of carbon. Photosynthesis is the chemical process by which solar energy is stored as chemical energy.

Catabolism—the breaking down of complex materials into simpler ones using enzymes and releasing energy.

Cellular basis of life—cells are the basis of life. There are two types of cells: prokaryotes and eukaryotes.

Chloroplasts—packets that contain *chlorophyll a* and other pigments.

Chrysolaminarin—the carbohydrates reserve in organisms of division **Chrysophyta**.

Climax community—the terminal stage of ecological succession in an area.

Cold-blooded animals—body temperatures change with the environment.

Companion cells—specialized cells in the phloem that load sugars into the sieve elements.

Competition—is a critical factor for organisms in any community. Animals and plants must compete successfully in the community to stay alive.

Compound—two or more elements chemically combined.

Cone of depression—as the water in a well is drawn down, the water near the well drains or flows into it. The water will drain further back from the top of the water table into the well as drawdown increases.

Confined aquifer—an aquifer that is surrounded by formations of less permeable or impermeable material.

Conidia—asexual spores that form on specialized hyphae called **conidiophores**. Large conidia are called **macroconidia** and small conidia are called **microconidia**.

Contamination—the introduction into water of toxic materials, bacteria, or other deleterious agents that make the water unfit for its intended use.

Cotyledons—leaf-like structure (sometimes referred to as seed leaf) that is present in the seeds of flowering plants.

Decomposers—heterotrophic organisms that break down dead protoplasm and use some of the products and release others for use by consumer organisms.

Diatoms—photosynthetic, circular, or oblong chrysophyte cells.

Dicot—one of the two min types of flowering plants; characterized by having two cotyledons.

Digestion—process by which food is broken down into simpler substances.

Dinoflagellates—unicellular, photosynthetic protistan algae.

Diploid—having two of each kind of chromosome (2n).

Dissolved oxygen (D.O.)—the amount of oxygen dissolved in a stream is an indication of the degree of health of the stream and its ability to support a balanced aquatic ecosystem.

DNA—the double helix of DNA is the unifying chemical of life; its linear sequence defines the diversity of living things.

Drainage basin—an area from which surface runoff or groundwater recharge is carried into a single drainage system. It is also called catchment area, watershed, and drainage area.

Drawdown—the distance or difference between the static level and the pumping level. When the drawdown for any particular capacity well and rate pump bowls is determined, the pumping level is known for that capacity. The pump bowls are located below the pumping level so that they will always be underwater. When the drawdown is fixed or remains steady, the well is then furnishing the same amount of water as is being pumped.

Ecological pyramids—three types of ecological pyramids are the pyramids of numbers, productivity, and energy. All of these pyramids are based on the fact that due to energy loss, fewer animals can be supported at each additional trophic level.

Ecosystem—a functioning unit of nature that combines biotic communities and the abiotic environments with which they interact. Ecosystems vary greatly in size and characteristics.

Ecotones—a transitional area between two (or more) distinct habitats or ecosystems, which may have characteristics of both or its own distinct characteristics. The edge of woodland, next to a field or lawn, is an ecotone, as are some savanna areas between forests and grasslands.

Elements—pure substance that cannot be broken down into simpler substances.

Energy—is the ability or capacity to do work. Energy is degraded from a higher to a lower state.

Enzymes—special types of proteins that regulate chemical activities.

Epitheca—the larger part of the frustule (diatoms).

Euglenoids—contain chlorophylls **a** and **b** in their chloroplasts; representative genus is *Euglena*.

Eukaryotes—these are cells with a nucleus; they are found in human and other multicellular organisms (plants and animals) and also in algae and protozoa.

Eutrophication—the natural ageing of a lake or land-locked body of water which results in organic material being produced in abundance due to a ready supply of nutrients accumulated over the years.

Evapotranspiration (plant water loss)—describes the process whereby plants lose water to the atmosphere during the exchange of gases necessary for photosynthesis. Water loss by evapotranspiration constitutes a major flux back to the atmosphere.

Evolution—the modification of species, it is the core theme of biology.

Excretion—process of getting rid of waste materials.

Extinction—the dying out of a species, or the condition of having no remaining living members; also the process of bringing about such a condition.

First Law of Thermodynamics—states energy is transformed from one form to another, but is neither created nor destroyed. Given this principle, we should be able to account for all the energy in a system in an energy budget, a diagrammatic representation of the energy that flows through an ecosystem.

Forest land—land at least 10 percent stocked by forest trees of any size, including land that formerly had such tree cover and that will be naturally or artificially regenerated. The minimum area for classification of forest land is 1 acre.

Fragmentation—a type of asexual algal reproduction in which the thallus breaks up and each fragmented part grows to form a new thallus.

Frustule—the distinctive two-piece wall of silica in diatoms.

Fungi—a group of organisms that lack chlorophyll and obtain nutrients from dead or living organic matter.

Gaining stream—typical of humid regions, where groundwater recharges the stream.

Groundwater—subsurface water occupying a saturated geological formation from which wells and springs are fed.

Guard cells—specialized epidermal cells that flank stomata and whose opening and closing regulate gas exchange and water loss.

Habitat—ecologists use the term habitat to mean the place where an organism lives.

Haploid—having only a single set of chromosomes (n).

Hardwood—a dicotyledonous tree, usually broad-leaved and deciduous.

Herbivore—an animal that feeds on plants.

Heterotroph (living organisms)—obtaining materials and energy by the breaking down of other biological material using digestive enzymes and then assimilating the usable by-products.

Hydrology—the applied science pertaining to properties, distribution, and behavior or water.

Hydrosphere—is the water covering the Earth's surface of which 80% is saltwater and 19% is groundwater.

Hypha (pl. **hyphae**)—a tubular cell that grows from the tip and may form many branches. Probably the best-known example of how extensive fungal hyphae can become is demonstrated in an individual honey fungus, *Armalloria ostoyae*, which was discovered in 1992 in Washington State. This particular fungus has been identified as the world's largest living thing; it covers almost 1,500 acres. Estimates have also been made about its individual network of hyphae: it is estimated to be 500 to 1,000 years old.

Impermeable—a material or substance water will not pass through.

Infiltration capacity—the maximum rate soil can absorb rainfall.

Ingestion—taking in food or producing food.

Inorganic compounds—may or may not contain carbon.

Invertebrates—organisms without a backbone.

Lake—may be defined as a body of standing surface-water runoff (and maybe some groundwater seepage) occupying a depression in the land.

Laminar flow—in a stream where parallel layers of water shear over one another vertically.

Larva (larvae plural)—juvenile form of many insects and other organisms that become different in form when changed into adults.

Lava cascade—a cascade of water is a small waterfall formed as water descends over rocks. In similar fashion, a lava cascade refers to the rush of descent of lava over a cliff. In Hawaii, lava cascades typically occur when lava spills over the edge of a crater, a fault scarp, or a sea cliff into the ocean.

Lava channels—are narrow, curved or straight open pathways through which lava moves on the surface of a volcano. The volume of lava moving down a channel fluctuates so that the channel may be full or overflowing at times and nearly empty at other times. During overflow, some of the lava congeals and cools along the banks to form natural levees that may eventually enable the lava channel to build a few meters above the surround ground.

Lava drapery—is the cooled, congealed rock on the face of a cliff, crater, or fissure formed by lava pouring or cascading over their edges.

Lava flow—Lava flows are associated with volcanoes and others are the result of fissure flow. These masses of molten rock pour onto the Earth's surface during an effusive eruption. Both moving lava and the resulting solidified deposit are referred to as lava flows. Because of the wide range in (1) viscosity of the different lava types (basalt, andesite, dacite, and rhyolite); (2) lava discharge during eruptions; and (3) characteristics of the erupting vent and topography over which lava travels, lava flows come in a great variety of shapes and sizes.

Lava spillways—are confined lava channels on the sides of a volcanic cone or shield that form when lava overflows the rim of the vent.

Lava surge—intermittent surges or accelerations in the forward advance of lava can occur when the supply of lava to a flow front suddenly increases or a flow front gives way. The supply of lava may increase as a consequence of a higher discharge of lava from the vent, a sudden change in the vent geometry so that a great volume of lava escapes (e.g., the collapse of a vent wall), or by the escape of ponded lava from along a channel. Lava surges may be accompanied by thin, short-lived breakouts of fluid lava from the main channel and flow front.

Life span—maximum length of time an organism can be expected to live.

Limiting factor—is a necessary material that is in short supply and because of the lack of it an organism cannot reach its full potential.

Limnology—is the study of freshwater ecology, which is divided into 2 classes: *lentic* and *lotic*. The *lentic class* (calm zone) is composed of lakes, ponds, and swamps; within this class are four zones: littoral, limnetic, profundal, and benthic zones. *Lotic class* (washed) consists of rivers and streams and is composed of two zones: rapids and pools.

Lipids—energy-rich compounds made of C, O, and H.

Lithosphere—is the solid components of the Earth's surface such as rocks and weathered soil.

Losing stream—typical of arid regions, where streams can recharge groundwater.

Macroinvertebrates—animals that have no backbone and are visible without magnification.

Macrophytes—big plants.

Meandering—stream condition whereby flow follows a winding and turning course.

Meristem—group of plant cells that can divide indefinitely provides new cells for the plant.

Microhabitat—describes very local habitats.

Monocots—one of two main types of flowering plants; characterized by having a single cotyledon.

Mortality—the volume of sound wood in growing stock trees that died from natural causes during a specified year.

Mycelium—consists of many branched hypha and can become large enough to be seen with the naked eye.

National forest—an ownership class of Federal lands, designated by Executive order or statue as national forests of purchase units, and other lands under the administration of the USDA Forest Service.

Neustonic—algae that live at the water-atmosphere interface.

Niche—the role that an organism plays in its natural ecosystem, including its activities, resource use, and interaction with other organisms.

Nitrogen—is required for construction of proteins and nucleic acids; the major source is the atmosphere.

Nonseptate or **aseptate**—when crosswalls are not present.

Oogonia—vegetative cells that function as female sexual structures in the algal reproductive system.

Organisms—require 20–40 elements for survival.

Overland flow—the movement of water on and just under the Earth's surface.

Pellicle—a *Euglena* structure that allows for turning and flexing of the cell.

Perennial stream—a type of stream in which flow continues during periods of no rainfall.

Periderm—a layer of plant tissue derived from the cork cambium, and then secondary tissue, replacing the epidermis.

Permeable—a material or substance that water can pass through.

Phloem—complex vascular tissue that transports carbohydrates throughout the plant.

Phosphorus cycle—is very inefficient; the greatest source is the lithosphere. Humans have greatly speeded up this cycle through mining.

Phytoplankton—made up of algae and small plants.

Plankton—free-floating, mostly microscopic aquatic organisms.

Planktonic—algae suspended in water as opposed to attached and living on the bottom (benthic).

Point source—source of pollutants that involves discharge of pollutants from an identifiable point, such as a smokestack or sewage treatment plant.

Pollution—an averse alteration to the environment by a pollutant.

Population—a group of organisms of a single species that inhabit a certain region at a particular time.

Porosity—the ratio of pore space to total volume. That portion of a cubic foot of soil that is air space and could therefore contain moisture.

Precipitation—the process by which atmospheric moisture is discharged onto the earth's crust. Precipitation takes the form of rain, snow, hail, and sleet.

Prokaryotes—are cells without a nucleus. These include bacteria and cyanophytes (blue green algae). The genetic material is a single circular DNA and is contained in the cytoplasm, since there is no nucleus.

Proteins—used to build and repair cells; made of amino acids.

Protothecosis—a disease in humans and animals caused by the green algae, *Prototheca moriformis*.

Pumping level—the level at which the water stands when the pump is operating.

Radius of influence—the distance from the well to the edge of the cone of depression, the radius of a circle around the well from which water flows into the well.

Raw water—the untreated water to be used after treatment for drinking water.

Recharge area—an area from which precipitation flows into underground water sources.

Riffles—refer to shallow, high velocity flow over mixed gravel-cobble (bar-like) substrate.

Second Law of Thermodynamics—asserts that energy is only available due to degradation of energy from a concentrated to a dispersed form. This indicates that energy becomes more and more dissipated (randomly arranged) as it is transformed from one form to another or moved from one place to another. It also suggests that any transformation of energy will be less than 100% efficient (i.e., the transfers of energy from one trophic level to another are not perfect); some energy is dissipated during each transfer.

Septate hyphae—when a filament has crosswalls.

Sexual spores—In the fungi division Amastigomycota, four subdivisions are separated on the basis of type of sexual reproductive spores present.

Sieve cells—conducting cells in the phloem of vascular plants.

Silviculture—management of forest land for timber.

Sinuosity—the bending or curving shape of a stream course.

Softwood—a coniferous tree, usually evergreen, having needles or scale-like leaves.

Specific yield—the geologist's method for determining the capacity of a given well and the production of a given water-bearing formation, it is expressed as gallons per minute per foot of drawdown.

Sporangiospores—spores that form within a sac called a sporangium. The sporangia are attached to stalks called sporangiophores.

Spore—reproductive stage of the fungi.

Spring—a surface feature where without the help of man, water issues from rock or soil onto the land or into a body of water, the place of issuance being relatively restricted in size.

Static level—the height to which the water will rise in the well when the pump is not operating.

Stimulus—signal that causes an organism to react.

Stomata—pores on the underside of leaves that can be opened of closed to control gas exchange and water loss.

Structure and function—at all levels of organization, biological structures are shaped by natural selection to maximize their ability to perform their functions.

Subalpine—describing the region, the climate, the vegetation, or all three found just below alpine regions, usually on mountainsides at 1300 to 1800 meters in elevation. Subalpine vegetation is that just below the treeline, often dominated by pine or spruce trees.

Surface runoff—the amount of rainfall that passes over the surface of the earth.

Surface water—the water on the earth's surface as distinguished from water underground (groundwater).

Thallus—the vegetative body of algae; main plant body, not differentiated into a stem or leaves.

Thalweg—line of maximum water of channel depth in a stream.

Timber land—forest land that is capable of producing crops of industrial wood and not withdrawn from timber utilization by statute or administrative regulation. (*Note:* Areas qualifying as timber land are capable of producing in excess of 20 cubic feet per acre per year of industrial wood in natural stands.)

Trophic level—the feeding position occupied by a given organism in a food chain, measured by the number of steps removed from the producers.

Tropism—plant behavior; controlling the direction of plant growth.

Turbulent flow—in a stream where complex mixing is the result.

Unconfined aquifer—an aquifer that sits on an impervious layer, but is open on the top to local infiltration. The recharge for an unconfined aquifer is local. It is also called a water table aquifer.

Unity in diversity—explained by evolution: all organisms linked to their ancestors through evolution; scientists classify life on earth into groups related by ancestry; related organisms descended from a common ancestor and having certain similar characteristics.

Vascular tissue—tissues found in the bodies of vascular plants that transport water, nutrients, and carbohydrates. The two major kinds are xylem and phloem.

Volcanic domes—are rounded, steep-sided mounds built by very viscous magma, usually either dacite or rhyolite. Such magmas are typically too viscous (resistant to flow) to move far from the vent before cooling and crystallizing. Domes may consist of one of more individual lava flows. Volcanic domes are also referred to as lava domes.

Warm-blooded animals—maintain a constant body temperature.

Water rights—the rights, acquired under the law, to use the water accruing in surface or groundwater, for a specified purpose in a given manner and usually within the limits of a given time period.

Watershed—a drainage basin from which surface water is obtained.

Water table—the average depth or elevation of the groundwater over a selected area. The upper surface of the zone of saturation, except where that surface is formed by an impermeable body.

Xylem—vascular tissue of plants that transports water and dissolved minerals from the roots upward to other parts of the plant. Xylem often also provides mechanical support against gravity.

Index

abrasion, 81
abyssal plain, 194
air masses, 140–141
air motion, 125–133
albatross, 332
Albedo, 148
albino rainbow, 117, 120–121
alderflies, 391–392
algae, 297–302; blue-green, 255; brown, 298, 302; characteristics of, 299; golden-brown, 298, 301–302; green, 298, 300
alligators, 361–362
American badgers, 352–353
amoebae, 296
amphibians, 354, 359–365
angiosperms, 310
animal cell, 275
animal group, 321–435
animalia, 288
animals, 315–320
annelids, 317
aphanitic, 55
aplite, 56
Apollo Project, 16
aquatic insects, 380–394

aquifers, 186
arêtes, 77
arthropods, 318
arthrospores, 306
asexual reproduction, 315
atmosphere, 43, 95–106
atmospheric moisture, 102
atmospheric water phenomena, 107
atoms, 228
autotrophic, 241
auxins, 313

bacteria, 29, 219, 233, 242, 255, 259, 276, 288–294, 381, 397–398, 400; arrangements of, 289; shapes of, 289
bacterial cell, 276–277
Barchan dunes, 83
bars, 206
basalt, 54
basaltic formations, 17
batholiths, 60
beargrass, 25
beetles, 389–390
Bergeron process, 110–111
binomial system of nomenclature, 286

bioelements, 244
biogeochemical cycles, 241–259; types of, 246
biomass, 225, 265
biomass pyramids, 265
biome, 229
biosphere, 43, 225–237
biotic factors, 243
biotic indices, 380–394
birds, 321–342
black bear, 347–348
black-footed ferrets, 350–352
blastospores, 306
blowouts, 81
bluebunch wheatgrass, 24
Bowen's Reaction Series, 54–57
breccias, 56
Bryce Canyon, 70–71
bryophytes, 401–402
butterflies, 377–380
butterfly effect, 136

caddisflies, 386–387
calderas, 61
California condor, 336
calorific flow, 260
Candide, 62
capillary action, 176
capsule, 277
carbon cycle, 250–252
Carson, R., 239
Castle Lake, 30
Cavendish, H., 172
cell genome, 277
cell junctions, 283
cell nucleus, 279
cells, 228, 273–284; structure of, 278–284; types of, 276–278
cell wall, 277, 282
centrosome, 282
chemical weathering, 69, 71–72; types of, 72; water impact, 72
chemically stratified lakes, 221
chloroplast, 277–278, 283, 311

chlorophyll, 311
chordate, 318
chromatin, 279
cilia, 283
ciliophora, 295–296
cinder cones, 17, 24, 57
cirque, 77
cirrus, 105
classification, 286
climate, 135–152; division of, 143; subdivisions of, 144–146
cloud classification system, 104
cloud formation, 102–105; types of, 103–105
cloudless rain, 114
coalescence, 110
coastal erosion, 198–199
cold front, 141
Coldwater Lake, 30
colonial-nesting waterbirds, 335–336
Colorado River fish, 387–388
colored rain, 113–114
Columbia River, 26
commensalism, 266
community, 229
competition, 267
composite volcanoes, 57–58
concordant intrusions, 60
confined aquifer, 188
continental crust, 44
continental drift, 66
continental rise, 194
continental shelf, 194
continental slope, 194
Continuous Reaction Series, 56–57
contrails, 106
convectional precipitation, 112
convective clouds, 105
convergent boundaries, 67
conversion efficiency, 264
Coriolis effect, 124
Coriolis force, 129
cork cambium, 313
coulees, 59

Cowlitz River, 26
Craters of the Moon National Monument and Preserve, 16–24
cristae, 281
crop rotation, 253
crustal plates, 66–68; boundaries of, 67; major, 66; minor, 67
Cryptosporidia, 166, 294
crystal growth, 69
cumulus, 104
currents, 196–197
cutthroat trout, 370–372
cysts, 294
cytoplasm, 282
cytoskeleton, 282

dabbling ducks, 328
damselflies, 238, 392–394
deep ancient lakes, 221
deflation, 80
denitrification, 253
desertification, 86–87
desert pavement, 81
desert salt lakes, 221
desert tortoises, 362–363
desert varnish, 81
desmosomes, 283
dew point, 108
diatom, 298
dicots, 310
dikes, 60
dinoflagellates, 298
discontinuous reaction series, 56
discordant intrusions, 60
dissolution, 214
distilled water, 171
diurnal oxygen curve, 268
divergent boundaries, 67
diversity, 285–320
dobsonflies, 391–392
doldrums, 125
domain, 288
dragonflies, 238, 392–394
dry climates, 143

Dry Falls, 32–33
ducks, 327–331
duff, 12
dunes, 82–83
dystrophic, 220–221

eagles, 336–337
earth, 41
earth core, 44
earthflows, 86
earth geology, 43–44
earthquakes, 52, 62–66; intensity of, 65; magnitude of, 64–65
earth scientists, 48
earth structure, 44–50
echinoderms, 318
ecological pyramids, 264–266; types of, 264
ecosystems, 229, 237–243; types of, 242
elastic rebound, 63
elk, 354
El Niño, 137–138
endogenic forces, 52
endoplasmic reticulum (ER), 279
energy flow, 259–270
energy pyramids, 265
entrainment, 87–88, 210
epicenter, 63
erratic, 78
eskers, 80
estimated net productivity, 269–270
ethylene, 313
Euglena, 301
eukaryotic cells, 277–284
eutrophication, 219–220, 298
eutrophic lake, 220
evaporation, 115
evapotranspiration, 114–117, 249; process of, 117
evolution, 319–320
exogenic forces, 52, 68–89
extrusion, 55
extrusive rocks, 53–54

Fata Morgana, 117–121
feldspar, 55
felsic, 55
felsites, 54
Ferrel cell, 123
fire regimes, 430–433
fish, 365–373
fish populations, managed, 368–369
fjords, 78
flagella, 277, 283
floodplain, 206–207
food chain, 261–264
food chain efficiency, 263
food web, 262
forested wetlands, 428–430
fresh water, 153; location of, 182; sources
     of, 180–190
freshwater fish, 367–368
friction, 131
fringe-toed lizards, 364–365
frontal precipitation, 113
fronts, 141
frost wedging, 70
fumaroles, 61
fungi, 287, 303–307, 397–398

gabbro, 54
gap junction, 284
geese, 326–327
Generic Mapping Tools (GMT), 42
genome, 292
genus, 286–289
geological processes, 44–46
geosphere, 42
geysers, 61
Giardia lamblia, 189, 294, 297
gibberellins, 313
glacial erosion, 77–80
Glacial Lake Missoula, 32–33
glaciations, 72–80
glaciers, 73–80; characteristics of, 76;
     deposits of, 78; types of, 73–76
global air movement, 122–123
Global Positioning System (GPS), 42
global wind patterns, 124–125

glory, 118
golgi apparatus, 280
granite, 54
grasshoppers, 376–377
gravitropism, 314
gray wolves, 346–347
Great Rift, 24
green turtle, 357
grizzly bear, 348–349
gross primary productivity, 267
groundwater, 186–190, 211–214; geo-
     logic activity of, 214
GUDISW (groundwater under direct
     influence of surface water), 189
guyots, 195
gymnosperms, 310

Hadley cell, 123
hanging valleys, 78
haploid cell, 300
Harmony Lodge, 25
Hawaiian-type eruption, 59
hawksbill, 358
headlands, 198
heterotrophic, 241
highlands, 144
hoodoos, 71
horns, 77
hot springs, 61
huckleberries, 25
hydraulic action, 198
hydrogen oxide, 170
hydrosphere, 43
hygroscopic nuclei, 110
hypha, 306
hypocenter, 63

Ice Age, 32
icebergs, 79
ice caps, 78
ice sheets, 78
impoundments, 221
Indiana bats, 345–346
infrared light, 100
inhibitors, 313

insects, 376–396
intermediate filaments, 282
intrusions, 60
intrusive rocks, 53–54
invertebrates, 316–319
isobars, 127

Jacobs Lake, 10–15
jet streams, 131
Johnston, D. A., 25
joints, 69

Kaibab Plateau, 10–16
Kaibab Squirrel, 10
kames, 79
karst topography, 214
Kemp's Ridley, 357–358
kettle holes, 79
kingdoms, 287
kinins, 313
kipukas, 20
Kock's postulates, 234
Koppen climates, 145–146

laccoliths, 60
lacustrine, 372
lag gravels, 81
lahar, 6, 88–89
lancelets, 319
land organism, 423, 438
landslides, 85
lava domes, 58
lava plains, 61
lava tubes, 17
Lavoisier, A., 173
leaves, 310–311
lentic habitat, 216–219
levels of organization, 228–229
lichens, 23, 49, 401
life, 226–228
limiting nutrient, 253
lithosphere, 46–50, 93
living, 228
lodgpole pine, 432–433
loess, 83

loggerhead, 357
longitudinal dunes, 83
lopoliths, 60
lyse, 292
lysosomes, 280

macroinvertebrates, 384–394
macromolecule, 228
mafic, 55
magma, 53, 55; eruption of, 57–59
mammals, 343–354
mantle, 44
marine deposition, 199
marine mammals, 344–345
marine transportation, 199
marine turtles, 357–359
marl lakes, 221
marmots, 24
mass movements, 85–86
mass wasting, 83–86; by freezing and
    thawing, 84; by gravity, 83–84; by or-
    ganic activities, 84; by shock waves,
    84; by undercutting, 84; by water, 84
mastigophora, 295
mayflies, 241, 384–385
mean residence time (MRT), 245
meniscus, 170
Mercalli Intensity Scale, 66
mesotrophic lake, 220
meteorology, 136
Michigan forest, 434
microclimate, 146–152
microfilaments, 282
microfungi, 398–399
microtubules, 282
microvilli, 279
mineral resources, 42
Mississippi sandhill cranes, 340–341
mitochondria, 280
mixing ratio, 108
molecules, 228
mollusks, 316–317
monera, 287
monocots, 310
moraines, 78

moths, 377–380
motile, 290
motility, 298
mourning doves, 338–339
Mt. Freemont Trail, 3–8
Mt. Rainier, 3–8
Mt. St. Helens, 6–7, 24–31
mudflows, 85
mule deer, 24
mussel, 394–395
mutualism, 266
mycelia, 304
mycology, 303

naiads, 374
native rancid frogs, 362
native vascular plants, 403
natural disaster, 8
natural events, 8
nature, 3–34
needlegrass, 24
net community productivity, 267
net primary productivity, 267
net yield, 268
neutralism, 267
nitrobacter, 255
nitrogen cycle, 252–256
nitrogen fixation, 253
numbers pyramids, 265
nutrient cycles, 243–259

obligate, 266
obsidian, 51, 56
occluded front, 141
ocean floors, 194–195
oceans, 192–200
oligotrophic lakes, 220
olive ridley sea turtle, 358
one-factor control hypothesis, 258
one problem/one solution syndrome, 258
organ, 229
organelles, 228
organisms, 229
organ system, 229
orographic precipitation, 112

outwash plains, 79
outwash terraces, 79
oxygen sag curve, 381

parabolic dunes, 83
Paramecium, 295
parasites, 321
parasitism, 267
parent material, 48
particulate matter, 100–101
Pelean-type eruption, 59
pellicle, 296
peridolite, 54
permatite, 56
peroxisomes, 281
phage, 292
phaneritic, 55
phloem, 310
phosphorus cycle, 256–258
photosynthesis, 311–312
phototropism, 314
physical weathering, 69–71; by heat, 70;
    by biological activities, 70
pili, 277
piping plovers, 341
Plantae, 288
plant cell, 274, 309
plant hormones, 313
plant kingdom, 307
plant reproduction, 31
plants, 306–31
plant succession, 20
plasma membrane, 277, 279
plateau basalts, 61
plate tectonics, 52, 66–68
plinian eruption, 59
plutons, 60
pocket mouse, 24
polar cell, 123
polar climates, 143
polar lakes, 221
polysome, 281
pond ecosystem, 242
ponderosa pine, 432
pools, 206

population, 229
population growth, 267
porphyry, 56
potable water, 181
prairie grass, 379
precesses, 41
precipitation, 108; types of, 109, 112–113
productivity, 267–270
prokaryotic cells, 276–284
Protista, 287
protists, 292
protozoa, 276, 293–297
pumice, 54–56
P-waves, 64

raindrop, 111
raptors, 336–337
raven, 339–340
red alder, 4, 30
red-cockaded woodpeckers, 341–342
relative humidity, 108
reptiles, 354–359
reservoirs, 244
residence times, 250
ribosome, 281
Richter magnitude, 65
riffles, 206
rock cycle, 45–46
rock-forming material, 56
rock type and structure, 68
rodents, 24
roots, 312
running waters, 201–214
run-off, 178–179

sagebrush, 24
sagebrush-grass, 431
sagebrush steppe, 17
sage scrub, 434
salinity, 193
saltation, 80
saprophytes, 303
sarcondinea, 296
scablands, 10
scientific method, 229–236; use of, 230

scoria, 56
scree, 198
sea arches, 199
seabirds, 332–335
sea cliffs, 198
seamounts, 194
sea stacks, 199
second productivity, 267
Sediment Retention Structure (SRS), 26
seismic activity, 63
seismic waves, 64
seismograph, 63
seismology, 63–66
setae, 317
shear, 84
shield volcanoes, 58
shorebirds, 331–332
sills, 60
sinkhole, 214
sinuosity, 205–206
Sitka forest, 30
slash/burn agriculture, 253
slime, 302–303
slope, 69
slumps, 85
Snake River Plain, 24
soil, 46–50
soil creep, 86
soil engineers, 48
soil formation, 48–50
soil scientists, 47–48
solifluction, 86
songbirds, 323–325
southern pinelands, 433
southwestern willow flycatcher, 342
spatter cones, 17
species, 229
specific gravity, 56
Spirit Lake, 25, 30–31
sporangia, 306
sporozoans, 296
starch, 306–307
still waters, 159–161, 216–222
stocks, 60
stoneflies, 385–386

stratified drift, 79
stratosphere, 99–103
stratus, 103–104
stream channels, 204–207
streamflow, 207–208
stream genesis, 201–203
stream profiles, 204–205
streams, 203–214
stream water discharge, 208–209
strombolian-type eruption, 59
sulfur cycle, 258–259
sun, 138–140
surface tension, 169
surface water, 181–186, 249
surface waves, 64
S-waves, 64
syenite, 54
symbiosis, 266
system ecology, 241

tall-grass prairie, 43
Tarahumara frog, 365
tarn, 77
tests, 296
thermal areas, 61–62
thermal inversions, 141–142
thermosaline circulation, 197
thignotropism, 314
Thiobacillus thiooxidans, 259
tidal currents, 196
tides, 195–196
tight junctions, 283
till, 78
tissue, 228
tombolos, 189
Toutle River, 25
trade winds, 124
transform boundaries, 67
translation, 281
transpiration, 115, 247
transport of load, 87
transport of material, 209
transverse dunes, 83
trophic level, 261
tropical climates, 143

tropisms, 313–314
troposphere, 98–99
true flies, 387–388
Truman, H. R., 25
tubuline, 282
tuff, 56
tunicotes, 319
turtles, 356–357

unconfined aquifer, 187
universal solvent, 157
urea, 255
U.S. Forest Resources, 425–428
Utah desert, 435

vacuoles, 281
vascular cambium, 312
vascular plants, 309–315; growth in, 312
ventifacts, 81
vertebrates, 319
Vesuvian eruption, 59
viruses, 291
visible light, 100
vital air, 173
volcanic craters, 61
volcanic lakes, 221
volcanic landforms, 61
volcanic mountains, 61
volcanism, 52–62
vulcanian-type eruption, 59

wading birds, 335–336
Warm Fire, 10–16
warm front, 141
water balance, 115
water cycle, 176–178, 247–250
water flow, 87–88
water fluxes, 249
waterfowl, 325
water molds, 302–303
water movements, 178–180
water penny, 390
water reservoir, 249
watershed, 249
water strider, 391

water vapor, 99, 248
waves, 197
wave-cut bench, 198
weather, 135–152
weathering, 68
weed, 286
Weeping Rock, 155–156
western juniper, 431–432
whirligig beetle, 290
whitebark pine trees, 436–437
white sturgeon, 372–373
white-tailed deer, 353–354
wilderness, 421–438

Wildland Use Fire (WUF), 12–16
wild turkey, 337–338
wind, 124
wind chill, 132
wind-driven erosion, 80
wind erosion, 80–83
Wisconsin oak savanna, 437–438

xylem, 309–310

yardangs, 82

zebra mussel, 375

About the Authors

Frank R. Spellman is author of more than seventy published texts covering many areas of environmental science. He is also a consultant to the U.S. Department of Justice on environmental-health issues related to the U.S. military services. He has a BS in management and an MBA, along with an MS and PhD in environmental engineering.

Joni Price-Bayer is a speech language pathologist with Norfolk Public Schools. She has degrees in both English and education and is a professional member of the American Speech-Language Health Association. She coauthored *In Defense of Science* with Frank R. Spellman.